WITHDRAWN

To Ralph Case & Virginia
in gratitude to you for opening up
home to me & opening my horizons to
West Texas & in admiration & friendship

Carleton Gajdusek   April 1981

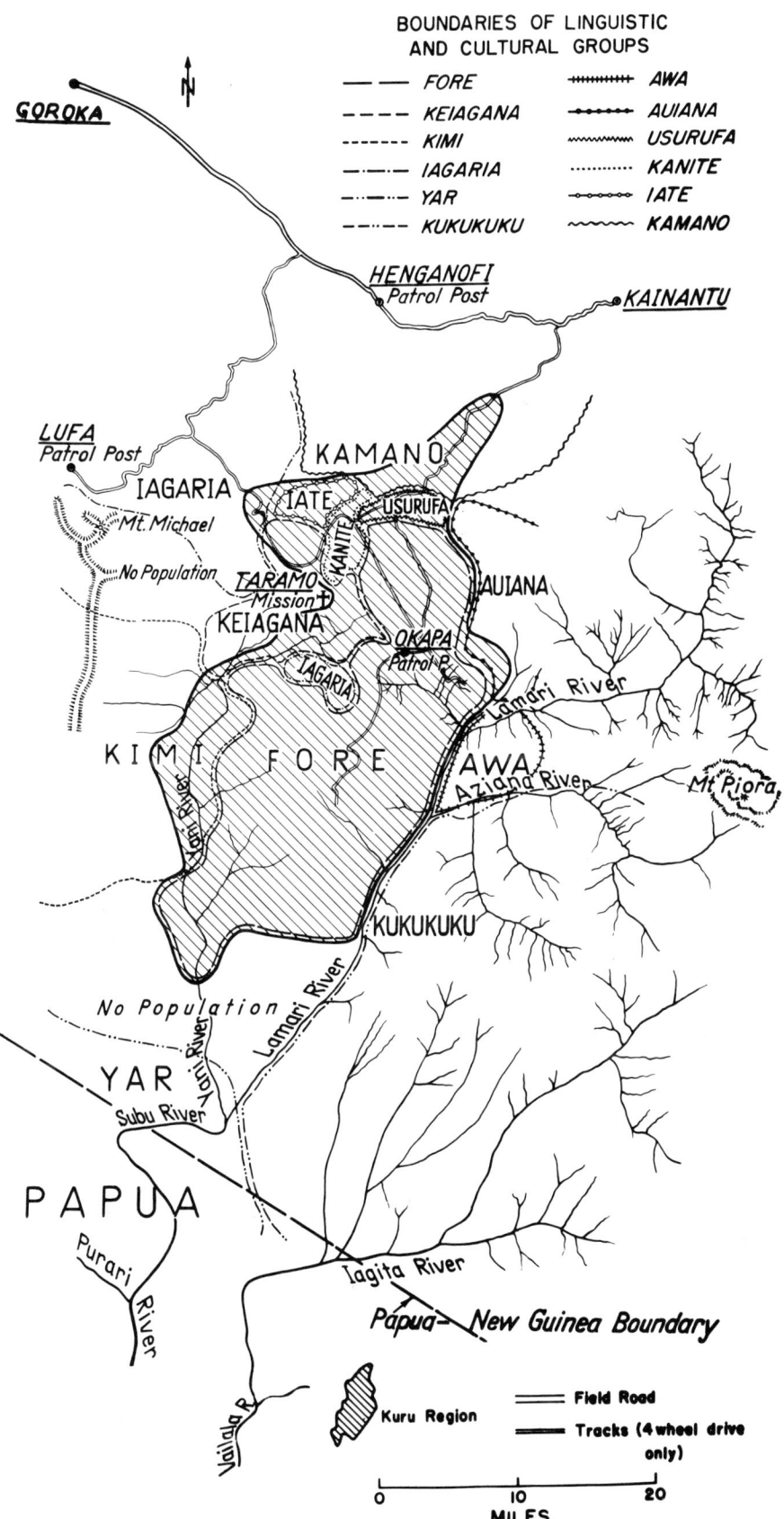

RC
394
.K8
K87
1981

KURU

*Early Letters and Field-Notes from
the Collection of D. Carleton Gajdusek*

Edited by

Judith Farquhar   D. Carleton Gajdusek

Raven Press · New York

Raven Press, 1140 Avenue of the Americas, New York, New York 10036

© 1981 by Raven Press Books, Ltd. All rights reserved. This book is protected by copyright. No part of it may be reproduced, stored in a retrieval system, or transmitted, in any form or by any means, electronic, mechanical, photocopying, recording, or otherwise, without the prior written permission of the publisher.

Made in the United States of America

Library of Congress Cataloging in Publication Data

Main entry under title:

Kuru : early letters and field-notes from the collection of D. Carleton Gajdusek.

   Includes index.
   1. Kuru. 2. Fore (New Guinea people)—Diseases.
3. Virologists—United States—Correspondence.
4. Neurologists—United States—Correspondence.
5. Gajdusek, Daniel Carleton, 1923–   I. Farquhar, Judith. II. Gajdusek, Daniel Carleton, 1923–

[DNLM: 1. Kuru—Correspondence. 2. Research—Correspondence. WC540 K96]
RC394.K8K87   616.8′3   78-68642
ISBN 0-89004-359-0

Table of Contents

List of Contents		vii
Preface—Judith Farquhar		xv
Editorial Note		xvii
Introduction—D. Carleton Gajdusek		xxi
I.	General Correspondence *26 December 1956–30 August 1957*	1
II.	Bush Correspondence *29 August–1 September 1957*	125
III.	Excerpts from the Field-Journals Kuru Epidemiological Patrols *3–14 September 1957*	133
IV.	Bush Correspondence *6–11 September 1957*	147
V.	General Correspondence *4 September–4 October 1957*	153
VI.	Excerpts from the Field-Journals Kuru Epidemiological Patrols *26 September–9 November 1957*	175
VII.	Bush Correspondence *28 September–16 October 1957*	219
VIII.	General Correspondence *4 October 1957–28 January 1958*	225
Appendixes:	1. Sorcery Among the South Fore, with Special Reference to Kuru (Charles Julius)	281
	2. Kuru [An Administrative Report to the NIH] (H. A. Imus)	289
	3. Report on the Kuru Disease (H. N. Robson, S. Sunderland, and J. C. Eccles)	291
	4. A Chronology of the Kuru Area	293
	5. Letters from Lois Larkin	296

Notes	305
Correspondents—Writers and Recipients	321
Glossary of Indigenous and Local Terms	323
References	325
Index	329

List of Contents

The order of the materials included in the body of this volume is chronological by date of writing except where classification into General and Bush Correspondence chapters required some minor shifts. D. Carleton Gajdusek (DCG) is designated by his initials only. Those letters and field-notes that have appeared in *Correspondence on the Discovery and Original Investigations on Kuru, Smadel-Gajdusek Correspondence* (C) or in *Kuru Epidemiological Patrols from the New Guinea Highlands to Papua, 1957* (J) are noted in the Reference column with the number of the letter (in C) or the page number (in J).

Page	Date	From	To	Reference
		General Correspondence		
1	26 Dec 1956	Zigas	Gunther	
	1957			
2	12 Feb	Burnet	Gunther	
2	15 Feb	Gunther	Burnet	
3	[late Feb]	Gunther	Burnet	
3	5 Mar	Scragg	Anderson	
4	11 Mar	I. Burnet	F.M. Burnet and family	
5	13 Mar	DCG	Burnet, Wood, Anderson	
8	15 Mar	DCG	Smadel	C 14
10	15 Mar	DCG, Zigas	Burnet, Anderson, Wood	C 14a
13	[mid Mar]	DCG	Scragg	
17	19 Mar	Scragg	Burnet	
17	20 Mar	Burnet	Scragg	
18	20 Mar	DCG	Burnet, Wood, Anderson	C 15a
21	20 Mar	DCG	Scragg	
23	22 Mar	Anderson	Zigas	
23	22 Mar	Anderson	Zigas	
24	22 Mar	Anderson	Scragg	
24	26 Mar	Burnet	Scragg	
25	26 Mar	Burnet	DCG	

Page	Date	From	To	Reference
26	28 Mar	Scragg	DCG	
26	29 Mar	Burnet (summary of conversation with Scragg)		
27	29 Mar	Smadel	Morris	C 16c
27	30 Mar	Scragg	DCG	
28	n.d.	DCG	Scragg	
28	3 Apr	DCG	Gunther	
29	3 Apr	DCG	Smadel	C 17
31	5 Apr	Burnet (summary of conversation with Scragg)		
32	5 Apr	Burnet	Scragg	
33	6 Apr	DCG	Scragg	
35	[early Apr]	DCG	Burnet	
38	9 Apr	Burnet	DCG	
38	9 Apr	Gunther	Burnet	
39	9 Apr	Gunther	DCG	
40	9 Apr	Burnet	Scragg	
40	9 Apr	Smadel	Rivers	C 18a
41	[mid Apr]	Burnet	Gunther	
42	16 Apr	Gunther	DCG	
42	[20 Apr]	DCG	Smadel	C 19
43	20 Apr	DCG	Burnet	
46	20 Apr	DCG	Gunther	
48	24 Apr	Gunther	Burnet	
49	3 May	Gunther	DCG	
49	5 May	DCG	Wood	
51	5 May	Anderson	Burnet	
51	7 May	Anderson	Zigas	
52	14 May	Burnet	Scragg	
52	n.d.	C. Cleland	Dept. of Territories, Canberra	
54	[mid May]	DCG	Smadel	C 23
55	16 May	Anderson	Zigas, DCG	
56	17 May	Anderson	DCG, Zigas	
57	19 May	DCG	Burnet, Anderson	
59	20 May	Simmons	DCG	
60	21 May	Rivers	DCG	
61	21 May	Rivers	Burnet	
61	23 May	Curtain	DCG	
62	23 May	Wood	DCG	
62	24 May	Burnet	DCG	
63	27 May	Burnet	Rivers	
64	28 May	DCG	Smadel	C 24
66	[late May]	DCG	Smadel	C 25
66	[late May]	DCG	Smadel	C 26
68	[late May]	DCG, Zigas	Price	C 26a

LIST OF CONTENTS

Page	Date	From	To	Reference
70	29 May	Smadel	DCG	C 27
70	[30 May]	DCG	Smadel	C 28
71	4 June	DCG	Burnet, Anderson	C 29a
74	4 June	Abbott	DCG	
74	7 June	DCG	Smadel	C 29
75	7 June	Anderson	Zigas	
75	14 June	Anderson	Zigas	
76	19 June	Anderson	DCG	
76	20 June	Anderson	DCG	
77	25 June	Simmons	DCG	
77	25 June	Smadel	DCG	C 30
78	29 June	DCG	Smadel	C 31
80	30 June	DCG	Simmons	
81	[early July]	DCG	Curtain	
83	4 July	DCG	Simmons	
84	4 July	DCG, Zigas	Abbott	C 33a
87	8 July	DCG	Smadel	C 33
88	9 July	Berndt	Scragg	C 35a
90	10 July	DCG	Smadel	C 34
91	12 July	DCG	Anderson	
92	15 July	Curtain	DCG	
93	17 July	Rees	Southern	C 64a
93	19 July	Anderson	Southern	C 64b
94	19 July	Simmons	DCG	
94	24 July	Scragg	DCG	
95	25 July	DCG	Smadel	
97	26 July	Anderson	Zigas, DCG	
98	26 July	C. G. Baker	DCG	C 36
99	5 Aug	Rivers	DCG	C 42a
100	[6 Aug]	DCG	Smadel	C 38
104	6 Aug	DCG	Burnet, Anderson, Wood	
106	6 Aug	DCG	Price	
108	[c. 6 Aug]	DCG	Wilson	
109	8 Aug	Imus	Smadel	C 40b
110	9 Aug	Scragg	Zigas, DCG	
110	15 Aug	Scragg	DCG	
111	15 Aug	Garland	Smadel	C 43d
111	[mid Aug]	DCG	Anderson	
112	15 Aug	Klatzo	Smadel	C 40a
113	16 Aug	Smadel	DCG	C 40
114	16 Aug	Robertson	Scragg	C 62c
115	16 Aug	DCG	Smadel	C 41
117	20 Aug	Kurland	Imus	C 43b
118	21 Aug	DCG	Anderson	
119	22 Aug	DCG	Anderson	

Page	Date	From	To	Reference
120	25 Aug	DCG	Smadel	C 42
121	26 Aug	Imus	Smadel	C 43a
122	28 Aug	Simmons	DCG	
122	29 Aug	Smadel	Ingalls	C 43e
123	30 Aug	Smadel	DCG	C 43

Bush Correspondence

Page	Date	From	To	Reference
125	29 Aug	DCG	Zigas, J. Baker	J 27
128	30 Aug	DCG	J. Baker	J 31
131	1,2 Sept	DCG	Zigas, J. Baker	J 40

Excerpts from Field-Journals

Page	Date	From	Reference
133	3 Sept	Paiti village, South Fore	J 1
134	4 Sept	Misapi village, South Kimi	J 2
135	6 Sept	Amusa Rest House, South Kimi	J 4
136	6 Sept	Uvai Rest House, South Kimi	J 5
137	7 Sept	Mani Rest House, South Kimi	J 9
138	8 Sept	Mani Rest House, South Kimi	J 10
140	9 Sept	Hegeteru village, Kimi	J 13
140	10 Sept	Gono Rest House and Mission, Lufa Kimi	J 15
142	11 Sept	Lufa Patrol Post	J 18
142	12 Sept	Lufa Patrol Post	J 19
144	14 Sept	Numpuru Rest House, Frigano area	J 23

Bush Correspondence

Page	Date	From	To	Reference
147	6 Sept	I. Burnet	DCG	J 44
147	[7 Sept]	Hamilton	DCG	J 44
148	7 Sept	J. Baker	DCG, Berkin	J 46
149	8 Sept	DCG	J. Baker, Zigas, Hamilton	J 47
150	11 Sept	DCG	Berkin	J 50

General Correspondence

Page	Date	From	To	Reference
153	4 Sept	Anderson	DCG	
153	10 Sept	Burnet	DCG	
154	12 Sept	Smadel	DCG	C 44
155	13 Sept	Klatzo	DCG	C 44b
156	14 Sept	Barrau	DCG	C 44c
156	18 Sept	DCG	Smadel	C 46
160	19 Sept	DCG	Winton	C 47a
160	21 Sept	DCG	Smadel	C 48
162	22 Sept	DCG	Simmons	C 49b
163	23 Sept	DCG	Burnet	

LIST OF CONTENTS

Page	Date	From	To	Reference
166	24 Sept	DCG	Scragg	
167	25 Sept	Burnet	Scragg	
168	25 Sept	DCG	Ingalls	C 49c
169	27 Sept	DCG	Smadel	C 49
171	27 Sept	Burnet	DCG	
171	27 Sept	Curtain	DCG	
172	4 Oct	Scragg	Burnet	

Excerpts from Field-Journals

Page	Date	Location	Reference
175	26 Sept	Okapa Rest House, North Fore	J 54
176	27 Sept	Yakeia village, Awa	J 56
177	28 Sept	Yakeia village, Awa	J 58
178	30 Sept	Mobutasa village, Awa	J 61
179	1 Oct	Agamusei village, Awa	J 63
180	2 Oct	Agamusei village, Awa	J 65
181	2 Oct	Auroga village, Kukukuku	J 67
182	4 Oct	Auroga village, Kukukuku	J 71
184	5 Oct	Auroga village camp No. 2, Kukukuku	J 74
185	6 Oct	Tchaiorogoro hamlet, Auroga village, Kukukuku	J 78
187	7 Oct	Camp No. 3: Tchaiorogoro village, Auroga Kukukuku	J 81
189	8 Oct	Camp No. 4: Iwane hamlet, Simbari Kukukuku	J 86
190	8 Oct	Camp No. 5: below Kaiguanbi hamlet, Simbari Kukukuku	J 90
192	9 Oct	Camp No. 5: below Kaiguanbi hamlet, Simbari Kukukuku	J 97
193	10 Oct	Camp No. 6: Maiguanga village, Muniri Kukukuku	J 99
195	11 Oct	Camp No. 6: near Maiguanga hamlet, Muniri Kukukuku	J 106
195	12 Oct	Camp No. 7: above Kataramapinti hamlet, Moraei Kukukuku	J 106
196	13 Oct	Camp No. 8: near Anjapte and Watcheramapinti hamlets, Moraei Kukukuku	J 110
197	14 Oct	Camp No. 8: near Anjapte and Watcheramapinti hamlets, Moraei Kukukuku	J 114
198	15 Oct	Camp No. 8: Moraei Kukukuku	J 118
200	16 Oct	Bush camp No. 1, on the shores of the Kataramunga River, en route to the Yar people	J 123

Page	Date	From	To	Reference
201	17 Oct	Bush camp No. 2 en route to the Yar people, at the site of deserted Kuraripinti village		J 127
202	18 Oct	Beside the Tu River, below Kuagat-nunga hamlets		J 129
204	19 Oct	Weme village (Papua), Yar Pawaiian people		J 136
206	20 Oct	Weme village, Yar Pawaiian		J 140
208	21 Oct	Weme village, Yar Pawaiian		J 144
209	22 Oct	Weme village, Yar Pawaiian		J 148
209	23 Oct	Weme village, Yar Pawaiian		J 150
210	24 Oct	Bush camp en route to So'o from Weme, Yar Pawaiian		J 155
210	25 Oct	Bush camp No. 2 between Weme and So'o, Pawaiian		J 155
212	26 Oct	So'o village, Pawaiian		J 158
214	27 Oct	Kapuna, London Mission Society Hospital site on the Wame River		J 165
216	31 Oct	Beara Government Station, Gulf District		J 170
216	8 Nov	Lae		J 174
218	9 Nov	Lae		J 179

Bush Correspondence

Page	Date	From	To	Reference
219	28 Sept	Hamilton	J. Baker, DCG	J 190
219	28 Sept	DCG	Hamilton, Berkin, Zigas	J 190
220	30 Sept	Hamilton	DCG, J. Baker	J 192
221	1 Oct	Hamilton	DCG	J 195
221	14 Oct	DCG	Berkin, Zigas, Hamilton	J 195
222	15 Oct	Berkin	J. Baker, DCG	J 201
222	15 Oct	Hamilton	DCG, J. Baker	J 201
223	16 Oct	DCG	Berkin, Zigas, Hamilton	J 202

General Correspondence

Page	Date	From	To	Reference
225	4 Oct	Scragg	DCG	
225	8 Oct	Berndt	Burnet	
226	14 Oct	Burnet	Berndt	
226	15 Oct	Hamilton	Smadel	C 49d
227	18 Oct	Kurland	Smadel	C 52a
228	21 Oct	Smadel	DCG, Hamilton	C 50
228	21 Oct	Smadel	DCG	C 51

LIST OF CONTENTS   xiii

Page	Date	From	To	Reference
229	31 Oct	Smadel	Hamilton	C 49e
229	31 Oct	Rees	Southern	C 64c
229	31 Oct	Robertson	Greenfield	C 62b
230	4 Nov	Curtain	DCG	
231	4 Nov	DCG	Smadel	C 54
235	5 Nov	Eccles	Bailey	C 54a
235	7 Nov	Smadel	DCG	C 56
236	7 Nov	Anderson	Rees	
236	8 Nov	Burnet	Winton	
236	10 Nov	DCG	Smadel	C 57
240	12 Nov	Bailey	Smadel	C 62a
240	12 Nov	DCG	Simmons	C 59a
243	13 Nov	DCG	Curtain	C 59b
245	13 Nov	DCG	Robertson	C 59c
246	14 Nov	DCG	Smadel	C 59
248	15 Nov	DCG	Curtain	
249	15 Nov	DCG	Winton	C 59d
250	16 Nov	DCG	Burnet, Anderson	C 60a
253	16 Nov	DCG	Smadel	C 60
253	17 Nov	DCG	Smadel	C 61
256	20 Nov	Davis	Garland	C 64d
256	21 Nov	Smadel	DCG	C 62
257	22 Nov	Simmons	DCG	
257	24 Nov	DCG	Simmons	
259	24 Nov	DCG	Smadel	C 63
261	27 Nov	Burnet	DCG	
261	2 Dec	Simmons	DCG	
262	[early Dec]	DCG	Smadel	C 65
263	6 Dec	Masland	C. G. Baker	C 66a
264	6 Dec	Burnet	Rees	
264	7 Dec	DCG	Smadel	C 67
267	8 Dec	DCG	Simmons	
268	11 Dec	DCG	Editor, *Klinische Wochenschrift*	C 69a
268	19 Dec	DCG	Smadel	C 68
270	24 Dec	Masland	Smadel	C 63a
270	24 Dec	DCG	Smadel	C 69
271	24 Dec	DCG	Robertson	C 70a
273	31 Dec	DCG	Smadel	C 70
	1958			
274	10 Jan	Smadel	Masland	C 71a
274	12 Jan	DCG	Smadel	C 72
277	24 Jan	DCG	Smadel	C 73
279	28 Jan	DCG	Smadel	C 74

Preface

In 1957 kuru was a newly discovered fatal disease of the central nervous system that was occurring in epidemic proportions among the inhabitants of stone-age settlements in the central highlands of New Guinea. The disease has since nearly disappeared, and the affected peoples now live in consolidated villages as peasant societies with a cash economy. But the study of kuru has had a scientific impact far beyond the confines of New Guinea and the field of tropical medicine. It was the first chronic disease of man proved to be a slow virus infection, and the investigation of its atypical properties has resulted in the description of a class of unconventional pathogens that appear to be far smaller, and simpler in structure, than any previously known virus. The discovery of the virus etiology of kuru was the first indication that chronic noninflammatory diseases of the CNS—even those occurring in a heredofamilial pattern—could be infectious. For his work in the biology of kuru and other slow virus diseases, D. Carleton Gajdusek was awarded the 1976 Nobel Prize in Physiology or Medicine.

Kuru and its scientific history are unusual in many respects. Seldom has a major medical investigation begun so suddenly in such exotic surroundings under the direction of such exceptional scientific minds. Furthermore, the inception of history-making research is most often recorded in retrospect, subject to the revisions of memory and a shortage of written records. But since necessity dictated that Gajdusek and his colleagues conduct their intellectual discourse by mail, many ideas that might otherwise have been forgotten or abandoned are preserved in this correspondence. The materials collected in this book thus provide a rare opportunity to observe the operations of scientific thought in grappling with a new and bizarre clinical entity. In them, then-attractive hypotheses that later turned out to be costly dead-ends are entertained along with prescient wild guesses that have now become widely accepted explanations.

But these letters record more than the medical thinking and the field and laboratory investigations that formed the foundation of slow virus studies. They also betray something of the professional and national territoriality that often affects the course of scientific research. The medical importance of kuru escaped no one's notice, and much of the early 1957 correspondence directly concerns the allocation of research rights over the problem. The tacit rules that govern the scientific division of labor and the competitive pursuit of significant research findings can be seen at work in these pages, though some would say that they are evident more in the breach than in the observance.

Kuru's dramatic setting in the Eastern Highlands District of the Territory of Papua and New Guinea—now a province in the new nation of Papua New Guinea—is a major component in the story that unfolds herein. In many respects the initial investigation of kuru was high adventure, taking place as it did among people who had only recently begun to have contact with Europeans and who lived in fortified mountaintop hamlets connected only by difficult footpaths. These people of the kuru area welcomed and assisted researchers with enthusiasm, and their continuing friendship has enabled such excellent kuru epidemiology that every death since 1956 has been recorded and most cases examined. The major epidemiological task in 1957, however, was the determination of kuru's real prevalence and the delineation of the area within which it occurred. This work required numerous expeditions on foot into territory previously unexplored and very difficult of access, visiting settlements that in some cases had never before been seen by Europeans. These "kuru epidemiological patrols" were undertaken by Gajdusek, Vin Zigas, and Jack Baker, sometimes with other Kuru Research Center personnel. Gajdusek recorded his observations and experiences during some of these patrols in an extensive personal field-journal, excerpts from which have been included in this volume to convey a sense of the challenges and seductions of the field situation.

Our intention in assembling this collection of letters and field-notes from the first year of kuru study is to provide a relatively objective and complete record of that dramatic and scientifically fruitful period. Though this book has been edited to be useful primarily to students of the history and philosophy of science, these materials have intrinsic interest as products of an unprecedented encounter between Western science and neolithic New Guinea, and of the tragic disease that brought them together.

Judith Farquhar

Editorial Note

In compiling this volume, we have tried to present all the available written records we consider to have substantive importance for an understanding of the first year of kuru investigation. Our selection of letters has excluded only those that were purely logistical or administrative (e.g., DCG's lists of needed supplies, or receipts for books borrowed from the Public Health Department library). Specimen-lists and tables of laboratory results have been excluded as well, since these data are summarized in the letters. Routine exchanges about such matters as the details of publishing the first three scientific papers on kuru have not been included in the main collection but have been summarized in the Notes. All of the written records of that period that we have been able to locate, no matter how mundane, have been preserved; and the entire corpus from which this book has been assembled is to be deposited at the Library of the American Philosophical Society, where it can be made available to scholars whose research requires access to the original materials.

We have not attempted to provide a completely representative or exhaustive picture of the contents of DCG's field-journal; it has, rather, been excerpted for this book with a particular focus on the ethnological and medical background of early kuru epidemiology. The journal has been published in its entirety in limited edition as *Kuru Epidemiological Patrols from the New Guinea Highlands to Papua, 1957*. (Page and line citations that accompany the excerpts included herein refer to that publication.) This field-journal is by no means a complete record of the kuru epidemiological patrols that were undertaken in 1957, since it was not begun until September of that year, after more than a dozen other patrols in and around the kuru area had already been made. But it does serve to provide a richer picture of the kuru area and its people than can be gleaned from the scientific correspondence.

Some of the letters have been previously published as well, in *Correspondence on the Discovery and Original Investigations on Kuru: Smadel–Gajdusek Correspondence 1955–1958*, which was offset-reproduced at the NIH in April 1975. The present book was originally intended to be a revised edition of that publication; in the process of preparing the manuscript, however, so many previously unpublished letters came to light that, with the excerpts from the field-journal, this volume little resembles its precursor. As has been pointed out in the Introduction, some of the additional materials were kindly provided by Sir Mac Burnet and Sir John Gunther; others surfaced as we made more thorough searches of long-abandoned correspondence files.

The time-period covered herein begins with Vin Zigas's 26 December 1956 report to Gunther (who was then Director of the P.H.D.) on "a form of encephalitis amongst the Okapa people," and ends with DCG's last letter to Joe Smadel from New Guinea at the end of January 1958. These letters serve to naturally delimit the first year of kuru investigation. The total corpus of original correspondence actually includes much more that is of interest in the history of medical research in Australia and New Guinea, and in the history of kuru, before and after the period covered here; these additional materials provide important context to the kuru story, but for obvious reasons of space limitation it has been impossible to include them in this publication.

The major editorial intervention we have made is to excise from some of the letters material that is egregiously redundant of previous letters, or so banal as to interfere with the flow of significant information; in these cases the excised portions have been transferred (in most cases in their entirety) to the Notes. Thus, those who care to mentally restore a letter to its redundant entirety may readily do so by referring to the Notes, which are cited numerically in square brackets. Occasionally—where these excisions consisted of long lists of patients' names, specimen numbers, and the like—they have been abstracted or summarized. Inside addresses have not been reproduced here, since the locations of all writers may be found in the List of Correspondents. DCG's letters are from the Kuru Research Center at Moke (Okapa) unless otherwise noted. Otherwise, we have preserved the letters as they were written, modifying punctuation and spelling for clarity only. More extensive editing was required with DCG's writings than with those of the other correspondents, but nowhere has the substance of a document been altered.

We have standardized the spellings of local place-names to those in general use at the end of 1957. Some names have since come to be spelled in a different way, but for the sake of historical accuracy we have elected to retain the 1957 spellings. Thus, for example, in the early period of government contact the Gimi linguistic group was known by government officials as the "Kimi"; early kuru publications used this spelling, as did most of the letters from the kuru region. Frequently a particular name was spelled several different ways, a predictable situation where non-literate languages are being transcribed by a variety of people. DCG's writings, especially the field-journal, contain many parenthetical variations in the spelling of local names, as well as considerable discussion of their correct pronunciation and orthography. Some examples of these alternate spellings and discussions of nomenclatures can be found in the present volume; but since this problem is only peripheral to the focus of this book, most names have been given in the standardized form only.

The cultural and linguistic groups in the New Guinea highlands were not true tribes; social organization in the Eastern Highlands seldom extended beyond the unit that we have most often called a "village." Residents of these isolated groups of hamlets were rarely aware of the full extent and boundaries of the lands inhabited by those sharing their language and culture. This situation has resulted in the assignment of linguistic-group names by outsiders, which names can refer to a people, the regions in which they live, or the language they speak—often without clear contextual distinction being made among them. The word *Fore*, for example, can refer not only to the language but to its speakers, their lands, and their cultural groupings. In the excerpts from the field-journals included in this volume, village names are often followed by the name of the cultural and linguistic group to which their inhabitants belonged.

Our use of the term *Kukukuku* (and the shortened form *Kuk*) for the linguistic groups in the Eastern Highlands and Morobe Provinces, east and southeast of the

EDITORIAL NOTE

kuru region, requires special explanation. This term was universally used for these groups in 1957, but due to its pejorative connotations it has been replaced since the mid-1960s by the name *Anga*. We have not altered its earlier usage here, however, in keeping with our convention of preserving the usages of the period in which these documents were written.

In 1957, the Territory of Papua and New Guinea was governed from Port Moresby by the Australian government through its Minister for Territories in Canberra. After World War II, the Territory of New Guinea was delegated to Australia as a trust territory by the United Nations; its administration had been previously ceded to them after World War I by the League of Nations, and they had been governing it since then as the Mandated Territory of New Guinea. By the early 1950s, the administration of the Territory of Papua (annexed by Australia in 1884) had been combined with that of the Territory of New Guinea. Thus, in 1957 Australia was administering the two territories jointly from Port Moresby; but separate annual reports were issued, and the distinction between them was not entirely lost. References herein to "the Papuan side" reflect these historical differences; but in this book wherever the word "Administration" is found, it refers to the consolidated Territorial administration in Port Moresby.

At the time that kuru research began, the Okapa Patrol Post (founded in 1956 by John Coleman) was staffed only by a contingent of the Royal Papua and New Guinea Constabulary. This was a period during which Australia was trying to bring its officially designated Uncontrolled Areas under its administrative aegis, suppressing intervillage warfare and, where it was practiced, cannibalism. This was brought about—usually without recourse to force—by the introduction of Australian territorial courts of law and British-style justice, as well as by the strict use of annual censuses of every village. Most of the New Guinea highland groups found these newly introduced institutions relatively congenial. The North Fore area was declared a Controlled Area in late 1957 and was opened up ("derestricted") to missionaries, government personnel, miners, and other Europeans; but the South Fore and Gimi remained Uncontrolled Areas until they were "derestricted" late in 1958. Thus, all our work in the kuru area required special government permission in 1957.

The cine and still photographic records mentioned in the letters and in the journal have been preserved and cataloged in the Archive for the Study of Child Growth and Development and Disease Patterns in Primitive Cultures under DCG's direction at NIH. The photographs in this volume have been selected from that collection and are identified by their Archival numbers. Those still photographs identified by a number starting with 57- or DCG- were taken by DCG, while those identified by a number starting with LH- were taken by Lucy Hamilton, now Mrs. Jack Ried. The 16mm cine footage of kuru has been compiled into a Research Special Film that we described in a 1970 paper in *Brain*; it is on deposit at the National Library of Medicine.

Materials are arranged chronologically according to date of writing, with minor deviations occurring as a result of separating the letters into "General" and "Bush" correspondence and in keeping the field-journal excerpts separate from the letters that were being written in other parts of the world while DCG was on patrol. The List of Contents should serve as a guide in locating particular materials.

We wish to thank Marta Nicholas, Steve Ono, John Runman, and Raul Zaretsky for their unstinting assistance in preparing the manuscript.

J. F.
D. C. G.

Introduction

D. C. Gajdusek

With the first report—in 1963—of the transmission of kuru to chimpanzees from cell-free suspensions of human brain, kuru became the first human disease proved to be a slow virus infection, thereby stimulating the search for such etiology in other chronic human diseases. By 1967 a worldwide form of presenile dementia, Creutzfeldt-Jakob disease (CJD), was proved to be caused by a similar virus which remains to this day indistinguishable from the virus of kuru except for unstable biological properties of host-range, incubation period, and virulence. The viruses causing both diseases have since been found to have unusual properties similar to those of the scrapie virus of sheep and goats.

With the disappearance of kuru from the youngest age-group in 1964, and then progressively from older age-groups, cannibalism of those who had died of kuru was established as the unique means of dissemination of the disease—involving the contamination particularly of women and children, who preferentially participated in mourning rites and the associated cannibalism. Transmission being interrupted by the cessation of cannibalism during the 1950s, no one born since then has developed kuru. Kuru should thus disappear completely in the next two decades.

The usual mode of transmission of CJD remains unknown. Iatrogenic infection from contaminated surgical instruments (on which the CJD virus has resisted usual means of sterilization), and transmission by corneal transplantation from a CJD-infected donor, have been established. Furthermore, the discovery of dozens of families in which members in three or even four generations have developed CJD at approximately the same age has again presented the problem of genetically determined susceptibility to infection with these viruses, which has remained a problem in the study of kuru and scrapie.

Finally, the viruses of kuru, CJD, and scrapie have been found to be alike and to possess very different properties from any other group of viruses—so different that many virologists prefer to consider them a new category of microbial agent rather than true viruses. Their intriguing atypical physical and chemical properties have already demonstrated that these are viruses with smaller genetic information than that of any other microorganism, and that they are the only animal viruses lacking non–host-specified protein sub-units. Thus a new group of investigators interested in the basic molecular biology of the replication of macromolecules has entered the field.

This volume does not attempt to be a history of the study of slow virus infections or of the study of the biology and structure of these unconventional viruses. Instead, it allows the reader to trace the diverse origins and meandering paths of ideas, recording something of the personality conflicts and compatibilities and the varied and divergent devotions that led to the discovery and first investigations of kuru. The events recorded in these letters and field-notes inspired and shaped all future work on slow virus infections of man, and led to a major breakthrough in the study of chronic and heredofamilial degenerative diseases of the human central nervous system.

I have always been intrigued with the social organization of science, the derivation and interdependence of ideas, and the conflicting motives and methods of the investigators involved in this shared quest for intellectual understanding and professional recognition. Early in the history of kuru investigations, we at Okapa and the National Institutes of Health (NIH) began to speak of "the disease people caught from studying kuru," and jokingly described it as more interesting than kuru itself. I thus recognized the desirability of preserving much of the written record arising from the discovery of this strange disease and its "secondary infection" of involvement-in-kuru-investigation.

The letters and field-notes compiled herein document, in great detail, the origins and pursuance of the first year of study of kuru. They speak for themselves, and the informed reader will see that many important subsequent developments are prefigured in the scientific exchanges of this period. These records should, moreover, dispel any misconceptions about who did what—as well as answer many questions about how, when, and why the early field and laboratory work was undertaken.

I first entered the kuru region in March of 1957 with Dr. Vincent Zigas, the Medical Officer at Kainantu in the Eastern Highlands District of the Territory of Papua and New Guinea. We were transporting back to their homes two kuru patients whom Dr. Zigas had earlier brought to his sub-district hospital at Kainantu, north of the kuru-affected region. These were the first two kuru patients I had seen. While I was a guest for a night at Dr. Zigas's home, he convinced me to alter my plans and to travel with him south from Kainantu to the Okapa Patrol Post. Though I had intended to meet Patrol Officer Ian Burnet at Lufa Patrol Post on the western shoulder of Mt. Michael (a 12,500-foot peak dominating the area south of Goroka, Henganofi, and Kainantu—the three administrative centers on the main Highland Road), I abandoned those plans to accompany Zigas to Okapa, east of Mt. Michael.

I had heard of kuru in Port Moresby from the new Director of Public Health, Dr. Roy Scragg. From the first accounts of the disease Dr. Zigas had sent in 1956 to Director of the Public Health Department (P.H.D.) Dr. John Gunther (who early in 1957 became Assistant Administrator of the Trust Territory), it appeared to be an epidemic of encephalitis-like disease. We were thus thinking of infection, and particularly virus infection, before I had ever seen the disease.

I had just left the Walter and Eliza Hall Institute of Medical Research (WEHIMR) in Melbourne, Australia, where, as a visiting investigator, I had gone (from previous training in the laboratories of Linus Pauling and Max Delbrück at Cal Tech, John Enders at Harvard, and Joseph Smadel at Walter Reed) to work with Dr. Frank MacFarlane Burnet on influenza-virus genetics and infectious hepatitis virus. Instead of doing much work on flu or hepatitis viruses, during my one-and-a-half years at the Hall Institute I discovered auto-immune antibodies in sera from chronic hepatitis, disseminated lupus erythematosis, and multiple myeloma patients—which antibodies were individually specific in each patient. I had submitted for publication three papers on the auto-immune complement fixation (AICF) test which I had used to estimate the auto-antibodies; Sir MacFarlane Burnet subsequently used these data in formulating his clonal selection hypothesis of antibody formation. I was on my way back to Harvard via the Territory of Papua and New Guinea, where I hoped to continue my studies of child growth and development and disease patterns in primitive cultures; this project had already taken me to visit several groups of aborigines in Australia as well as Melanesian peoples in New Guinea and New Britain in 1955 and 1956. On hearing of kuru from Dr. Scragg, I decided to go through Kainantu and visit Dr. Zigas, who had first alerted the medical service to its existence.

Vincent Zigas's enthusiasm and Eastern European extravagance matched well with my own. From what he told me and from what I saw of the two kuru patients at his Kainantu Hospital, it appeared that kuru was most likely to be an epidemic of encephalitic disease, or at least a post-infectious phenomenon akin to the post-encephalitic parkinsonism that had appeared following the von Economo's encephalitis epidemic after World War I. Before I ever entered the kuru region with him, our thinking was prejudiced in favor of an infectious etiology. We even delayed our departure to obtain buffered glycerine in which to store autopsy tissues for virus studies—the classical way of attempting to preserve a virus in tissues in the absence of refrigeration or of animals, chick embryos, and tissue cultures that could be immediately inoculated. A few months earlier, Zigas had succeeded in getting a brain specimen of a kuru victim at autopsy, and had sent it and 26 serum specimens from kuru victims to the Hall Institute for virus studies. There S. G. Anderson obtained no virus on chorioallantoic inoculation of chick embryos and on intracerebral inoculation of suckling and adult mice. When we entered the kuru region, we brought with us equipment to do further autopsies and to collect specimens for extensive microbiological studies, especially serological and virological.

The first months of kuru research, and the first clinical and laboratory investigations and autopsies, were carried out with the hope and expectation of finding an infectious pathogen. It was the scientific study of kuru that made this first assumption seem less and less tenable. In the mid-1950s, we knew of no brain infections with helminths or protozoa, bacteria, spirochetes, fungi, rickettsiae, or viruses, which did not produce inflammatory change in the cerebrospinal fluid (CSF) as evidenced by a pleocytosis and an elevated protein, and which failed to evoke a febrile response and an associated neuropathology suggestive of infection. We soon found that none of these phenomena occurred in kuru. Furthermore, an exhaustive search early in 1957—with the aid of S. G. Anderson, Lois Larkin, and others at the Hall Institute—for a transmissible agent through inoculation of suckling or adult mice and of rats, guinea pigs, chick embryos, and tissue cultures, produced no positive results. We reported this early negative experience in our first paper on kuru in the *Medical Journal of Australia* (Zigas and Gajdusek) in 1957 and again in *Nature* (Gajdusek and Gibbs) in 1964. This search for a conventional virus agent, however, preceded even the first publication of a description of kuru.

We are often asked when and how we first came upon the idea that cannibalism was involved in the spread of the disease. It is useless to speculate about the origin of this idea; I know of few Europeans who did not arrive at such a conjecture. All the missionaries, traders, miners, and government workers and their families in the Eastern Highlands knew that most of the indigenous peoples in that area had been cannibals—mostly endocannibals, eating their close relatives in mourning rites—prior to and at the time of first missionary and government contact. Missionaries and government workers had engaged in a vigorous campaign to discourage the practice, and the government often imprisoned people for continuing to eat their dead kin after administrative control had been established. Europeans thus often suggested that "cannibalism probably spread that disease."

Under these circumstances, we certainly did not think it was necessary to write about it, having no evidence that the disease was infectious (despite our initial expectations) when we first described it. It was obvious that were kuru infectious, the opening of the bodies and close contact with them would be a source of infection. Even today, however, it seems unlikely that eating the bodies caused the spread of the disease; feeding huge doses of kuru and CJD viruses to experimental chimpanzees has not yet produced disease. It is, rather, hand contamination and the consequent inoculation through direct contact with mucous membranes and minor skin lesions that probably caused most infections in women (who were those chiefly involved in opening and cooking the bodies) and in the infants and small children in their care. Most laymen who knew that the people were cannibals, seeing us in close contact with the kuru patients and watching us do autopsies, regularly speculated that we might also become infected. Laymen often make such conjectures regarding cancer and other diseases that are not demonstrably infectious; and doctors regularly dismiss these suggestions as unlikely as long as there is no evidence that a pathogenic microorganism is the cause.

It is, however, strange that modern historians and scientists interested in the origin of ideas should ask how such a hypothesis arose. Such ideas are universal property today, since we have the idea of infectious disease and the knowledge of infectious microbes. But even centuries before Pasteur, a less precise concept of contagion led people to flee from plague and leprosy victims, and to burn their houses from fear of contamination.

The real question is not how we came to speculate that the disease might be infectious or that cannibalism might be involved in its spread; these were facile conjectures for which there was no scientific support at the time. Much more compelling for us than popular speculations about cannibalism was Dr. William Hadlow's suggestion, in 1959, that scrapie, an infectious chronic disease of sheep and goats, was clinically and pathologically quite similar to kuru; this suggestion led us to consider scrapie as a model for the kuru situation. When we became aware in this way that viruses could produce a chronic neurological disease, we obviously pursued this lead experimentally with animal models. If kuru were infectious, it was obvious to us that cannibalism would be implicated as a mode of transmission; the real question, however, was whether it was infectious in such a way that standard indices of infectivity were inadequate to demonstrate its transmissibility. All our early investigations of the infection hypothesis had given negative results, and the phenomenon of cannibalism in many regions of New Guinea, including areas adjacent to the kuru region, seems to have been medically innocuous. The dreadful and decisive accident appears now to have been the rare intrusion of a Creutzfeldt-Jakob–type virus into a community practicing this sort of cannibalism. Had the incubation period been short—a few days or weeks, as with most viruses—the people themselves would undoubtedly have become aware of the relationship and successfully avoided both the practice and the disease. The long silent incubation period of the virus made the correlation of exposure to human tissue with the occurrence of the disease a problem for them and for us.

It is more interesting to inquire how, in the course of cautious medical investigation, scientific inquiry led us away from the obvious initial expectation regarding infectious etiology, even though in the end that expectation was borne out. I have explained that initial scientific study of kuru had dissuaded us from the infectious hypothesis because of the absence of fever, the absence of pleocytosis or elevated protein in the CSF at any stage of the disease, and neuropathology that revealed no significant inflammatory response such as perivascular cuffing with mononuclear cells. In the absence of any evidence of infection, it became very difficult to see how cannibalism might be involved. At one rather casual stage of hypothesizing, we wondered whether cannibalism without infection might be involved by a mechanism of hypersensitivity—perhaps in children who, after consuming human brain via the immature infant gut, might then develop a late reaction of allergic encephalitis on subsequent reexposure to human brain. Since this was not an immediately evident hypothesis, I published it in a lecture; anthropologist Robert Glasse has quoted this as the only mention of cannibalism he found in our published writings. That this should have been the case, however, indicates not the prior absence of the idea, but rather its complete obviousness.

Slow nonvirulent infections of the brain remained on our minds as gnawing possibilities (particularly mycobacterial, yeast, fungal, and protozoan infections); but in such infections, inflammatory response is usually severe and the microorganisms readily visualized. Nor could we dismiss a process similar to general paresis or post-encephalitic parkinsonism. In the former, the *Treponema pallidum* was often difficult to find; and in the latter, no virus had been demonstrated. Dr. Webb Haymaker of the Armed Forces Institute of Pathology had in 1958 impressed upon me that we still had to consider the possibility of a parasitic infection with an unusually long incubation period; the parasite could be difficult to see in the central nervous system, and might provoke only minimal inflammatory response. A case in point was African sleeping sickness, caused by trypanosomes that had only rarely been visualized in neuropathological material. For this reason, in 1958 at the NIH we were already seeking more carefully collected frozen tissue specimens to inoculate animals again, but this time for long-term observation; and so we were trying to secure chimpanzees and other primates for the work.

Bill Hadlow's suggestion of an analogy between kuru and scrapie came as a surprise to me and to all of us in kuru work. In 1960 I visited Hadlow and W. S. Gordon at Compton, England; J. T. Stamp at the Moredun Institute in Edinburgh; and Pall Pállson at Keldur, Iceland—all of whom were studying slow and latent viruses of sheep and goats. I came back with the conviction that we had to urgently pursue our planned inoculations and long-term observations of animals, especially primates. Thus, in 1960, we returned enthusiastically to the infectious disease hypothesis which, although never completely abandoned, had seemed increasingly untenable. At the Tenth Pacific Science Congress (Honolulu, 1961) I presented a paper entitled "Kuru: An Appraisal of Five Years of Investigation, With a Discussion of the Still Undiscardable Possibility of Infectious Etiology." In that paper I stated clearly the trend that our work had taken in response to the impetus of

Hadlow's observations. The lecture was reproduced in mimeographed form and later published obscurely; the last few paragraphs are thus worth quoting here as a record of our thinking at that crucial juncture:

> In spite of all the genetic evidence, both the pathological picture and the epidemiological peculiarities of the disease persistently suggest that some yet-overlooked, chronic, slowly progressive, microbial infection may be involved in kuru pathogenesis. Similar suspicion prevails in our current etiological thinking about a number of less exotic and less rare chronic progressive degenerative diseases of the central nervous system of man. Thus, while we still prefer to consider multiple sclerosis probably an example of a hypersensitivity phenomenon associated with a disturbance in the immune mechanism, amyotrophic lateral sclerosis, Schilder's disease, leucoencephalitis, Kozhevnikov's epilepsy syndrome in the Soviet Union, the Jakob-Creutzfeldt syndromes, acute and chronic cerebellitis, and even many forms of parkinsonism (especially the parkinsonism-dementia encountered among the Chamorro population in Guam) continue to suggest the possibility that in man there may be infections analogous to the slow infections of the nervous system of animals which were intensively studied by Bjorn Sigurdsson, the Icelandic investigator who formulated the concept of "slow virus infections." *Scrapie* in sheep in England and Scotland, and *visna* in sheep in Iceland are the two most dramatic and well-known such infections—although *rida* (an Icelandic form of scrapie), *maedi* (a lung disease of sheep in Iceland), and the infectious adenomatosis of sheep's lungs are additional examples of transmissible slow infections. In human virology we have not established a counterpart for these phenomena, unless it be the slowly progressive infection of lymphogranuloma venereum or of the common wart.
>
> Latency in virus infections is a well-known phenomenon from the lysogenic coliphages through Herpes simplex virus in the mucosae in man. The slow infections are different in character from other viral diseases in that the viral agent continually proliferates slowly—causing a progressive pathological lesion, either without eliciting an antibody response or in spite of such a response. It is thus remarkable that Hadlow noted an extreme similarity between the epidemiology, the clinical and the histopathological picture of kuru in man and scrapie in sheep (or in its transmitted form to goats). Since this at first seemingly highly unlikely analogy was drawn, the clinical and pathological and epidemiological similarities have become even more apparent. It is now established that scrapie behaves as a genetic trait in sheep in the British Isles. Likewise, it is now established that a very unusual filterable, transmissible, remarkably heat-stable agent—which must be accepted as a virus—is capable of producing the disease when inoculated into the central nervous system or peripherally into sheep or goats. Marked differences in genetic susceptibility of different breeds of sheep to transmissible scrapie have been clearly demonstrated with susceptibility varying from 0 to over 80%; the goat is almost 100% susceptible. Thus, we have the strange phenomenon of a transmissible agent, clearly fitting the definition of a virus, producing a disease which in nature behaves as a recessive genetic trait. In recent months the work of Gordon, Pattison, and Chandler with scrapie has resulted in the establishment of what appears to be the same infection in laboratory mice—with a much shorter incubation-period than the 1 to 5 years encountered in sheep, or the 7 to 12 months in goats. Thus, a laboratory model for study is now available to us. Likewise, in visna, the viral and etiological agent has now been isolated and cultivated by Thormar in tissue culture of ovine choroid plexus cells, or ovine and bovine kidney cells, in which it behaves with a rapid, cytopathological effect.
>
> In view of these stimulating successes in veterinary virology, we no longer can ignore the possibility that kuru (or for that matter, the other syndromes mentioned above) might not be analogous slow infections—this, even in spite of the convincing genetic model which can be constructed. With this in mind, we have launched an intensive program aimed at exploring this possibility. As with the cancer-producing viruses, the possibility of similar agents involved in human disease must be considered. In kuru, as in a number of other chronic, progressive degenerative diseases of the central nervous system, we are conducting long-term transmission experiments to higher primates and simultaneously pursuing attempts at isolation of a transmissible agent in tissue cultures,

embryonated hens' eggs, and small laboratory animals. We are, furthermore, studying the virus causing slow nervous-system infections in animals as models in our laboratory.

By 1963 we had inoculated chimpanzees and monkeys with tissues from kuru and from many other subacute and chronic degenerative diseases of the central nervous system, but had as yet seen no illness develop in the animals under observation. Our work on scrapie and other slow viruses had progressed so far, however, and our conviction that kuru was going to prove to be a scrapie-like infection had advanced to such a degree, that I and my NIH collaborators Dr. Clarence J. Gibbs and Dr. Michael Alpers organized an international symposium on slow virus infections. Papers from this meeting were published as a monograph in 1965. Between the dates of the meeting and the publication of the monograph, the first two inoculated chimpanzees developed clinical kuru in our laboratories; this news was published as an addendum to our symposium paper.

Since that time, research on slow virus infections has grown, and it now involves numerous investigators and laboratories. This is not the place to trace the full history of that work nor to assess my own role and that of my colleagues at NIH and collaborators around the world in the evolving awareness and understanding of slow virus infections. This book does, however, provide an opportunity to acknowledge the valuable assistance and encouragement that Vin Zigas and I received from many quarters in the first year of kuru study, and perhaps to repay some of those old debts by giving credit where credit is due for considerable hard work in frequently inconvenient circumstances. As for the subsequent history of kuru and slow viruses, the copious and still-growing literature in the field makes the substantial and indispensable participation of its many authors very clear.

Sir Frank Macfarlane Burnet and Sir John Gunther have made an enormous contribution to this volume by permitting us to use letters from their own files, including some letters that I wrote but of which I had no copies. Lady Burnet located many of the missing letters in the Burnet files on kuru deposited in the Hazel Jenkins Library, Department of Microbiology, University of Melbourne, and we are grateful to Professor F. O. White for access to these letters and permission to use them. As is evident here, I am indebted to Sir Mac for his generous intermittent encouragement and support during the first years of kuru investigation and for his continued interest and friendship in subsequent years. Dr. Gunther, after an initial embarrassment at having a foreign "interloper" working on the kuru problem, gave me considerable support and encouragement thereafter and continued to play a key role in the organization of kuru research in New Guinea for some time.

Since the chief concern of this book is to provide a record of the original investigations of kuru, and since it confines itself to the period from February 1957 to January 1958, I wish to render special thanks to the few people whose help in that year was very great but who have for one reason or another received scant notice in the pages that follow:

Miss Lois Larkin, microbiological technician at the Hall Institute, worked and patrolled with me in the kuru region for over a month in 1957 (and as a result of her visit to Okapa, became Mrs. Jack Baker); for most of the year she coordinated the receipt and handling of many blood and other tissue specimens in Melbourne. Miss Elizabeth Jackson managed all correspondence and receipt of specimens and field data at NIH throughout 1957. Mrs. Gloria Zigas provided for me (and for many of our visitors to the kuru region) accommodations, food, friendship, and support during our many passages through Kainantu. The late Dr. Bill Smythe joined us for several weeks to help with the collection of pathological specimens and with problems of linguistics in the area. The late Assistant District Officer Michael Foley of Kainantu—later District Commissioner—provided important logistic and administrative support to Vin Zigas and Jack Baker, and friendship and encouragement to us all, which greatly facilitated our work.

I am particularly indebted to the team of young New Guinean boys who lived and worked with me untiringly for the whole first year of kuru investigations, serving as translators to their various languages while they learned Pidgin faster than I progressed with their native tongues. They served as my field assistants and became our hospital orderlies, assisting with every task at our kuru research base at Okapa and on patrols through all parts of the vast region (some of which they explored for the first time with us). These invaluable helpers and friends include Wanevi and Tiu from Agakamatasa village and Anua from Ketabi-Purosa village, South Fore linguistic group; Masasa and Taka from Aga Yagusa, Tosetnam from Miarasa, and Aurika from Moke, North Fore linguistic group; Evesa from Henegaru village, Keiagana linguistic group; Aroia and Asoi from Hepavina village, Gimi (Kimi) linguistic group; Morieto from Arora village, Auiana linguistic group; and Waiajeke (Haus Kapa) and Agurio from the Moraei (Dunkwi) subgroup of the Simbari Anga linguistic group. Waiajeke, Wanevi, and Tosetnam are now dead after having made 15, 20, and 23 years, respectively, of further contributions to kuru fieldwork. They and the others became and have remained close life-long friends.

Native Medical Assistants Sinoko, Liklik, Oiya, and Tarangau contributed enormously to our construction and management of the Okapa Kuru Research Center and Hospital, and assisted us in every aspect of the investigations. Sinoko's and Liklik's wives were our first two female nurses for the kuru victims. Kosinto (from Ke'efu village) was translator at the Okapa station for the Keiagana, Gimi (Kimi), and Fore languages, and the late Muriso (from Pusarasa village) was translator for Fore; both of them were enormously helpful while we were struggling with the problems of the many new languages that confronted us. Finally, the officer in charge of the detachment of the Royal Papua and New Guinea Constabulary at Okapa, Sergeant Malekor, and his intrepid assistant Corporal Homeguei were very helpful in teaching me the problems and techniques of dealing with peoples with little or no previous contact with outsiders.

As is readily discernible from the pages that follow, Vin Zigas and Jack Baker were the sustaining pivots of kuru field-investigations, and their friendship and wisdom made possible my familiarity with the region and its people and the continuation of kuru investigations on a long-term basis. Above all, the publication of these letters is a tribute to the late Dr. Joseph E. Smadel, who had so much to do with the first year of our kuru investigation and who later provided me with the laboratory at the National Institutes of Health in which we were able to demonstrate the slow virus etiology of kuru.

I.
General Correspondence
26 December 1956 – 30 August 1957

ZIGAS to GUNTHER[1] Kainantu
26 December 1956

I have to inform you that on 22/10/56 I left station for Moke area to investigate a form of encephalitis amongst the Okapa people–returned on 12/11/56 with preliminary brief report as reads below.

A number of people were found suffering from a probably new form of encephalitis attributed by inhabitants to sorcery and called 'Kuru', the prominent clinical symptoms of which are as follows:–

Disease originally started with fever, somnolence, muscular pain and weakness, headache (mostly occipital), vertigo, occasional vomiting. As course progresses condition becomes far more pronounced giving the well-known condition of Parkinsonism with its mask-like face, flexed arms and wrists, unsteady walk, ocular disorders such as diplopia, strabismus, nystagmus, tremors of fingers and hands giving a cigarette rolling movement. In the late stage–no control of sphincters, increased W.B.C. and C.S.F. under increased pressure. Duration of disease approximately from seven to nine months, slowly progressive and usually ends in death.

During my stay at Moke no more than 27 cases were under my close observation for three weeks, during which time two cases died in coma. I also visited surrounding villages and found another 11 cases. I observed that no age is immune, female more affected than male and I would say in ratio 3 to 1. I sent 22 samples of blood sera and a brain to Dr. Anderson of Eliza Hall Institute in Melbourne and am anxiously awaiting the results of tests.

The present evidence that this disease is caused by virus is not absolute, only suggestive. It is also of interest to note that the disease is strictly endemic involving as area appr. 13 miles in radius. A speculation would arise as to cause of transmission whether it is direct or by insect and birds which are numerous in this particular area, especially bird of paradise and cockatoos.

I do hope, Sir, you will agree from the above mentioned it is worth further close investigation, especially when I was approached by a large number of influential natives begging me to rid them of this killer. I have to note that they are just in a stage of evolution and attributing a disease to sorcery. We just succeeded to clean

---

[1] This letter is quoted as it appears in Gunther's historical and autobiographical manuscript "Australia, Kuru, and a Nobel Prize."

this particular area from yaws which was prevalent and which they always thought was brought about by Hoodoo. They, in their primitive minds, now think that the doctor has power to beat their sorcerers.

I would like to ask your authority to give me the task of investigating this disease and I feel strongly that given time and close co-operation of Dr. Anderson that this disease could be eradicated [and] by that gain their confidence.

BURNET to GUNTHER   12 February 1957
Dear Dr. Gunther,

I have naturally been in close touch with Anderson in regard to the possibility of his undertaking an epidemiological investigation of the Okapa outbreak described by Zigas.

I may say at once that the Institute board and I are agreeable to Anderson doing this, the only qualification being in regard to the possible dangers from hostile native reaction. He is aware of the normal dangers of any such work but I feel that if, in the opinion of those knowing the region and the people, the particular situation is likely to give rise to serious additional danger from hostile action, I should not be justified in consenting to the project.

In any case I think the work would need to be done in association with someone on the Native Affairs side who was competent to deal with any situation that promised to become sticky. If the opinion is that no abnormal risk will be entailed and Anderson undertakes the work, I should like to know whether the Administration will accept any responsibility either directly or through insurance for death or serious disability incurred as a result. The Institute will, as in the past, take out an air-travel policy; but other risks should also be covered.

Yours sincerely,
F. M. Burnet

GUNTHER to BURNET   15 February 1957
Dear Sir Macfarlane,

On Monday, I am going to Kainantu where Dr. Zigas has at least one case to show me. I have had long discussions with our government anthropologist who has done special work among the Fore people, and he has no doubt that kuru is a clinical disease. These people apparently attribute all disease to "sorcery," and give us credit with having "magic" superior to theirs for conditions other than kuru. He tells me the area is closely controlled, and they are some of the friendliest people in the Territory and are greatly cooperative. They are intensely interested in getting rid of kuru. I specifically asked if they would submit to having blood samples taken and faeces examined etc.; he assures me they will. I could surely, therefore, allay any fears Mrs. Anderson may have. Nevertheless, I have sent your query to the Secretary for Law.

I will write again in about a week, after seeing the patients at Kainantu and discussing the problem on the spot.

Would you please tell Dr. Anderson I will reply also to him then.

With warm regards.

Yours sincerely,
J. T. Gunther

GUNTHER to BURNET [late February 1957]

Dear Sir Mac,

I referred your queries concerning compensation in the case of hostile attack and disease to our Secretary for Law. His reply to me is enclosed. [1]

2. You will see that the reply is not very satisfactory. However, in his final paragraph he points out that a special contract or a special insurance policy could be taken out by the Government in the event of Dr. Anderson investigating kuru.

3. You will probably be aware that I have now left the position of Director of Public Health and have handed over the follow-up of this disease to my successor, Dr. Scragg. I think Dr. Scragg is in touch with Dr. Anderson in the matter. The more we hear about it, the more fascinating the problem is.

4. I understand Gajdusek has gone to the highlands. Unfortunately I did not see him in Port Moresby. No doubt, though, he will be hot on the trail if he gets the opportunity. It is particularly interesting as your son is at Lufa which, of course, is next door to our problem.

With very warm regards,

Yours sincerely,
J. T. Gunther

SCRAGG to ANDERSON 5 March 1957

Dear Dr. Anderson,

Dr. Gunther, following his promotion to Assistant Administrator, has requested me to reply to your letter of the 12th February.

Since seeing your letter, I have paid a visit to Kainantu and examined fully three cases of the condition *kuru*, as well as having discussed the whole problem with Dr. Zigas. The clinical details are as follows:

CASE I was a woman aged over 40 who had been sick for more than four months. At this stage, her speech was already unintelligible to others who should have understood it. She had a fixed stare and some repetitive nodding of her head. There was no pill-rolling. There was slight spasticity but no cogwheel rigidity. The reflexes were slightly increased and the plantars were both flexed. Her gait had a wide base, with her arms fixed in a semi-flexed position, and she was extremely unsteady and only with difficulty could be encouraged to take even a few steps. There were no signs of any neurological condition other than the extrapyramidal. A lumbar puncture was done. The CSF pressure was not raised. There was no significant increase in cells, being only less than 1 per mm^3 of leucocytes. Sugar was present in 30–40 mgs% and protein was less than 100 mgs%. The haemoglobin was 82 sahli.

CASES II AND III were both women aged over 50 who had had the disease for about six months. Examination revealed the same as above, but both were bedridden in addition. Lumbar punctures were not done.

COURSE OF DISEASE: In response to questions, the following was elucidated. There is no evidence as yet of an acute onset. According to the native story, poison is worked and four months later the symptoms develop with headache, vertigo, and occasional falling over. They then gradually lose control with dysarthria, ataxia, parkinsonism, athetosis, and opisthotonos. Eventually rigidity becomes complete and, over the last two weeks of life, coma develops and they lose control of their sphincters. From the first symptom to death takes approximately eight months.

EPIDEMIOLOGY: The disease only occurs within 15 miles of Okapa, the population of this area being approximately 15,000. There have been 25 deaths in the last six months, and Dr. Zigas considers that there are five new cases and five deaths amongst old cases in each month. No cases have been known to recover. There are about twice as many cases amongst women as compared with men. The youngest case was aged about five and the oldest approximately 55. The disease has been in the area for about 30 years, and the natives say it comes from the

Kukukuku area, which is to the east over impassable mountainous terrain. However, in Goroka I was informed that a similar case in a child had been seen at Lufa (south of Goroka and separated from Okapa by Mount Michael, one of the highest mountains in the area). It was also stated that the same disease, under the name of *opa-fi*, was known in this Lufa area.

No person entering the Okapa area has ever contracted this disease. Dr. Zigas and Patrol Officer Coleman both requested that the native community work the poison against them, and this was done with obviously no ill effects. Recently, since the absence of the Patrol Officer, the body of at least one person dying of the disease has been cut into small portions and eaten (particularly by the children)—in their minds, as a prophylaxis.

In my opinion, the aetiology of the condition is one of the following:
(1) post-encephalitic syndrome,
(2) copper or manganese deficiency, or
(3) associated with some plant toxin similar to lathyrism.

In all, this is the most interesting condition that I have ever encountered in the Territory; and in view of the high mortality and the readily available postmortem material, I consider that no effort should be spared in determining if possible the aetiology and ascertaining whether any cure is available.

With regards to your visit, Dr. Zigas is overdue for leave but naturally it is essential that he be in the area at the same time as you are there. He suggests that the month of April, when the wet season has started to abate, would be the best time; but as soon thereafter as possible for you would be convenient. A house is available at Fore, and Dr. Zigas—who has the complete confidence of the people in the area and would be able to take the brunt of any possible resentment (which, however, he does not think would occur)—desires to stay with you all the time you are in the field.

A new Patrol Officer has also arrived there. Patrol Officer John Coleman, who has lived in the area for two years and is the man most in touch with the anthropological and epidemiological aspects of the problem, is at present in Sydney doing a course at the Australian School of Pacific Administration, Mosman. Dr. Gunther has stated that if you would like to consult with him, approval will be given, on reference to us, for Mr. Coleman to visit you in Melbourne.

I have enclosed the report of Mr. Charles Julius, the Territory Anthropologist, on his investigation into Sorcery in the Fore area.   [See Appendix 1.]

Dr. Gunther has requested legal advice on the possibilities of attack by tribesmen and will be writing personally to Sir Macfarlane Burnet on this matter.

Yours sincerely,
R. F. R. Scragg
Acting Director of Public Health

I. BURNET to F. M. BURNET and family 11 March 1957
Dear Family,

I have just returned from my first patrol, and have been very busy catching up on affairs on the station and getting ready for another short patrol near Goroka.

I received Pa's letter and the follow-up request for Gajdusek's picture. I have been out of contact with Goroka for over a month now, and so I don't know whether I have missed him or not. If not, all will be well because I now possess a super Paxette with interchangeable lenses, built-in rangefinder, etc. I have nearly finished a reel of Kodachrome and can't wait to see the results.

The patrol was rather uneventful. I was accompanied by a European Medical Assistant and an Agricultural Assistant. The EMA supervised the giving of about 12,000 injections of penicillin as part of a yaws eradication campaign. We saw only about 20 active cases, which made the whole thing seem pretty futile. Dr. Symes came out from Goroka and enquired after cases of Parkinson's disease, which he said were currently thought to be the aftermath of encephalitis. Is it? In any case, we found about three cases, which was significant because Parkinson's disease had only been found in the Okapa area before.

I was sorry I did not have the camera on this patrol, because there was some real photographic potential. We had a line of over a hundred carriers, spread out over nearly half a mile. At each rest house, an offering of food was made. This usually consisted of about half a ton of *kaukau*, potatoes, tomatoes, maize, etc.; also about six chooks [chickens] and an occasional pig. As a result, on the station I now have about 60 chooks and four pigs and a goat. The cost of living here is not high.

The natives in this area are about the most advanced in my control, but by Manus standards they are pretty bushy. Many still hide from census and have to be rounded up by police-*bois*. When an old man was jailed, they usually offered a stand-in for him as bribery by way of a large pig.

Apart from census, the main point of the patrol was to encourage the progress of a new road from Kami (about 15 miles from Lufa on the road to Goroka) to Okapa. In Manus, one would have gone round making speeches and achieving very little; up here it was merely a matter of marking the route and putting a Corporal and two police-*bois* on the spot. The population is so dense that most of the 12 miles of road was completed in the first eight weeks. As most of the road had to be dug out of the side of the hills, that is pretty good going.

I have finally received my radio and other effects which I had sent up from Melbourne. Did Pa ever figure on Guest, or is his effort still to come?

Love,
Ian

DCG to BURNET, WOOD, and ANDERSON   Kainantu
                                                     13 March 1957

Dear Dr. Burnet, Dr. Wood, and Dr. Anderson:

What I write concerns you all and thus I address you all. I arrived in Port Moresby and started to plan out sites for potential child study and mentioned the possibility of seeing Ian Burnet at Lufa just at the time when Dr. Scragg (now acting Director of Health since Dr. Gunther is now promoted to Assistant Administrator) was dealing with correspondence on kuru, the disease about which Anderson and the Department have been corresponding at length. He told me the story, and I soon got all details of the anthropological study of the region and kuru sorcery from the State Anthropologist, Julius, and agreed to have a look at the Okapa area as well as Lufa as possible regions for long-term child study of the sort I am interested in as well as looking at kuru cases—especially those reported in children, which immediately place the disease outside of all the possibilities of vascular accidents, degenerative processes, genetic-controlled neuropathies, since most cases of these latter are adults. The more I read of the literature and correspondence, the more obvious was it that I had no intention of stepping into your project, but likewise I was somewhat disappointed that I had heard no word about kuru and its interesting and intriguing problems.

In an attempt to be of some help and service to you and the Administration, I have agreed to look into it at length; and I am rushing you a note now just after my arrival in Kainantu in order that it can get back to Goroka and on to you promptly, since any delay is apt to find this flooded airstrip closed for a week or more. Already in Moresby I voiced the opinion that standing beside a group of kuru patients I would be little better off than Dr. Zigas; I can give a thorough clinical pediatric study of the younger cases, check eye-grounds in children rather well, cerebrospinal fluids etc. a bit more expertly, perhaps, than the local men. But real study of such a progressive neurological disease cannot be properly done in Kainantu, let alone Okapa, and Goroka itself and Moresby are hardly much better. In the long run, from all I have heard, electroencephalography (for a post-encephalitic–type pattern which is rather definite), pneumoencephalography (which I have myself often done and would gladly do in Moresby but *not* Goroka), ventricular dye studies, etc.—along with prompt cell counts and chemistry on fresh CSF immediately examined, and such clinical chemistry as glucose-tolerance and endocrine-function tests—are *all* indicated in a complete workup of such an unusual new disease. In the long run, such studies could not be reliably or profitably done even in modern Lae or Moresby.

I thus already suggested in Moresby the possibility of sending an ideal case to Brisbane, Sydney, or Melbourne for study in a unit such as Dr. Wood's Clinical Research unit. This would yield, in the long run, far more information and far more reliable results at a far smaller expense than all sorts of half-hearted efforts at getting experts and equipment into the highlands—which will never really work out in the study of a progressive neurological disorder, at that! Naturally, the immense problem of shifting a patient to Australia was a daring suggestion I made only cautiously; but already Dr. Scragg, the Assistant District Officer, and the District Commissioner in Goroka, and other administrators were quite in accord with it, although some top-level decision would have to be made. Since arriving here an hour ago and talking at length with Dr. Zigas, my opinion is greatly strengthened.

He tells me that *all* (yes, astoundingly, all 100%) of 28 cases collected two to four months ago are now dead! Furthermore, cases dying out of Okapa in the hospital in Kainantu are removed from the culture but have so many relatives about that autopsy is refused (although in Okapa he can talk them into it), and a case really shifted out of the region would neither bother them nor would autopsy in the event of death disturb them unduly. In the community (where facilities are mighty poor) or in Goroka, Moresby, or Melbourne, autopsy might be less resented back home than it would be just outside the community in Kainantu.

Now, I am not suggesting accepting a classical early case in Melbourne on the Clinical Unit ward for autopsy purposes, but rather for clinical study and evaluation. [2]

My suggestion was to find, if possible, an early case in a young adult or older child (Zigas has had several fatal cases in 7-, 12-, 13-, 15-, 20-year-olds) and send this patient with perhaps one Pidgin English–native language interpreter, who could be much more easily handled in a Melbourne hospital than an older adult patient and who would adapt to it (beds, etc.) much more easily than the fixed, older adults. I hardly dare mention this suggestion, but just breathing the suggestion, Dr. Zigas jumped upon it with enthusiasm, and he is convinced that:

1. He can convince the community of the need for, and the family into willing acquiescence to, such a trip for a young adult patient without a family retinue required.

2. He can easily supply a young Pidgin English–speaking schoolboy or young man who can interpret Okapa language well in a Melbourne hospital.

3. The duet could be easily handled in Melbourne hospital if they were youths, young adults, or older children, as I suggest.

4. Repercussions would be insignificant back here in Okapa, since the disease is uniformly fatal sorcery to the natives now and *anything* which might counteract this sorcery he can persuade them to accept.

He was so pleased at the very idea, so sure that it could be worked from here locally, that the real crux of the matter now is you people in Melbourne. Could you give a Clinical Unit room to patient and interpreter? Would you be in accord with such an approach, and could you arrange intensive research-study and therapy (if any could be found) for a classical, typical case in a young adult or older child if all legal permissions, family and tribal assent, and Territorial sponsorship were arranged? Once that is known, then Zigas here—with my help for a week or two—can act, and it is quite probable that Moresby Public Health and Administration would be in full accord and arrange the matter. However, it would require at the outset information about your readiness to accept a case and study it, your interest in the possibility, and your desire for this, the only sensible clinical approach! Now, from the point of view of seeing the disease *in situ,* working out its epidemiology, and collecting blood, plants, possible vectors, etc., etc., a trip here to the site by Anderson, if he is interested, would certainly be in order. However, none of this is going to give the clinical leads that a good clinical workup will give, and I thus bring it to your consideration.

If this approach is of interest to you, let the Administration know immediately and—since I am now on the site and leaving for Okapa tomorrow—I can really hunt up the appropriate early case for either brief or long-term study in Melbourne. Even one or two days' study in Melbourne might yield more than a more costly study attempted in the highlands based on flying up all equipment, etc. I shall proceed with plans to bring a classical young patient, if one is found, with me to Goroka and Moresby; but the possibility of further transport to Melbourne with a companion interpreter is for you to decide. Should you answer *no* (a reply I fully expect, given the radicalness of the proposal), it changes nothing; should it be *yes*, Dr. Zigas and I can really get to work to find the proper case to rush to you promptly. Let me know care of the District Commissioner, Goroka, or through Public Health, Moresby; and initiate the formal request to Public Health from yourselves. They are fully prepared for it, and know of my presence on the site ready to select a proper case.

Toxic poisoning must be seriously considered; cases are overwhelmingly found in females—with fewer and fewer males being found, Dr. Zigas says. They are all dying, he says. I am going to see two in the hospital now, and then will proceed to Okapa to hunt up new cases. If we should decide to send a case *on your suggestion,* rest assured I will not send a dud—only a good early, promising case. Nothing I have heard sounds like an acute infectious disease or a virus disease. It sounds more like fatal anorexia nervosa plus hysteria to me; but the neurological symptoms, I am told, are *real*. I shall now have a look, and write further after I see. But this decision about whether or not you want a case for a few days only or for real long-term observation—either could be worked—is what you must make.

Further assurance. Aside from trying to help Dr. Zigas collect proper fixed pathological specimens, proper sera and bloods etc., I plan to do nothing but study the region for possible child studies of the type I am interested in. If you can use a case, I shall then do all I can with Dr. Zigas to get a proper one to you, and send nothing if no really good early case in a young person with full permission, etc., turns up.

Excuse the rush. There is no time for rereading. I have typed without looking at the paper or planning my remarks. The letter must go off now.

Sincerely,
Carleton

DCG to SMADEL

Moke, at the Okapa Patrol Post.
Fore, Kimi, Keiagana linguistic areas
Territory of Papua and New Guinea
15 March 1957

Dear Joe,

Enclosed is a somewhat self-explanatory letter [to Burnet, Anderson, and Wood; 15 March]. I am in one of the most remote, recently opened regions of New Guinea (in the Eastern Highlands), in the center of tribal groups of cannibals only contacted in the last ten years and controlled for five years—still spearing each other as of a few days ago, and only a few weeks ago cooking and feeding the children the body of a kuru case, the disease I am studying. This is a sorcery-induced disease, according to the local people; and that it has been the major disease problem of the region, as well as a social problem for the past five years, is certain. It is so astonishing an illness that clinical description can only be read with skepticism; and I was highly skeptical until two days ago, when I arrived and began to see the cases on every side. Classical advancing "parkinsonism" involving every age—found overwhelmingly in females although many boys and a few men also have had it—is a mighty strange syndrome. To see whole groups of well-nourished healthy young adults dancing about, with athetoid tremors which look far more hysterical than organic, is a real sight. But to see them, however, regularly progress to neurological degeneration in three to six months (usually three) and to death is another matter and cannot be shrugged off. If psychological, we have the very best epidemic psychosis for study in the world; and the Institute of Mental Health should be most interested. If a heredofamilial disease, we have the highest incidence-concentrated "epidemic" ever seen; and the geneticists should be on the ball. If a post-encephalitic or toxic or allergic encephalitic epidemic of basal gangliar disease is here, then all of medicine should be interested.

The Public Health Department in New Guinea is interested; and Anderson and the others at the Hall Institute have been corresponding with Dr. Zigas here, who has found the disorder over the past two years[2] and has seen it consume to a fatal outcome the child and adult alike—it is no chronic senile neurological degeneration!

I am now on the spot—in being in the Territories—and in the process of setting up further child growth studies in primitive cultures. Obviously, I have the study on my hands at the moment and, in any event, will be a collaborator in whatever goes on in the future. However, Joe, I stopped all my polio-fund salary when I left Melbourne in February, and since January will have no income until I finally get back to the States. I am now living off my small savings and have no grant at the moment, with plenty to do in a dozen fields, on my return; however, this is nothing to put off and nothing to drop. I write first to let you know about it, and second to ask whether there might be a grant you could dig up. I am in no position to make a more formal application, being devoid of mail, radio, or telephone and all else in the bush where I am working on the disease—but with astoundingly good native hospital, clean mountaintop jungle-surrounded native buildings to keep the sick, and sometimes even a serviceable jeep-track to import supplies and rations from the government. All I need, however, is enough to insure my keeping alive once I get out, if this runs to several months (as it certainly should if I can afford to stay). With no initial expenses of equipping an expedition or shipping off supplies or personnel, might there not be some Public Health Service grant or any other you could find to give me some sort of salary while I work here? I could already give you a set of papers on a truly new disease. If all this is impossible, OK.

May I send to you specimens and blood serum, etc., if it proves feasible? Most are now going to Melbourne. I would not presuppose this to be an infectious disease problem. Briefly, some fifty people (children included) have died of the disease in a small community within two to three years, all with the same neurological picture, afebrile and progressive; some four doctors beside myself have now seen it. We are all impressed that it is a new epidemic syndrome of some sort of parkinsonism or basal gangliar disorder (toxic, allergic, heredofamilial, post-infectious, infectious?? —or psychiatric?).

Although the people are currently still warriors and cannibals, they are well "under control" and very cooperative; language is no problem, for Pidgin English translators are available for each language and I am already at work studying the languages.

My auto-immune complement fixation test paper in *Nature* should be out any day now. The definitive paper was submitted to *JAMA* before I left, and a second paper with clinical correlations will now go in. A third, on the auto-antibody nature of some macroglobulins and on the characterization of organ specificity of antibodies and antigens in certain diseased sera of high titer, will soon be ready. It is work that I must go on with; but since I left, Sir Mac has converted most of the Institute work to this type of auto-immune study, and I believe (and, to be frank, fear) that my AICF test will soon be followed by a barrage of Hall Institute publications. I have in manuscript form a paper on skull-distortion in infancy and

---

[2]In fact, Zigas had not been in the kuru region long enough to have been observing kuru for two years. His 26 December 1956 report to Gunther appears to have been the first official notice he gave to his superiors of having seen the disease.

its effects on development, as studied last year among the Mamusi and Arawe "savages" in New Britain—and sundry esoterica. The current disease study on kuru is no esotericum, however.

[3]

Joe, I had planned to write about much and at leisure during my New Guinea fieldwork, but this has me now rushed beyond all hopes. Since I would like to be able to stick to it, I must request your consideration for any possible (even moderate) financial support while it goes on. The last two months have been without pay; I am by no means broke as yet, and—with or without help—I intend to stick this one out a bit. [4]

Sincerely and gratefully yours,
Carleton

DCG and ZIGAS to BURNET, ANDERSON, and WOOD

Moke (incorrectly called Okapa)
15 March 1957 (in the pouring rain)

Dear Doctors Burnet, Anderson, and Wood,

I realize that my last letter, written rather spastically in a great rush, calls for prompt follow-up. Since we are trying to get a jeep out (Dr. Zigas will go and I will stay) tomorrow, I am writing now, although landslides may have now blocked the road and made exit impossible for a few days.

At Kainantu there were two old women from Moke (called Okapa by mistake, although Okapa is really another settlement) who had kuru. They had been brought to Kainantu for Dr. Scragg to see when he visited Kainantu (he did not come to Okapa), and had classical advanced disease. When I saw them they were no longer ambulatory; and the tremors, athetoid movements, and blurred speech all pointed to a chronic neurological disorder unassociated with any acute infectious disease at onset (or even in the months or years before onset) which was dramatic enough to be recalled by reliable informants in their community. They were rational, but articulation of speech was very poor. Silly smiles, with grimacing, were prominent. Fixed and pained facies and slow, clumsy, voluntary motion (apparently in an attempt to overcome tremors and athetoid movement) were prominent also. They were carried to our jeep and managed to hang on, sitting up for the rough four-hour drive here. No fever during the course has been noted, and the story of headache and "*skin i ot*" is very variable and uncertain at onset. I was certainly not able to say anything but (?) parkinsonism—a rather apparent disturbance of the midbrain and basal ganglia in the older woman, with no real cause for believing it post-encephalitic.

Now that I am here, the picture is shaping up better in my mind. We have been collecting all the cases we can in the entire region. I am getting detailed anamneses, which requires a half-day of work on interviewing each case; and it is already apparent that the majority of cases do not have acute onset, and no acute febrile disease suggesting encephalitis in any way has been present in the region, as seen by patrol officers or by Dr. Zigas or recognized by the natives or early patients. As Dr. Zigas had noted, the picture is astoundingly uniform and distinct. Native diagnosis of kuru is as reliable as any modern medical appraisal would be. They know what they are talking about; and once you have seen a few cases, you also know. However, nothing in the course suggests microbiological studies, I fear. Instead,

Sydenham's chorea comes to mind first; secondly, various types of hysteria and motor aspects of psychoses, such as behavior of certain schizophrenics—but the hysterical analogy is closer, for rationality is usually good early in the disease, as well as late. As it progresses, neurological damage appears to be rather certain, but it is most difficult to pin down, now that I have been doing very complete neurological examinations. In fact, cerebellar function tests are very mildly disturbed, if at all, and the whole picture would be mesencephalic or basal gangliar.

The predominance in women is most marked: a ratio of between five and ten to one, perhaps higher. This is so certain that, coupled with the picture, one must think of heredofamilial neurological degeneration of which texts are filled, with dozens of eponymically designated syndromes. The incidence is spread over two linguistic groups—clans which often intermarry. A third linguistic group also bordering on this region of tri-linguistic junction may have had a few cases in villages directly adjacent to the kuru region, which appears to be rather small; but further study is required. The fact that parkinsonism has been seen in natives in Goroka was revealed to me at Goroka by Dr. Symes; but the story of kuru at Lufa is new to Dr. Zigas and he questions it, since he was there himself.

On arrival here, I concentrated on two classical cases in middle-aged women with all the features which have been described to you. This morning I did thorough physicals and neurologicals and found no truly abnormal reflex patterns: sometimes slightly hyperactive, other times slightly hypoactive—triceps, biceps, radial periosteal, tibial; no abnormal reflexes. One of the severe cases seen yesterday, unable to walk and with extreme tremors and athetosis, has been walking about most of the day without a sign of disease, and it is thus certain that rather advanced symptoms can reverse overnight at this stage. One naturally keeps thinking of hysteria or anorexia nervosa–type disorders (in this there is no anorexia); but the eventual outcome to real debilitation—tremor and athetosis, and speech disturbances—beyond reversibility is likewise certain. Thus, I prefer to believe that the overnight reversal I have so clearly seen in one typical case is part of the exacerbation and remission of a basal gangliar–type disturbance, with different levels of emotional excitation; the "remitted" patient still had tremors at one time during the afternoon.

Naturally, all this makes me fix my attention on my own domain, pediatrics, in which I consider myself rather competent as a neurologist. There are childhood cases which end fatally, I am told, and I have tried to see them. One child, a boy of about seven, had been carried here; he is obviously unlikely to survive for long, can no longer walk, has hardly distinguishable speech, and urinates and defecates in the house (although he is not dribbling or incontinent). As with all cases, he has to be fed now, having lost the ability to bring food to his mouth. It is hard to believe that he is a recent case—a boy previously of normal intelligence and physical development—but multiple reliable informants testify that he was walking, running, and playing normally only three months ago. His speech and growth and his development were apparently normal. The course is that described for classical kuru: i.e., one month of unsteady gait, followed by tremors and athetosis and blurred speech in the second month, and now in the third month almost complete incapacitation. He is a nice cooperative fellow—a bit too far gone, I fear, to consider taking to Melbourne, unless a doctor accompanies him.

Today a ten-year-old girl, intelligent and cooperative, was brought in in her second month, having marked tremors and instability which was marked with slight athetoid movements. I examined her completely. She had no adiadokokinesis, only slight disturbance of finger-to-nose, finger-to-finger, or heel-shin cerebellar function tests and negative Romberg. Her speech sounds okay, and natives assure me that it is fully rational though blurred. Optic fundi are normal, as are those of all seven typical cases I examined thoroughly today, and there is no papilledema, retinal exudate, cupping or blurring of disc, and no optic atrophy; no nystagmus or extraocular motion abnormality in this girl (nor in the other six); pupillary reflexes to light or accommodation are normal. Lumbar puncture on the girl, with immediate and repeated cell counts: RBC 40–50/mm^3 (fresh from slight trauma of tap); WBC 6/mm^3 (fluid clear and colorless); WBC checked by acid solution of RBCs 5/mm^3; Pandy's negative; heating (boil) and then Pandy's reagent, slight trace of protein. Thus, normal WBC in peripheral blood about 6000/mm^3; slight but not marked anemia. I shall try to get urine, etc. tomorrow. Before fluid was removed, dynamics were checked and found to be normal; and pressure was also normal.

This case is now in the second month. A most typical case, and as with most others there is no sign of encephalitic onset—although by questioning for hours, we have elicited the story of an acute stomatitis in the seven-year-old boy just before his onset (? herpes simplex), and in the girl now in question "some headache" and "hot skin" (which never really kept her from play) at onset (? did we put it there by the intensity of interrogation). That no one has seen an acute febrile disease they associate with the onset of kuru, and that we have found none, is now certain; there has not been an antecedent epidemic of any sort, and although rubella was said to have gone through the region, no evidence that these cases were preceded by rubella was pointed out, although it could have been. Thus, it seems further apparent that microbiological field epidemiology, although perhaps indicated, is not the primary nor most important line of attack at the moment. Good clinical and laboratory pathology is the current need in deciding what this is. Thus, I still urge high-grade hospital workup with thorough neurological consultation and long-term follow-up (a total course is three to six months to death in most cases, although I have now located "cured" cases of the once-classical kuru). I can do as much field neurology as any neurologist, I feel; but further accurate chemical, encephalographic and X-ray studies, and serial neurological observation with good neuropathology on autopsy material are the primary needs.

We shall do our best to secure whole brain, fixed in 10% formol-saline, on the next death. There is little to suggest pathology in any other system, but we shall try to get other fixed viscera also. Virus isolation material will also be sent, but it would appear highly dubious that anything microbiological would be found in these afebrile, chronic, progressive neurological degenerative conditions. Neuropathology, using every available technique, is essential; since most cases die, and those now on our hands probably will also, please send detailed and exact formulae and handling methods for brains, as determined by an expert neuropathologist. Many prefer long-term fixation before cutting of brains, fixation being on the whole brain. For silver impregnation, special fixation may be required. Please be specific, since reference works to check the formulae are not here. Should we find anything that even remotely suggests a possible kuru-antecedent infectious disease, I shall rush to you all possible tissue and washings for isolation studies. A year or two of residence might be required to settle the matter.

We have already located familial cases occurring years apart, with a mother and sibling (older sister) of the 10-year-old girl I have written about having died of classical disease five and three years ago, respectively. Please consider carefully the possibility of accepting in Melbourne this 10-to-12-year-old girl (her age could easily be stretched to 13 or 15 years for the purposes of Royal Melbourne). We think (?) we could persuade the parent (father) and the village to let her go, and we could send with her a very intelligent 16-to-18-year-old lad who is native Fore in speech, fluent in Pidgin and a good translator, and who wears clothes well. The girl herself is ambulatory, cooperative, and a good patient who could still travel easily. All evidence points to a progression beyond this favorable point within one to two months, if she remains here. Thus, faced with a fatal disease and with a young cooperative patient on hand, with about as early a case as one can have with kuru, this is the case for you if ever you want one. Please send us word directly; and if it is *yes,* start immediate arrangements with Drs. Gunther and Scragg, telling them of our progress, since you are better informed now than they. I shall write a brief report to the Public Health Department in Moresby now. Our work at the moment consists of plotting out the lineage and location of cases, the study of cases on hand and the hunt for new cases, as well as therapeutic attempts on those on hand: long-term doses of vitamin B complex, ascorbic acid, and chloramphenicol—truly a shot in the dark! If we can get it, we may try ACTH and cortisone. We have not recorded a single even slightly elevated temperature; no sedimentation rates have been done as yet.

Excuse this disjointed report. Dr. Zigas and I are rushed, and he states that he can remain only a few more months in the region before taking leave (i.e., at Kainantu). No kuru at Kainantu, but we shall certainly bring cases out and head for Moresby, if possible; best of all would be Melbourne, for where in Moresby are we going to get reliable pneumoencephalograms and especially electroencephalography (which can be of value in suggesting post-encephalitic damage), reliable metabolic studies, or endocrine studies? What you need is good neurology, laboratory pathology, clinical pathology, and neuropsychiatry. It would be cheaper in the long run to send you a good case; and in a child, it would be manageable—we are trying now to get dresses on her. The P.H.D. and Dr. Zigas would like to know of your plans. We can undertake anything you desire at the moment; and should any of you want to see the disease *in situ,* come up promptly.

Sincerely,
Carleton Gajdusek
Vin Zigas

DCG to SCRAGG [mid-March 1957]
Dear Dr. Scragg,

Dr. Zigas transported me promptly to Moke (Okapa Patrol Post) after showing me the two old women at Kainantu whom you also saw. I started out hunting vigorously for cases in children, since these are the most promising for study, I believe. Being a qualified pediatrician, I know much more of what I am about clinically and consider myself an adequate neurologist in the years below adulthood. This approach has already yielded us three classical cases in children, which we are studying intensively. We have at the moment 11 cases at the Moke Hospital.

I am trying to keep them all here on various controlled hit-or-miss therapeutic regimens, using whatever leads pharmacology might suggest.

Thus, the three children and two early cases in adult women are on high doses of chloramphenicol supplemented by good doses of ascorbic acid and vitamin B complex with folic acid. This is purely empirical, but it is a vigorous approach which must be tried. A second series of cases are on good doses of phenobarbital for its anticonvulsive effect—again a purely empirical trial. We have rounded up word of another dozen cases, and I am trying to assemble these also. What is most important is to observe these completely and fully throughout their course; and we are doing our best to keep the confidence of the Fore, Keiagana, and Kimi—the three linguistic groups involved. The full evolution of the disease must be fully observed, and thus we are holding all patients under these clinical therapeutic trials.

Other suggested courses of therapy which I should like to try, and will if medications become available, are:

1) Any of the modern tranquilizers used in psychoses nowadays, with dosage schedule on whatever can be supplied. Some six patients, at least, should be given these.
2) ACTH and/or cortisone, perhaps with antihistamines.
3) Testosterone—considering predominance (which now appears closer to 10:1 than 2:1) in women.
4) Other anticonvulsants:
   −Tridione is most wanted, since the 2–3 second frequency of the tremors approaches the 2–3 second rate of the petit mal triad which is specifically suppressed by Tridione.
   −Dilantin and/or Mesantoin.
   −Phenergan (in spite of its moderate toxicity), since it has been moderately active in basal gangliar, midbrain types of epilepsies, as I recall.

Although we have elicited a history of acute stomatitis (or yaws, but more likely herpetic) at the onset of one case and a few cases of "skin i ot" and headache as antecedent illnesses, we have *not* found this at all regularly, and have been—as you and Dr. Zigas were—totally unable to find any evidence of epidemic encephalitis, meningoencephalitis, or any exanthematous or acute infectious disease which might lead to a post-infectious encephalitic phenomenon. Rubella is said to be in the region, but we have not found any association with the cases as yet. Thus, I am most dubious about any microbiological studies which can now be made and about field epidemiology of a microbiological nature, although such studies should supplement other approaches. As I am most at home in microbiological fieldwork, I have written this opinion to Drs. Burnet, Anderson, and Ian Wood (at the Royal Melbourne).

As a first approach, I have strongly suggested that an attempt be made to bring one or more typical early cases to a good investigative center for further study. Electroencephalography, which is certainly indicated, is available only in Melbourne and Sydney and cannot be transported easily to the Territories. Pneumoencephalography is also indicated; and although I have often done it in the field when good stereoscopic skull films could be made available, I think it would be a waste of time lacking good facilities for stereos of the skull. Finally, what is most urgently needed is expert neuropathology of an entire, well-fixed brain, with all types of staining techniques and long-term intensive study. This will establish whether post-encephalitic changes, a demyelinating process, or some other toxic or degenerative process is involved—or perhaps that *no* pathology is visible at all. With this in mind, we are waiting for such material and will do our best to get such a brain. I would like word—and I have asked Anderson and Wood to consult a competent neuropathologist and secure the data—of *exactly* (with exact formulae, etc.) how brain material should be prepared for such study. Usually whole-brain fixation in 10% formol-saline is required before sectioning, if I recall correctly. However, for myelin sheath staining or silver impregnation other fixatives may be desired also.

There is no evidence as yet of disease of other systems, but we shall try also to secure specimens for visceral histopathology. Can you supply Zenker's and Bouin's, or some other such fixatives to us (besides the formalin, which we have)? Specimens for virus isolation studies will be collected for shipment to the Hall Institute; but in view of the definitely afebrile course, normal white blood cell counts, and normal cerebrospinal fluid examinations (I am doing lumbar punctures on all patients, with careful cell counts), it is hard to conceive of an infectious process operating at death (other than secondary).

[5] We are keeping full hospital charts of the Boston type as closely as possible on each patient; this may be gilding the lily a bit. Thorough ophthalmoscopy of all patients has failed to reveal any retinal changes.

At the moment we have three children on hand with the disease. One is far advanced and no longer ambulatory (a cooperative boy of about 7), while the other two (girls of about 10 and 8 who are in the second and first months of their illness, respectively) are completely cooperative and friendly and easily manageable. These two girls, along with a Fore-speaking lad of 16–17 who is a good interpreter, are my suggestion for the Royal Melbourne Hospital if you, Dr. Gunther, and the Administration would back it and could swing it. Dr. Zigas thinks that we can easily talk the Fore parents, guardians, and villagers into permitting the trip. The Administration, however, would have to settle the legal aspects, about which I know little. I have written to Drs. Burnet, Anderson, and Wood (director of the Clinical Research Unit of the Hall Institute at Royal Melbourne, which would be primarily concerned) about this possibility. I fully expect them to back out, faced with such an immediate prospect. However, these two children are *ideal* cases—the like of which may be difficult to obtain again—and easily manageable, so that the whole wild scheme might really be feasible after all. Native adults—so fixed in a pattern of life and behavior, and unadaptable—would obviously be an almost unmanageable problem on a Melbourne ward. These children would not. I have had them in our house here as patients. If this scheme is practical, please suggest it to Dr. Gunther and voice your accord to Drs. Burnet, Anderson, and Wood in Melbourne, who may shortly be writing to you about my letters to them. If it is not practical at the moment, I shall follow these patients here carefully, for I think we can do as much here as in Kainantu or Goroka, or even perhaps Moresby or Lae. However, if you would like thé patients in Moresby, I can bring them along when I come. Most important, however, is *haste* if the Melbourne scheme is feasible, for one case is now in the *earliest recognizable stages*—a really hard thing to find. We are now tracking down further cases, have records of some dozen more cases which we hope to see soon; and I am preparing a village-by-village, hamlet-by-hamlet list of current and past disease—with even an attempt at lineages, for the possibility of an inbred heredofamilial degenerative process must be considered. (High sex-preference, high family-incidence over the years—not by direct contact!—and much of the clinical picture strongly suggest this intriguing possibility.) We have not yet located any evidence of poisoning or deficiency.

Please excuse this rapidly typed, disjointed report. I am typing at Moke before making ward rounds. Dr. Zigas is due back from Kainantu today or tomorrow with further supplies and medicines, I believe. I shall try to get this progress note out to you as fast as possible—but when??

Sincerely,
Carleton Gajdusek

P.S. Dr. Zigas has returned. We now have thirteen cases under careful study and controlled medication; four are in children (aged 7, 8, 10, and 13), three of whom are ideal for Melbourne—and one or all of them could go, we believe. Dr. Zigas is convinced we could talk the local people into the matter without much difficulty. Seven are on chloramphenicol; full doses for five days will be our trial routine, although I have *no* faith in it in this condition. This is supplemented with vitamin B complex and C and supplemented diet. Four are on high doses of phenobarbital. Two are on acetylsalicylic acid only—old cases, the two women you saw. We have now done six lumbar punctures here, and all spinal fluids have been scrupulously counted immediately within a few minutes of the tap. Pressures were all normal, cerebrospinal fluid all clear, and Pandys all negative. There has been no pleocytosis—and this includes all of our early cases, some only two to three weeks since first signs of illness.

I shall try to complete the lumbar punctures on the other seven tomorrow, and get white blood cell and red blood cell counts and sedimentation rates done along with urines on all of them shortly. In addition we have a number of other studies in mind; and I shall try to ship out cold specimens of all the spinal fluid for Wassermanns, gold sols, chlorides, and total protein in Melbourne—along with a new large serum collection sometime later when we can arrange good consecutive air transport.

I think that virus isolation studies are not the first approach, and thus am not concentrating on isolation materials. There is no seasonal preferential incidence, no febrile course, no evidence of cerebrospinal fluid protein or cells, and thus no reason to suspect infection in these cases; and death is the end of a chronic process. We have been plotting out epidemiology carefully, and I now believe that my impression is shifting toward a heredofamilial illness, strange as this may seem. There are many cases with a family history of death from kuru years ago in siblings or mother. For this reason, Dr. Zigas and I would like [F. R.] Ford's *Diseases of the Nervous System in Infancy, Childhood, and Adolescence*, along with whatever most complete neurological text you have on the genetic heredofamilial diseases (included in general neurology, probably). I am thinking of such things as Wilson's disease.

In the last two days we have elicited valid histories of rather miraculous complete recoveries. Since one seems to have been in an advanced patient seen previously by Dr. Zigas and other reliable informants, we shall pursue this tomorrow. If true, it may shift our thinking toward a psychosis or hysterical type of disorder akin to anorexia nervosa or depressions in manic-depressive psychosis—states in which many pseudo-neurological damage signs and symptoms may be found. This shift in our thinking I am reluctant to jump at, but in the adult patients it is ever in the back of our minds as we observe how exaggerated their symptoms become under attention and how, in one patient, completely severe symptoms have regressed to minimal ones right under our observation.

Obviously, I am musing along in these keys, for toxic phenomena have not been excluded at all as yet. The psychiatric and possible genetic factors, however, may only add interest to the disorder, which is certainly the most impressive medical matter in this region.

I have written to my American colleagues requesting a bit of a further delay in my return, and have postponed my arrival in Dutch New Guinea so that I am free to continue our study until something more definitive is on hand.

We shall send this letter out to Kainantu by foot-carrier today in the hope it will make the next plane to you. Should Melbourne be willing to accept a case, let us know promptly. If you can get any of the supplies I have listed above to us, we shall be most grateful. Especially the anticonvulsants. Any further lab equipment—such as those items listed above—would be most helpful. We have no urinary hygrometer, a poor hemocytometer, no rectal thermometers (and only rectal temperatures are really valid in kids!), uncertain Pandy's reagent and Fehling's solutions, and no good lab manual for reference. Anything else you can think of, or which you believe we should check, would be of value. The motion-picture camera and film we requested by cable is for photographing cerebellar function tests and reflex patterns, tremors, Rombergs, etc. Please send any necessary suggestions as to its use, for although I am well-versed in Leica still-photography and optics, I have done little cine work.

Finally, we beg that you bring these rambling observations to Dr. Gunther's attention, for Dr. Zigas informs me of his past extreme interest in this matter and he may be influential in convincing Dr. Ian Wood to accept a case.

SCRAGG to BURNET
[telegram]   Melbourne, 19 March 1957. Reference Kuru Patient: Patient coming to Port Moresby. Please advise if you consider essential patient go in person to Melbourne or if forwarding of necessary specimens from Moresby would be adequate.

BURNET to SCRAGG   20 March 1957
Dear Dr. Scragg,

I had some difficulty in hearing all that you said on the telephone, so, in addition to sending the gist of my remarks, I will add my understanding of your side.

After discussion with various people in Melbourne, particularly Dr. Anderson, we decided that the chief reasons for removing a typical patient from the environment are:

1. To see whether this in itself has a beneficial effect, e.g., by eliminating chronic poisoning or allowing relief of some psychosocial tension.

2. To have an expert neurological assessment of the clinical condition, and to carry out such laboratory tests as seem appropriate.

3. In the event of death of the patient, to have a fully adequate postmortem examination made.

To some extent this could be done in Moresby, and we felt that it would probably be both wise and administratively unavoidable to retain the patient for some days at least in Moresby. This would allow a full clinical examination, and if a full report could be sent as soon as possible we should be in a position to discuss with you whether it would be desirable or imperative to send the patient to a major medical centre in Australia.

If it appears desirable for Dr. Anderson to come to Papua, he will be ready to do so at short notice, so that I feel it is desirable that appropriate authority for his visit should be obtained now.

We should be very glad to receive the fullest clinical notes on the patient, and I will undertake that they will be discussed at once with the most appropriate people here in Melbourne. If it is felt that certain pathological examinations which cannot be done in Moresby should be undertaken immediately, we would be glad to receive and deal with the specimens concerned.

My summary of your remarks is that the patient from Okapa is expected in Moresby on Friday, that he/she will be kept in hospital for the time being in Moresby, and that full notes on his/her condition, etc., will be sent to us as soon as possible, that a Patrol Officer at present in Sydney will be authorised to come to Melbourne to discuss the tribal background with Dr. Anderson, and that you will approach the Minister for authority in regard to Dr. Anderson's probable visit.

I am most grateful for your help in the matter.

Yours sincerely,
F. M. Burnet

DCG to BURNET, WOOD, and ANDERSON 20 March 1957
Dear Drs. Burnet, Wood, and Anderson,

While we are awaiting word or suggestions from you, I thought it wise to avail myself of the opportunity to get mail out to Kainantu. We have now located some 27 cases of kuru, all new, which were not present in October 1956 when Dr. Zigas tried to round them all up and found 36 (and bled 22) in his three weeks in the region. All, or all but a few, of his 36 cases are now dead; but this check on deaths has not been fully completed as yet. Our 27 represented a distinct pediatric concentration and, thus, we have rounded up 8 cases in children who are now in the hospital. In addition, 9 adults are in our special kuru hospital, and we plan to keep the children at least throughout the course of their illness to recovery or death. It will be the latter, I fear, for some who are far gone already.

I have developed an extensive anamnesis technique and am finding the natives as accurate and as good informants and observers as any parents I ever encountered in civilized pediatric practice, if one takes into account the two-to-three–hour discussion a history requires. By this technique, it is now evident that kuru has been recognized here for at least two decades, perhaps longer. It has a traditional focus in a few "villages," and has apparently spread to a few adjacent ones in the last decade or so but this is not certain. There is no detectable seasonal incidence of onset, with cases appearing all through the year; but it will require more statistics before a curve to detect peak incidence can be plotted. I noted earlier the frequent accurate stories of kuru in the mothers of some of our patients. In checking this carefully, we found that independent observers agreed on this in their past history. By taking extended family histories, it is now evident that there are multiple cases of kuru within the families of many of our patients; and the histories are filled with such other descriptive causes of death as: spear wounds, arrow wounds, old age (*lapun tru*), skin ulcerations, capture and stab deaths, abdominal disease, and diarrhea with bloody stools. Since these descriptions are given consistently and accurately, and since recall of kuru seems to be fairly stable in the minds of the villagers to the extent of acting out tremors and late incapacity, we must accept these histories as accurate. They point to our classical patients and say "the very same thing" in describing past cases.

Of the two children admitted yesterday, one girl of seven years had both of her father's two siblings die of kuru, and her mother also died of kuru. The other, a girl of ten years, had had kuru in seven close relatives: although neither parent died of the disease, both of the two older siblings died of the disease in the past at the age of about ten years. Furthermore, one paternal uncle and three first cousins on the father's side also died of kuru, and one child of one of the paternal cousins also died of the illness. On the mother's side, we could find no cases of kuru. Now, whether this high family-incidence is spurious because of a high incidence of the disease in the entire community, I do not know—due to the time-consuming nature of necessary "control" histories. But that there is a rather astonishing family incidence is now clear. I am using Western relationship terms, disregarding native kinships. This can be done, since they understand well that we want the lists of "*pikaninis blong wonpella bel*" (i.e., the lists of all pregnancies and children born to each woman or man involved) and they keep this straight in their minds! In spite of the familial incidence (with an almost uniformly progressive course which likewise suggests one of the heredofamilial neurological diseases, as does the greater than 10:1 predominance in females) we are finding stories of fever, malaise, headache, and knee pains—without any stiff neck and loss of ambulation (this is definite)—in several patients, which might suggest an antecedent acute infectious disease. This is not, by any means, the case in all patients. Many had gradual onset without antecedent illness; none had incapacitating disease with sleep, coma, paralysis, meningeal neck stiffness, or any other symptoms which might have kept them off their feet. Thus, we have no suggestion of true encephalitis preceding the illness, such as the encephalitis lethargica which occurred after World War I in Vienna. In fact, rash illness, stomatitis (the one history thereof is unique), severe acute febrile disease, or any other such antecedent illness is most certainly lacking. Thus, although we are disturbed and still trying to hunt down the febrile diseases which might be prodromata to this syndrome, we have not as yet found any ourselves but will eventually make a house-to-house search during patrol in the villages.

Our hands are full in trying to keep charts and progress notes, records of medication (our clinical trials, in which I have little faith, but which we follow with true clinical investigative procedure), and feedings in the two cases advanced beyond self-feeding possibilities. We have now completed lumbar punctures on 14 cases, including all 8 children. In no case did we get more than a very slight cloudiness with four drops of Pandy's reagent, heated, in warmed cerebrospinal fluid. All pressures were in normal range, with normal dynamics (using 120–150 as normal). All fluids were clear and colorless; no pellicles formed. Careful cell counts on all fluids have been done, repeating the count completely three times on each fluid. None had over 10 WBC/mm^3 in early or advanced cases! Red cells from slight tap traumas have never exceeded 200 to 300/mm^3; and in any case with over 10 RBC/mm^3, fluid has been recounted for WBCs after dissolving the RBCs in acid. Our counts are valid, done promptly after the taps.

We shall be mailing to you all fourteen fluids shortly, perhaps with this letter. They have been kept at 4°C since they were done. I suggest egg and animal inoculations, but have little hopes for any such attempts—yet they should be done (? herpes, etc.). But I am most anxious to have the CSF chemistries and serology: colloidal gold sol curves, Wassermann, total protein, globulin (Pandy's or some other method), chloride, and sugar. I know that sugar is not valid on old fluid, if many cells are present. This has been refrigerated, and has no bacteria or cells to speak of. Thus, the sugars should not be too far off and should at least give us a minimum level. I shall also send some of our Pandy's solution for you to check on a known Pandy positive CSF, since Pandy's reactions and reagents tend to be fickle. Because of the prime interest in proteins, gold sol, etc., I am addressing the fluids to Lois or to the clinical unit, but hope that you or Anderson or French may be willing to passage them once. The shortage of tubes, etc. has made it necessary to pour off some fluid for cell counts, etc., so I hope we have not contaminated many. Fever charts reveal no temperature elevations thus far in any of the seventeen hospital patients.

We are waiting for detailed instructions on handling specimens for research neuropathology. We expect to be able to get a brain. Since many of the diseases of the corpus striatum and cerebellum require exquisite care to demonstrate their pathology, with the study of the globus pallidus, lenticular nuclei, caudate nuclei, putamen, etc., and the search for demyelinating lesions, we feel that such studies should take precedence—at least in our first case—over virus isolation attempts which might, in securing the proper tissue, mutilate pathological specimens. Will you take care of securing the expert neuropathological investigations? We hope to get new supplies of fixatives from Moresby soon.

In spite of the unlikelihood of any process which might be influenced by chemotherapy, we have given doses of chloramphenicol to eight patients, and will to three new patients, keeping them on a five-to-six day course. This includes all the children. Ascorbic acid, vitamin B complex, and folic acid supplements—together with improved diets and fish oil—have been the concomitant supportive regime for these patients. Others have received one grain of phenobarbital t.i.d. without evident effects, but we are now anxious to try tranquilizers, anticonvulsants (especially Tridione and Phenergan), cortisone and testosterone, on purely empirical grounds in controlled series. Have you any suggestions as to trial therapeutic regimes for our consideration?

The high incidence of Sydenham's chorea in girls, the similarity of the syndrome to progressive lenticular degeneration (Wilson's disease), and familial history such as that of the three siblings contracting the disease at about the same age—all point to a degenerative disease of the corpus striatum on a heredofamilial basis. By concentrating on the children, we may be able to exclude the possibility of previous visits to the low country, since some have never gone; and severe encephalitic illness in the past is more easily excluded, since past history is restricted at their age and more easily recalled. As I recall, the rather rapid and consistent progress to death is not characteristic of post-encephalitic parkinsonism, while it is of the illnesses of the Wilson's disease type. We have not found Babinski or nystagmus of Friedreich's ataxia, or the cord degenerations or usual features of disseminated sclerosis. Neither toxic reaction or deficiency, nor antecedent acute infectious disease of which this is a sequel, have yet been totally excluded. Enough of the philosophizing! This is no time for armchair medicine on our part, but any suggestions from you or gleanings from the heavy tomes devoted to familial neurological degenerations, toxic neurological disorders, etc. will be most appreciated.

We are telegraphing the P.H.D. for permission to construct an adequate treatment room and ward for child cases who will be long-term nursing problems, and have requested medical supplies and lab supplies for further studies and therapeutic trials. Fortunately we are here alone, with no other Europeans as yet, and thus have used the patrol officer's house as a treatment room for the severe patients here with us. Shortly the officer will return, however (a Mr. Baker, who is new), and we must evacuate. We have been busy lysol-ing down the house and the government office, both of which are our lumbar puncture stations at the moment. Whatever the outcome of this study, the disease remains most devastating—and fascinating. It looks as though there will be very little to do for the patients once neurological damage is evident, although stories of a few (very few) recovered cases are told and are apparently valid. Since it is apparently the major cause of death (in girls at least) in childhood during the present time and for our group of 27 patients—almost all younger and previously healthy natives with a high probability of death—we feel that it should take precedence over most other matters in this region and hope the Administration will agree and grant us the few new structures. We are in clouds most of the day and it rains fully half of each day.

Finally, we are anxiously awaiting word as to whether you will take a case or two for careful study, in Melbourne, for: stereos of the skull, pneumoencephalography, electroencephalography, and thorough neurological consultation; and we await word, furthermore, of any suggestions for study, treatment, and collection of specimens. Will Dr. Anderson be coming up to see any cases, and when?

Sincerely,
Carleton

DCG to SCRAGG   20 March 1957
Dear Dr. Scragg,

Dr. Zigas and I are completing one week in Okapa (Moke) now, and we are by no means ready to make any definitive progress report as yet. However, we have addressed rather garrulous letters to you and the Hall Institute crowd in the past—since these require less time than a definitive analysis of results, which we are not yet prepared to make. They serve to keep you informed as to what is going on and the trends of our thinking, so that any suggestions, corrections of our errors, or added assistance which you may think of can be brought to bear on the problem while the chance is in our hands.

I am enclosing a carbon-copy of the letter we have just sent off to Melbourne to my colleagues there apprising them of our progress in order to elicit their advice and suggestions, since we certainly are open to any. At the moment we are concerned chiefly with accurate documentation of what we have on hand, and a bit of epidemiological detective work in plotting out the natural history of the disease in these communities.

With the arrival today, or shortly, of Mr. Baker, who will be the Patrol Officer here, we cannot continue to use his home as a treatment center for LPs [lumbar punctures] and microscopy and careful examination of patients. The Administrative Office is likewise a hospital auxiliary, which we will soon have to relinquish in deference to the courts. What remains is a hopelessly inadequate hut of the medical *dokta bois* (without even light or a table for LPs) and hospital wards. Our immediate need is a treatment hut—which, if authorized, might be thrown up in a few days—with a good table for examinations and LPs, and a desk for typewriting and laboratory studies. Secondly, we now have eight child patients, and nine adult patients under surveillance in the largest ward, which we have taken over for kuru. With their guardians, this is very crowded; and since we are dealing in some cases with advanced neurological disease, space with platforms and washing facilities to control decubitus ulcer formation is essential. The existing wards are ample for the needs of other medical problems at Okapa for the moment, but we need a somewhat better special building for the patients we hope to follow on a long-term basis. Thus, these two building authorizations, along with benches in the kuru wards for the patients—rather than the floor they are now using—are our immediate request. If they are to be of any value, they are needed *immediately* while we are devoting time to the study of the disease. Should it turn out to be a virtually untreatable neurological degenerative disease, I must admit that it will not make much sense to keep patients hospitalized for long periods rather than in their homes. To study the disease, however, the home and village is hopeless.

[6]

We are living in the patrol officer's house, and our accommodations are okay as long as he permits us to have a bed in the corner somewhere.

[7]

Obviously we can now select from a series of eight patients between five and thirteen years of age, should you and the Administration decide that a case should be transported there. Two, however, are so advanced that transport would be a difficult problem.

Your telegram has just arrived to Dr. Zigas. We are rather well off in our study at the moment, and will, once therapeutic trials of drug are completed, send a patient or two to Moresby for pneumoencephalography and your opinion as to nature of illness with whatever neurological consultation you can get. Since we are primarily interested in electroencephalography (to see if there is any petit mal triad-type record, or anything resembling the slow wave–high voltage pattern of postencephalitis) and in laboratory studies not done except in clinical-investigative units, I fear that not much more can be done in Moresby than here. However, it might be good for doctors there to see a case and study it. We shall pick for this purpose a full-blown classical case on which thorough laboratory workup would be desirable, including cerebrospinal fluid chemistries repeated daily, white blood cell counts and differential counts for a few days, daily urines for a while, and whatever weird and unlikely tests you can work out—e.g., liver function tests (thinking of Wilson's disease–type illness), renal function tests, and hormone excretions. In dealing with such a new pattern, it is wise to spread oneself widely to pick up any clues one can.

Will you be able to supply any of the drugs and other things we have requested? We have several new patients just arrived since writing the first part of this letter—one a well-nourished adolescent *male* with full-blown kuru.

Sincerely,
Carleton Gajdusek

P.S. Dr. Zigas points out that a patient would have to be returned before progress of her disease to incapacitation, because of the grave local disturbance which her death outside the region would cause. It might jeopardize all further case-finding and cooperation, without which further investigation in the field is impossible. He would appreciate permission to have patient accompanied by himself or a medical assistant to Moresby, as well as one relative or guardian of patient.

ANDERSON to ZIGAS   22 March 1957
Dear Dr. Zigas,

I have just received the following reports on CSF which you kindly sent some days ago. The tests were done in Royal Melbourne Hospital.

Total protein:	40 mg %
Chloride:	752 mg %
Sugar:	0.03 mg %
Globulin: Nonne-Apelt / Pandy's	no increase
Cells:	no cells

So far we have not isolated any virus on the chorioallantoic membrane, nor in baby or adult mice nor in guinea pigs.

From the notes you and Gajdusek have sent I agree that this does not seem likely to be an infectious disease, although the possibility might perhaps still be kept in mind.

All the best,
Yours sincerely,
Gray Anderson

ANDERSON to ZIGAS   22 March 1957
Dear Dr. Zigas,

We have received the second long letter about kuru, which Dr. Gajdusek typed.

Sir Macfarlane Burnet has spoken to Dr. R. Scragg on the 'phone and has also written to Dr. Scragg. I understand the patient will be staying in Moresby for a week or so, and a decision will be then made regarding transfer to Melbourne.

We have discussed with the pathologist here the appropriate fixation procedure for brains. He advises perfusion of the brain with 200cc of 10% formalin in either water or saline, the perfusion to be done via the carotid artery or the circle of Willis. You could use either a tube and canula or a syringe and needle. Then please slice the brain in ½-inch–thick coronal slices, wrap in gauze to maintain correct orientation of the slices and suspend in 10% formalin. I would say at this stage not to take any brain specimens for infectivity studies.

No doubt you already have my previous letter regarding sections in formalin of all other organs, including spinal cord (complete cord) and two large peripheral nerves.

With kind regards and best luck,

Yours sincerely,
S. G. Anderson

ANDERSON to SCRAGG 22 March 1957

Dear Dr. Scragg,

I was very pleased to have your letter of 5th March, which arrived this afternoon, together with the attached report of Mr. Charles Julius.[3] Sir Macfarlane and I now have a clearer picture of the situation and, as you say, it is an extremely intriguing disease which interests us a lot.

Dr. Gajdusek will surely pursue local investigations vigorously and expertly. However, if I can still be of service when he leaves for America, I would be delighted to visit the Highlands and make what contribution I can. This could be in early April or any time after.

Would you please let me know when Dr. Gajdusek is about to leave, and also whether any local investigation then remains to be done. In the meantime, we will wait to hear from the Patrol Officer, Mr. Coleman. I would like to thank you for the suggestion that he should visit us in Melbourne.

It crossed my mind that Dr. C. E. M. Gunther reported an encephalitis in natives coming from the Kukukuku area in the *Medical Journal of Australia* 1955, vol. I, p. 715. Mr. Julius mentions the belief of the Fore people that kuru derived from the Kukukuku area.

About a week ago I had a sample of CSF and a sample of heparinized blood from Dr. Zigas. The CSF was normal biochemically, and no cells were seen in the portion which was formalinised. Passage of both samples through suckling and adult mice, and through two passages of chorioallantoic eggs, has not yielded any virus. I have sent this news to Dr. Zigas. I do not know the name of the patient.

Please excuse my poor typing. I did not want to wait till Monday for the typist.

With kind regards, I am,

Yours sincerely,
S. G. Anderson

BURNET to SCRAGG 26 March 1957

Dear Dr. Scragg,

We have received a series of long letters from Drs. Gajdusek and Zigas which indicate that they are working very hard on the kuru question, but which throw no light whatever on the administrative setup involved. For various reasons, I should like to get clear what Gajdusek's status is in regard to the Administration. He is no longer attached to the Institute, and I was most surprised to find him taking charge of an investigation that we had planned for Dr. Anderson.

---

[3]This appears to have been a condensed version of the report that is included in this volume as Appendix 1.

I believe that in many ways Gajdusek is particularly well-suited for such an investigation, and if you and others in the Administration feel that work well begun must be actively supported I shall wholeheartedly agree. I am also not unaware that Gajdusek has a tempestuous enthusiasm that may make him feel that he is doing both the Administration and the Institute a service in tackling the kuru problem and that the value of that service is more than enough to counterbalance the effect of his somewhat unorthodox intrusion into the problem. He has certainly been most conscientious in keeping us fully informed of developments.

The specimens of CSF are being tested as requested by Gajdusek and Zigas, and reports will be sent both to you and to Kainantu. We shall continue to make appropriate tests as requested, and await with interest a report on the patient sent to Moresby.

I should have preferred the whole investigation to have been an Australian affair, but the main thing is to get the investigation done as well as possible. Our cooperation in this matter was at the request of the Administration, and we shall continue to do our best to help in any matter within our capacity.

Yours sincerely,
F. M. Burnet

c.c.   Dr. D. C. Gajdusek

BURNET to DCG   26 March 1957
Dear Carleton,

Thank you for your three reports, which we have all found extremely interesting. The CSFs are being tested, and Anderson has sent the data requested re fixation to Zigas.

However, I would like to get the record straight, as your countrymen say, and I enclose a copy of a letter I have written to Dr. Scragg. In view of our present commitment in regard to kuru and further studies contemplated by the Institute in New Guinea, I would be very grateful if you could give me some idea
  (a) of when you are leaving Australian New Guinea,
  (b) of any plans you may have for epidemiological or child studies in New Guinea in the future, and by whom they will be supported.

Your help has undoubtedly been invaluable to Zigas, but it has badly disrupted our own plans. I don't like untidy situations developing, and before we plan ahead I should like to have as much information as possible.

All I have heard of kuru makes me think you are on the right track in looking for a familial CNS condition, and I shall be most interested to see some pedigrees if you can get them worked out under native mating arrangements. I found Julius's report most enlightening in removing a lot of my preconceived ideas about New Guinea sorcery. I presume you are alive to the possibility that the degree of expression of a genetic lesion might be influenced by various environmental factors.

With all good wishes,

Yours sincerely,
F. M. Burnet

c.c.   Dr. R. F. R. Scragg

SCRAGG to DCG   28 March 1957

Dear Dr. Gajdusek,

Thank you very much for your very interesting and informative letters on your current investigations into kuru.

I have contacted our dispensary and he is forwarding you supplies of those of the drugs that we have. These are cortisone, testosterone, Tridione, and Dilantin.

Dr. Price is forwarding you Zenker's, Benedict's, and Pandy's reagents; and a haemocytometer is also underway.

I am afraid we do not have a copy of Ford's *Diseases of The Nervous System in Infancy, Childhood, and Adolescence,* but I am forwarding two volumes on children's diseases and the latest edition of the *Yearbook of Neurology* etc.

I am afraid that we do not have a motion-picture camera that we can lend, and I think your best approach would be to endeavour to borrow one from someone locally.

We will be willing to receive a patient in Port Moresby as soon as you can see fit to transfer one or more. Approval is hereby given for Dr. Zigas to accompany the patient at our expense.

I agree very much with Dr. Zigas that no patient must be permitted to die out of Okapa area. Accordingly, cases you send should be in the early stages and should be returned to Kainantu before they progress too much.

We are willing to pay the fares of a patient and guardian to Melbourne, and Fairfield Hospital are willing to accept them.

We are capable of doing stereos and X-rays in Port Moresby but of necessity would have to forward the patient to Melbourne for electroencephalography and pneumoencephalography.

Recently, in Australia, I saw a new book on Kinnier Wilson's disease. However, the author of this tome slips my mind at the moment; no doubt Melbourne would be able to obtain it for you.

With your presence in the field, it will be hardly necessary for Dr. Anderson to visit the area; but I would welcome some idea as to how long you intend to stay at Okapa.

I will take up the matter of accommodation for patients with the Director of Works and endeavour to satisfy some of your demands in this respect.

Please keep me informed in this matter as in the past, for I will be only too glad to do anything I can to help.

Yours sincerely,
R. F. R. Scragg

P.S. Letters from Sir Mac Burnet will no doubt have reached you; dependent on these I think Dr. Anderson may visit [*illegible*].

BURNET   29 March 1957
                                                                    7:00 p.m.

[handwritten summary of a telephone conversation with Scragg]

Gajdusek was not authorized to undertake work in kuru area; he left with the intention expressed of going on patrol with Ian [Burnet] at Lufa. He will be advised tomorrow to leave the area, and Anderson will be contacted by telegram. If his [Anderson's] wife wishes, accommodations can be arranged for her.

Heard less clearly: Another American invasion—   definite arrangement with Anderson—   inform the Minister by radio—   Dr. May would be able to give you direct information—   surprise that Coleman the P.O. had not been to see me—   Messages from S. G. Anderson unanswered.

SMADEL to MORRIS   29 March 1957

Assistant Director of Professional Education
National Foundation for Infantile Paralysis

Dear Marion:

At the recent meeting of the Virus Committee you asked me what had happened about Gajdusek, since you had had no word from him since January. I have just received a letter from the wilds of New Guinea, indicating that Gajdusek is now among the cannibals studying a most important new disease with neurological manifestations.

I am enclosing copies of Gajdusek's letters. I attempted to reproduce them by Thermofax, but the result is almost illegible. When you have finished with the letters, please return them to me.

You will notice Gajdusek's plaintive plea for support. I would be delighted to find a means for supporting him here. In fact, we have already been negotiating about a regular staff appointment when he comes back to the United States. The main difficulty, as far as we are concerned in getting any kind of support for Gajdusek, is the red tape. With all the best intentions, we cannot put a man on a fellowship or on a staff appointment without having a warm body and a signature. These are rather difficult to get on short notice with Gajdusek at the Okapa Patrol Post.

This is to inquire about the possibility of extending Gajdusek's fellowship from the NFIP (which expired two months ago) for a period of three or four months. He hopes to remain for several months in the jungle. As he points out in his letter, he does not need any money there but he will need some when he gets out.

If the NFIP's red tape is as bad as ours and my suggestion appears entirely impractical, please ask Tom Rivers (who is an old red-tape cutter, too) if he has any ideas on the subject. If neither of you can think of anything, I'll take another crack at it.

Best wishes.

Sincerely yours,
Joseph E. Smadel, M.D.
Associate Director

SCRAGG to DCG
[radiogram]   30 March 1957. Burnet deeply concerned you taking over investigation planned for Anderson at request of Australian Government. Your interest in problem fully appreciated but my instructions request you look into Lufa area

only. Anderson now advises arriving within one week. Accordingly on ethical grounds request you consider discontinuing your investigations kuru. Will advise definite time of arrival Anderson and suggest rendezvous Goroka to discuss problem. Please advise me by radio your plans dependent on above.

DCG to SCRAGG
[undated reply handwritten at bottom of above radiogram]   Intensive investigation uninterruptible. Will remain at work with patients to whom we are responsible. Am in direct correspondence with Burnet. Will discuss plans with Anderson at Moke [Okapa]. Burnet advised.

DCG to GUNTHER   3 April 1957
Dear Dr. Gunther,

I just missed you in Sydney when I called on Drs. Walsh and Ward. I had just come from a week of work in Canberra (with Fenner, Cairns, and Fazekas)—to where you were heading, I was told in Sydney. I was most sorry. Arrived in Port Moresby, I again missed you since I was told that you were ill. I hope that I shall have a chance to see you when I return to Port Moresby.

For the past three weeks I have been here at Moke studying kuru rather intensely. Dr. Zigas and I have been working together at full speed on the illness and have dug up 41 cases already. We have a kuru hospital in function, with 25 cases (including 13 in children, since my specialty in pediatrics finds me more at home in clinical investigation in this field) in our wards this evening. Some 30 lumbar punctures have been done, and CSF Pandy's reactions and complete cell counts have failed to reveal any abnormalities. I have been learning the Fore language rapidly and have developed a fairly accurate history-taking technique which has already permitted the establishment of extensive family history of kuru in most of our patients. All search has failed to reveal any encephalitis in the region now or previously, and we cannot yet find any evidence of an antecedent acute infectious disease. I am placing my bets at the moment on a heredofamilial degenerative disorder of the corpus striatum (akin to paralysis agitans or Wilson's hepatolenticular degeneration) distinct from those yet described, restricted to the Fore and the adjacent intermarrying Kimi and Keiagana, and following a rather typical basal gangliar, corpus striatum degenerative disease pattern. Toxic, deficiency, infectious, and post-infectious hypotheses have not yet been substantiated.

We are running intensive therapeutic trials on what little evidence we have from pharmacological leads. Thus, broad-spectrum antibiotics (chloramphenicol and Aureomycin) used in high doses of 5–6 days do not affect the afebrile course, as we thought they would not. Phenobarbital used in high anticonvulsive doses has no effect thus far. Antihistamines are being studied now. Cortisone and testosterone will soon be used in controlled clinical investigative series.

We have just made a six-day patrol to the Lamari River, reaching the most distant South Fore villages from which we had never heard of kuru; in three of the most distant clan regions, with a total population well under 1000, we found 13 new cases now active! Six were in children. The female preponderance at better than ten to one in adults and three to one in children persists.

We have ticklish problems in trying to avoid any trace of coercion of the natives.

We have gained their full confidence around Moke. Dr. Zigas has repeatedly told me of your intense interest in kuru and enthusiastic encouragement of his quest in pursuit of the nature of this disease. Knowing of your interest in such problems, and myself being far too much intrigued by our current studies and careful clinical-investigative program to abandon them lightly, I write to urge you to consider the possibility of a brief—even a one-day—visit to our Kuru Center here at Moke. We can, in a few hours, show you well over two dozen fully documented cases, all with good laboratory data, complete "academic" medical charts and progress notes, LP findings, etc., etc. Your visit, if at all possible, would greatly boost our morale and lend encouragement to our study.

Dr. Zigas points out that if previously notified, he could arrange Land Rover transport from Kainantu to Moke and back to Kainantu in a given day.

Sincerely,
D. Carleton Gajdusek, M.D.

DCG to SMADEL   3 April 1957
Dear Joe:

I am writing to you just after returning from a patrol down to the southern fringe of the South Fore region, and to the Lamari River and Kukukuku and Awa linguistic areas. This is a little-visited region, uncontrolled at its extremity; and tribal wars, bow-and-arrow murders, raids, and cannibalism are still rather frequent—in fact, all three infringements of the supposed "control" have occurred here since my arrival. *Tukabu* (a traditional native way of throttling a sorcerer with strangulation, and stone-beating to break the femur, fracture the ribs, and pound the costovertebral angles) is a frequent source of "emergency" medical problems, as are the wicked arrow-wounds. The patrol was directed to this little-known region to determine the southern extent of the kuru disease which has so occupied me and is the subject again of this note.

Here in Moke, in a mat-floor hospital which was quickly built with native materials—in which we have a microscope, hemocytometer, a host of lab reagents and equipment, and all the diagnosis instruments that such a "bush" hospital would be expected to possess, including ophthalmoscope and tuning forks—Dr. Zigas (the Kainantu physician and only M.D. in the Eastern Highlands), who discovered the disease, and I have set up a kuru investigative station. Thus far it is operating largely on his and my own enthusiasm and his courage in daring to leave the European population relatively unattended while we work far out in this bush—that is, as long as the Administration which employs him will tolerate this "research" diversion.

We have now located 41 cases of kuru, a clear-cut central nervous system degenerative disease of rather rapid course, quite uniformly and progressively devastating and almost always fatal. We have examined these cases and are following them closely; 25 are now in our Kuru Hospital . . . , 13 children and 12 adults. Another 3 children and 13 adults are in their villages, about half of them having returned there after extensive hospital study; we intend to try to urge them back. I suspect that with 13 child cases of basal gangliar disorder (all relatively acute and most probably doomed to fatal outcome in less than a year), and with 3 other children we are trying to "bribe" in, and in view of the fact that the disease is restricted to the

Fore linguistic area and involves only neighboring (now intermarrying) linguistic groups of Kimi and Keiagana in the regions directly adjacent to the Fore tribe (a population of perhaps 15,000, scattered in some of the roughest mountain jungle in the world), it is probably the highest incidence of such a disease ever recorded in any population.

We know of another 40 cases (in addition to our current 41) who had the disease and died during the past eight months, all of whom were seen and documented by Dr. Zigas. We have accumulated histories of well over a hundred other cases that died during the past decade or so from the disease in families of our patients. It is undoubtedly the major disease problem among the Fore people, and beyond infancy—a period during which we have not yet diagnosed the disease—it is perhaps the major cause of death next to warfare wounds. We now have the people's confidence; they bring patients to us and permit us to live in their villages while we hunt down cases. Since our last week of patrol, we located 13 new cases in only three clan areas with a total population of well under a thousand; further patrolling to establish the geographic extent of the illness may roughly double the number of cases. Some 30 lumbar punctures have been negative; no elevation of cells and no Pandy stronger than a rare weakly positive, in any stage of illness. There is no evidence of encephalitis now or in the past, nor any convincing acute infectious disease picture antecedent to the insidious onset of gait instability, remarkable tremor that appears more hysterical than organic but which progresses to spasm, flexed posture, complete instability and inability to walk, loss of speech, and final and complete neurological invalidism and death—with ocular convergence, drooling, etc. late in its course. Many of our most severe cases are children of five to ten years of age, who have been ill for only three to ten months and are now near death.

Zigas and I are now preparing a paper for submission to one of the U.S. journals, with a title like "Kuru, A Localized 'Epidemic' of Chronic Progressive Degenerative Disease of the Central Nervous System in Children, Adolescents and Adults in the Eastern Highlands of Australian New Guinea." We both see clearly that unless we work out and publish our preliminary and very extensive studies, Zigas will be cheated out of anything by the administrative superstructure. Secondly, I suspect a good deal of jealousy from the Australian sources shortly, as word gets out. The fact is that besides Zigas and myself, no other medical man in the world has investigated or seen the disease, except for a few administrative M.D.s who saw some cases for a few hours when Zigas brought them out of the region to "civilization." Hall Institute's interests are here, but no one is doing a thing; and the microbiologic prospects look dimmer as I study the illness. I am shifting to further suspicions of one of the most remarkable incidences of a whole race inflicted with a heredofamilial genetic degenerative disorder of the central nervous system—specifically of the cerebellar structure. There could be no better setting in the world for the study of such a disorder; and in spite of the tremendous odds, we really have the confidence of these "cannibals" and "savages." I am rapidly learning the language, getting fine and extensive accurate family histories (which are cross-checked for validity) and a better controlled epidemiology than could ever be expected in less isolated "civilized" communities.

With this in mind, I now write to assure you that my last letter was no ruse; I have the "real thing" in my hands. I can and will continue to support myself and the project on my own funds and those that Zigas can wheedle from the Port Moresby

P.H.D. (now a rather complex bureaucracy, with many petty jealousies—far away from this remote mountain jungle post, but our lifeline). Will you let me know if you think even a temporary salary, which I shall throw into support of this project to pay for the axes, beads, tobacco, and other trading items with which we purchase bodies (with autopsy permission) and food for our patients—a normal U.S. salary will go mighty far here—could be arranged to permit me to carry out the work for several more months, at least long enough to watch the evolution of a number of current cases to their fatal outcome. If any added expense account could be arranged, all the better. For $100 worth of flown-in supplies, we hope to have a completely new building up, quite ample to house our 30-odd patients and another 10 or 20, all under good conditions; this would be built of largely native materials. Thus we are really launched on the study; a second such opportunity may never present itself. I am interested to know what your general opinion may be, particularly since I have not yet had a reply to my first letter. Actually, for two weeks we have had no contact with the outside world, and this letter is going out with Dr. Zigas on his return to Kainantu tomorrow.

Should you be able to make a grant, based on little more than your assurance of my integrity and medical judgment—I tell you Joe, this is no wild-goose chase, but a really big thing; everything in my medical training makes me confident—let me know, so that I can plan to extend my work here for at least several further months. On my own, I can hold out for one or two months and still have enough to get home via Europe. But I am fully aware of the unwisdom of trying to myself be both investigator and philanthropist—enough said. I stake my entire medical reputation on this matter.

In advance, however, I must tell you that we are probably dealing with a genetically determined CNS degeneration—with toxic deficiency, postinfection all failing to stand up. Since much suggests hysteria early in the disease, excepting the cases in children of five-to-ten years of age, I cannot get the psychosis idea out of my mind. But the advanced classical parkinsonism and basal gangliar disorder that results in eventual death cannot easily be linked with psychosis—in spite of the role this illness plays in local sorcery, murder, warfare, etc. (which gives one the consideration that it is a possible psychosis). This class of illness is not my specialty, but I can handle research on it as well as anyone. [8]

Autopsy material is most difficult to obtain and will require time and much persuasion, but we shall get it. We promised one brain to Melbourne, but if you can promise expert neuropathology, I shall get one off to you. [9]

If you really think I should not abandon this matter at any cost, let me know, since I shall stick to it even without any grant as long as possible. Should anyone in the States be wondering of my whereabouts, let them know, since my chances of keeping up much correspondence from here on anything but kuru is most remote.

Sincerely,
Carleton

BURNET 5 April 1957
[summary of telephone conversation with Scragg]

The Administration sent a telegram to Dr. Gajdusek, asking him to discontinue work on kuru and discuss the situation at Moresby. Gajdusek's reply was that he would continue working at Kainantu at least until Dr. Anderson arrived. Dr. Scragg has discussed the matter with Dr. Gunther, who is definite that no action can be taken to force Gajdusek to discontinue the investigation. In view of the fact that the present impression is that kuru is probably a familial disease rather than a result of encephalitis, he and Gunther consider that Gajdusek may be better equipped for the investigation than Anderson, although they fully agree in regard to the unjustified character of Gajdusek's intrusion into the problem.

Scragg has received authority from Canberra for Anderson to come to Moresby. In Scragg's opinion the best action is for Dr. Anderson to come to Moresby as arranged, on Tuesday the 9th; discuss the matter at Moresby; and probably proceed, with Scragg, to Kainantu for discussion as to the best method of continuing the investigation of kuru. Irrespective of whether Anderson agrees to work in association with Gajdusek, Scragg is very anxious that he should come to Moresby to discuss, in addition to kuru, matters concerning encephalitis and future activities in the Territory.

I said that I knew Anderson had strong objections to working with Gajdusek and that I supported these objections, but that I would urge him to go to Moresby as previously arranged in order to get a complete clarification of the situation.

BURNET to SCRAGG   5 April 1957
Dear Dr. Scragg,

Following our telephone conversation, I have had a long discussion with Dr. Anderson and also consulted one or two others whose advice would be of value.

I believe that the only way of resolving an unfortunate situation is for the Institute to withdraw from any responsibility for the kuru investigations, leaving them in the hands of Gajdusek and Zigas. I am still considerably irked by Gajdusek's actions, but there is little doubt that he has the technical competence to do a first-rate job. Since his position at Kainantu has now been accepted by the Administration our only course is to leave the entire investigation to him, and Anderson is of this opinion also. Under these circumstances, we feel there would be no point in Anderson visiting Moresby at the present time.

It seems reasonably clear that kuru is not of virus origin, so there is no particular reason why the laboratories of this Institute should be involved in the investigation. However, as I stated in my previous letter, we shall be prepared to help find ways of carrying out any investigations that are called for in Melbourne by the kuru work.

I considered seriously whether any useful purpose would be served by my making a short trip to New Guinea, but have concluded that there is nothing to justify it at the present time. The present difficulty will obviously solve itself in a few weeks, but I am quite concerned that it should not leave any impediment to a continuation of our association with you in regard to encephalitis or any other matters in our field which may emerge in the future. I look forward to hearing your reactions to the memorandum I sent to Canberra a few weeks ago in relation to Dr. Gunther's report on the possibilities for research on disease in infants and children. If for any reason it does seem important that I should come north for a brief visit, there is every prospect that I should be able to do so.

At a later date, perhaps in May, I hope it will be agreeable to you for Anderson to visit Port Moresby. This would allow him to discuss some of the implications of Dr. Gunther's report with you, and permit him to visit the areas of sentinel fowls and to see any cases of Murray Valley encephalitis which might have occurred in Port Moresby this year. I should appreciate your opinion in regard to this.

Enclosed is an abstract of findings with CSF sent to us over the last two weeks. The slips themselves have been sent to Kainantu.

I am sorry for all the trouble that you have been let in for. With kind regards,

Yours sincerely,
F. M. Burnet

DCG to SCRAGG 6 April 1957
Dear Dr. Scragg,

I am enclosing two complete history and record sheets on our two patients Manto and Asomeia, two classical cases of kuru—most representative of advanced but not late or near-terminal disease. These are the two patients picked for you in Moresby; and Dr. Zigas, now in Kainantu, will accompany them to Moresby. We have promised the local population to have them back as soon as possible, to not permit very prolonged absence at any cost, and to have them accompanied all the time. Their husbands have agreed and are going with them. There has been much wailing and uncertainty here in association with their departure, and only when they get back in good shape will our reputation with our patients be secured. The Fore are most resistant to having a death occur outside of the local village setting.

I am sorry that we have not done more work on these two women, and that our charts are thus incomplete. We have been concentrating on our child patients. However, I have dug up as complete a family history as possible; neither patient has the dramatic familial incidence of kuru that we have observed in some others. The history is as complete and as accurate as I can get at the present.

Manto has had a trial course of full doses of broad-spectrum antibiotics and supplementary vitamins, which was subsequently followed by a course of Benadryl. Asomeia has had a course of phenobarbital in good anticonvulsant doses, and this was followed by a clinical trial of Benadryl. None of these regimens have affected the course of the illness in these or our other patients.

The kuru research project continues at full speed, with an attempt to complete our current therapeutic trials. I hope that the equipment which you sent word of in your letter of March 28 will all be arriving shortly. We could particularly use the Dilantin, Tridione, cortisone, and testosterone. These will all be tried, as some corticotropin is now being tried (but our supply is nearly out) on several child patients, who require smaller therapeutic doses. I would also like as good a laboratory manual as possible, and any exact information on anticoagulant concentration for sedimentation rates; our sedimentation rates are thus far dependent upon the oxalate concentration of anticoagulant oxalate used, and this is thus useless. I have forgotten standard amounts of oxalate or citrate. Can you send information on these, plus the necessary anticoagulants and means in which to quantitatively supply the correct amount to blood-collecting tubes? We could also use further sed.-rate tubes. We have only two, and thus it takes days to run sed. rates on our patients.

We have still found only negative cerebrospinal fluids, with one exception: A case which is apparently spontaneously remitting—and which we did not see and examine thoroughly until a bit too late, and which is further complicated by current insistence that it is not kuru but a kuru-like illness without kuru tremors—in an adult woman has yielded a positive Pandy's and 30–35 cells, about 90% monos in the CSF. Thus this first intriguing CSF is complicated by all sorts of uncertainties and may not really be kuru at all. Advanced severe cases are thus far completely negative.

We have had a severe, nearly fatal, case of meningococcic meningitis (diagnosed by finding meningococci on smear and in cells)—now cured. This was in a girl of 13–14. I have just had a terminal case of pneumococcus meningitis in an infant of a few months who required six hours of artificial respiration, had complete cardiac arrest over two dozen times (for almost a minute on several occasions) and complete respiratory failure for several dozen times but has been kept going by an all-night manual respirator technique and is now rallying on the third day of intensive intramuscular and intrathecal therapy. CSF was loaded with pneumococci and cells even on the second day, in spite of intensive chemotherapy. It now appears that we may get by, but I shall soon have to be doing subdural taps—for in this age, subdural loculated accumulations of fluid are more the rule than the exception. I hope we get through. Moke is running 20–30 sick—nay, dying—infants a day, with acute upper-respiratory infection which has advanced to either lobar or intensive bronchopneumonia (hospital census); and we take in some 5–10 new cases a day. We have not yet had a death, but that is pure luck.

The [district boarding] school epidemic (involving almost 100% of the 106 schoolboys just assembled) gave us 60 admissions with prolonged fevers over 102°F. The illness resembled severe herpangina, but minute postpharyngeal vesicles rather than larger typical herpangitic lesions on soft palate were the rule. I think it was a virus, perhaps an adenovirus or a Coxsackie, but it is not worth the effort to find out. We think we prevented any complications by treating all mild or incubating cases chemoprophylactically and all febrile cases intensively. No bacterial complications other than one OMPA with discharge developed. I think the epidemic is about over.

This then, is the kind of thing that slows up our kuru study, but we are now going strong again. One patrol has established the high incidence of kuru south to the Lamari River in places we had not previously suspected it. We found 13 new cases, brought sick children back with us, and are now about to hunt further afield for more, while therapeutic trials are in progress here. I shall certainly welcome all of the equipment and supplies you spoke of in your letter, and anything else you can think of. Any lab supplies will certainly be used, for we are lab-minded.

We are anxiously awaiting Anderson. Field epidemiology and the tracking-down of every possible antecedent infectious disease lead is still a huge job which will keep a team of epidemiologists busy for months. Clinical study is also a matter for years, not months. We plan to move into a newly built kuru ward in a few weeks, but this building is purely on a local, emergency scale with Mr. Baker giving us every bit of his time and assistance as local Patrol Officer; anything that the Works Department may subsequently authorize will be most valuable to the future study of kuru.

My own plans call for continuing our clinical studies through on the current group of patients until we reach a static condition—which may be a few weeks, perhaps a month or more—and then to sit down and try to reevaluate the situation. We both, Dr. Zigas and I, anxiously await any and all collaborators and contributing visitors who may be able to show up. Our colleagues from the Hall Institute will still have their hands full when and if they arrive. We hope they do, soon.

Sincerely,
D. Carleton Gajdusek, M.D.

DCG to BURNET [early April 1957]
Dear Sir Mac,

I am grateful for your letter and the copy of that to Dr. Scragg. We received it just after returning from a lengthy, very rugged (the most rugged in Dr. Zigas's long experience and in that of the Patrol Officer Mr. Baker, who is helping us) patrol to the most southern extent of the Fore linguistic area to the Lamari River, where we were able to locate 13 new cases of kuru—all in regions more remote and previously unknown to us to harbor the disease. In fact, in three clan-areas with a total population of well under 1000 we found these 13 active cases, history of as many deaths, and evidence that kuru has been in this region for at least 20–30 years (although there is also legend of its moving in from the north). All informants—including some Kukukukus—agree that it does not exist in the Awa linguistic area and in the Kukukuku area but we could not get across the swollen Lamari to find out for sure, and this would be a most extensive patrol investigation requiring, I fear, months devoted to linguistics to be certain.

I am rapidly learning a smattering of Fore, and Kimi and Keiagana word-lists and medical lore are already mine. Fore language is near-essential to getting the histories on which our finding of strong family-incidence is based. The interview technique I have developed has given us in many instances full paternal and maternal family histories, but it requires hours to dig out with old men and women informants. Our patients keep us too busy to get everything done. We now have seen 41 active cases of kuru, and have had most in our Kuru Hospital to study the disease. We are building a native-material ward to house our growing collection of patients. It is well under way. A treatment room is now well constructed and I am back again at blood counts, CSF studies, differentials, urines, etc.

It is a full-time job for one or two trained clinical investigators to simply keep our charts up to date, let alone run therapeutic trials of various medications—taking the lead suggested by antihistamine efficacy in some types of paralysis agitans, etc.—and we have a new set of clinical trials, all controlled, being started using cortisone, testosterone and further antihistamine trial. That chloramphenicol did not in any way influence the afebrile course—we have established this with three-week temperature charts on over two dozen patients—by alleviating symptoms is certain. That we have seen now several rather remarkable remissions—but not complete except perhaps in one case—is also certain. That the majority of our severe patients have what will prove to be fatal corpus striatum disease is equally as certain; and that therapy is unlikely to influence these advanced cases, except to greatly prolong their life-span by supportive measures and improving their general well-being, seems sure.

We have sent many a complaining older patient home. At the moment we have 25 cases in the hospital, 13 in children. Of a total of 41 cases now active, we have found 16 children, of whom 13 are in our hands. I am concentrating on the child cases—which are more dramatic, most easily studied, and more easily handled (by me, at least). We may get the other 3 cases in.

Fear and suspicion have been ticklish problems, and every procedure is suspect. We have convinced many of our integrity, are using no coercion, and find that our entire study depends upon respecting their unwillingness to have a near-fatal case die away from home—or at any rate, here in Moke—unless we get the body back home for burial. We may be able, in a few cases of extreme confidence, to get around this matter. Cases not yet near-terminal could be sent out. Moresby indicated doubt about sending to Melbourne; and we are certain that we are doing more for the patients and for the study here than could be done in Moresby at the moment, with almost 24 hours a day devoted to them by two medical men, not medical assistants or technicians. Dr. Zigas must return to Kainantu now, but will stop back soon again. I shall keep at the kuru problem full time. I am now swamped trying to catch up on full new neurologicals on all the patients. We are sending in some new CSFs and new blood specimens today. If the serum could be used for a few renal function studies such as NPN and BUN and perhaps sugars, total protein, globulins, etc. with a few liver function studies as well—those which would still be valid—the data might be of importance in ruling out any thus-far undetected systemic associated ailment which we could have missed.

Now, Sir Mac, for the information you have requested.

1. I should like to remain in Australian New Guinea until I have exhausted what little I can contribute to this kuru problem on the spot. At the moment, I consider myself the most qualified pediatrician—both clinical and investigative—in the Territory, and the current problem is fundamentally clinical and pediatric in that we have selected pediatric material for concentration. With infant and child LPs, venepunctures, and other procedures more my scope than adult, I doubt that there is anyone around or likely to soon be around who can complete these studies any better than I. I therefore consider it a duty both to the kuru patients and to my intellectual curiosity to stick to it for a month or longer, as the matter works out.

2. Upon my return to the U.S., I hope to soon launch into intensive university-sponsored and foundation-sponsored child studies devoted to the entire problem of child growth and development in primitive cultures, from the point of view of everything from symbolism and language-learning to medical and epidemiological problems of every sort with emphasis upon whatever proves most profitable in a particular setting. There are a number of universities (including Harvard and Western Reserve) and a large number of financing sources (including U.S. Public Health, Ford, and Rockefeller) which have voiced an interest in the project. I have not made firm commitments as yet as to which to accept. However, I certainly do hope to do extensive such work in Australian New Guinea in the future; and knowing of no other group in the world set up for a specific study of "Child Growth and Development in Primitive Cultures," I think it would be depriving ourselves of one of the richest study areas in the world to chalk it off our list. Specifically, I have three studies already under way: West and Central Nakanai, Mamusi-Aveli of New Britain, and Orikaiva of Papua; I also have Cape York Aboriginal studies (with two visits behind me) already moving; and I hope to soon have, in addition to the kuru work, a third here on Keiagana-Kimi-Fore peoples.

I thank you, Sir Mac, for your frankness. I am being equally frank, and I beg that you understand this interest I have, legitimately, in New Guinea. The region is big enough, with enough tribal groups and peoples, to keep all the anthropologists in the world busy for decades; and all the medical research facilities of the U.S., of Australia, and of Britain would not crowd the field. I am not concerned with where the support for the work comes from, what nationality is behind it. Here on kuru we could use immediately a dozen workers—epidemiologists, microbiologists, and pathologists; two dozen would not hurt nor exhaust the problem, and the more who arrive the quicker, the better. It is certainly not a matter of limited problem, limited material; it is a matter in which collaboration from any and every source should be desired. Orthodoxy in intrusion into problems of medical investigation—insofar as ideas of others are not usurped without full acknowledgment—has never bothered me. The problem of medical investigation is an open field, and one which to me has always been noncompetitive.

Please, Sir Mac, accept my apologies for any disruption of plans which my work may have caused. I cannot see that this should be the case. We have found much further information; traced a probable heredofamilial basis for the disease; started to plot out as accurately as possible its clinical course, response to drugs, and geographical-ethnical distribution. This is a solid block of preparation for any further work at any time—immediately or later—which your group or any others might make on the disease. Furthermore, we have gained even further confidence of the people, brought medical attention to their affliction on a large scale—little though it may help them ultimately—and are still doing this.

Thus, Sir Mac, I beg your understanding in permitting my enthusiasm to have intruded a bit on your interests here. I believe the field is amply wide and that no matter how large a study or how many dozen fieldworkers you may have in the field, we could still find plenty of intriguing tribal groups with child development problems to study which you would not have touched. The time for such study is rapidly disappearing; one group will never finish it. And since my own personal interests are anchored to no one in particular but my own interest in study and work, I am fully prepared to have anything I do linked administratively with whatever superstructure you suggest—rather than creating any of my own in the U.S., should that be of some disturbance to Australian workers.

Thank you again for all your encouragement and assistance, and especially for your wonderful tolerance of all the time and facilities my studies in Melbourne required at the Hall Institute. I hope to see you in Europe this summer. I shall be in Sweden and Finland, once work here and in Dutch New Guinea is completed.

Sincerely and respectfully,
D. Carleton Gajdusek

P.S.   We are ready to send out to Moresby a classical young adult patient as soon as Zigas receives authorization to accompany the patient or send his medical assistant with her. The more complex removal of a child from the local situation we can accomplish later, when and if you in Melbourne can take the case and Moresby agrees; however, I think we are doing Moresby a service by sending a bit more typical, older patient, for they will have less difficulty with procedures they may wish to do. An adult without any civilization would be a problem in Melbourne—not in Moresby—whereas our children would not.

I trust, Sir Mac, that you would be willing to act as my reference from the

Australian side, should the Administration require such at any time; from the U.S., the U.S. Academy of Pediatrics, Harvard, and others will serve.

P.P.S. We still would feel, as at every previous phase of this study, greatly aided by the arrival of any of you from Melbourne—there is plenty of work.

BURNET to DCG   9 April 1957
Dear Carleton,

I was pleased to get your letter and hear how much you are getting done in the kuru area. I was particularly glad to get the administrative position from your side clear. Many thanks for your frankness.

My own difficulties, however, are not yet resolved. By all Australian standards, your taking over a job that had been allocated to Anderson was quite indefensible. Anderson has bitterly resented it, and I sympathize with him. On the other hand, as a realist I have got to admit that the Administration was wise in making use of someone uniquely fitted to make a contribution to the kuru problem. From what you have told us and from everything else I have heard, the likely answer is a genetic anomaly, possibly with some environmental factor determining its expression. As such, I agree completely that what is required is primarily a clinical study, plus an elucidation of family relationships which only someone with your linguistic abilities is likely to make a go of. It is also work that can only be done on the spot. As far as we are concerned, no encephalitis virus studies can be done since that is Anderson's province; but I shall be glad to do what I can to channel any other requests (e.g., for brain histology) to competent hands. The offer from Fairfield Hospital to take a case for study if requested by the Administration still holds, but they would naturally like reasonable notice. I look forward to hearing how the pedigrees emerge and hope you will be able to keep us *au fait* with how things progress, even if we can't claim to have any direct association with the project.

I am not making any more plans about eventual activities in New Guinea until things settle down. I agree that there is an infinite amount of interesting work to be done in New Guinea, and I hope Anderson or someone else will keep the Institute active in infectious disease problems there. I see no objection at all to your running child development projects in the primitive areas—there will be plenty of room for any other projects elsewhere.

With all good wishes,
Yours sincerely,
F. M. Burnet

c.c. Dr. Scragg

GUNTHER to BURNET   9 April 1957
My dear Sir Mac,

I am personally disturbed at the turn of events in the matter of kuru.

2. Dr. Gajdusek has been, to say the least of it, unethical. Some time ago I received a radiogram from Canberra, stating that a Dr. Gajdusek wished to come to this country, and would I sponsor him. Thinking he wanted to come and work, as he and I had discussed in your Institute when you were present, I had no hesitation

in recommending that a permit be issued. To this day I have not received a single word from Gajdusek, except that his mail is addressed care of myself.

3. Before seeking a permit to enter, he made no effort to contact me; and on his arrival in the Territory, whilst he saw Dr. Scragg, he failed to have the courtesy to call on me.

4. Dr. Scragg tells me that he stated that primarily he had come up here to patrol with your son [Ian] from Lufa. Scragg suggested that whilst he was there, he might like to investigate the report that there was kuru there. Instead of that—and in this I don't like to describe him as dishonest—he went straight to Kainantu, introduced himself to Dr. Zigas, represented himself as coming on behalf of the Institute and to seek advice for Dr. Anderson. Since then, apparently, he has lived as the personal guest of Dr. Zigas.

5. He was asked by radiogram from Dr. Scragg to leave the area, because arrangement has been made for Dr. Anderson to do research on behalf of the Government of the Commonwealth. He refused, saying that he had a responsibility to the people. He told Dr. Zigas that New Guinea was a Trust Territory, and he personally could do what he liked in it. He has told Zigas that there is plenty of work to be done and that he would welcome Anderson working in the area.

6. I frankly believe that Anderson should proceed. He will be given the full protection of the Administration. We will do what we can to prevent Gajdusek having contact with the people. I do hope you will persuade Dr. Anderson to come as soon as possible.

7. Dr. Scragg will accompany Dr. Anderson to Okapa, and try and persuade Gajdusek to leave the area.

With warm regards,

Yours sincerely,
J. T. Gunther

P.S. Enclosed is a copy of a letter to Gajdusek.

GUNTHER to DCG 9 April 1957
Dear Dr. Gajdusek,

Some time ago I received a letter from the Department of Territories, Canberra, asking that a permit be issued for your entry to the Territory. I had no correspondence from you in this matter, nor have I received any since.

2. From having met you and knowing something of your work, I recommended that a permit be issued. I was amazed, therefore, that you had the discourtesy not to call upon me or make some contact with me while you were in Port Moresby. I am also surprised, to put it mildly, that you did not write to me and have not written to me since of your intentions.

3. For a while we have been seeking the aid of Sir Macfarlane Burnet and his Institute, through the Government of the Commonwealth of Australia, to investigate a disease called kuru. Apparently conscious of these negotiations, and without sponsorship by Sir Macfarlane Burnet or his Institute, you have come to this Territory and are working in a field that we had proposed for Sir Macfarlane. Dr. Scragg tells me that you did not discuss the possibility of coming to Okapa, but rather that you were coming to Lufa.

4. Dr. Zigas now informs me that you told him that you are representing the Institute and would send advice to Dr. Anderson. This certainly does not fit in with the facts now available to me from Sir Macfarlane.

5. Whilst I agree that there may be scope for more and more research within this area, I believe it was grossly unethical for you to enter the area, as you have done, without the approval of either Sir Macfarlane, Dr. Scragg, or myself.

6. No one appreciates the need for improving our services and relieving the suffering of these people more than I do. However, we must proceed steadily and according to plan, so that suddenly no damage is done.

7. I cannot believe for one moment that two or three weeks' delay in the investigation of this disease was of great consequence. Waiting for Dr. Anderson would have given a programmed approach to the problem, and the work done would have been work in which I would have faith.

Yours faithfully,
J. T. Gunther

BURNET to SCRAGG   9 April 1957
Dear Dr. Scragg,

I have had another long letter from Gajdusek, in which he discussed his plans for the future very frankly. I thought that it would be advisable, in order to keep you completely in touch with what is going on, to send a copy of my reply to Gajdusek. You will see that my whole endeavour is to allow the situation to settle down as amicably as possible. I hope very much that Anderson will have a chance to make the next step in the encephalitis work some time in May.

With kind regards,

Yours sincerely,
F. M. Burnet

SMADEL to RIVERS   9 April 1957
Dear Tom:

It was good of you to phone this morning to discuss Gajdusek's problems. Of course, I was sorry to learn that it was not possible to extend Gajdusek's fellowship for several months. However, I was delighted to learn that it might be possible for you to use the administrative authority vested in the Medical Director to make available certain funds which would take care of Gajdusek during the remainder of his stay in the New Guinea jungle and permit him to buy a ticket to come home.

This is to assure you formally that we have begun negotiations with Gajdusek for employment at the NIH when he returns. Furthermore, we are most anxious to have him join our staff. The main reason for not bringing the matter to fruition is Gajdusek's casualness about such mundane affairs as jobs, money, etc. If he ever gets back here, I believe we can hold him down long enough to get the necessary signatures to put him on the payroll.

With the above assurance, may I suggest that you allocate something between $700 and $1000 to Gajdusek for his support for several months to continue his work on the new neurological disease (kuru) among the New Guinea cannibals, and for passage back to Washington. If you find this possible, I would suggest that you send an airmail letter to Gajdusek at: Okapa Patrol Post, Kainantu Subdistrict ADO, Eastern Highlands District, Territory of New Guinea.

Furthermore, it would be a good idea to send the money and a covering note to Sir Macfarlane Burnet at Melbourne. Thanks a lot, Tom, for your help in getting Gajdusek out of the jungle. By this time, Burnet is well aware of Gajdusek's casualness so will probably not be surprised at such an arrangement; if necessary, he can be expected to find Gajdusek to deliver the money.

Sincerely yours,
Joseph E. Smadel, M.D.
Associate Director

BURNET to GUNTHER [mid-April 1957]
[handwritten draft]

I was very pleased to get your letter clarifying your attitude towards Gajdusek's rather extraordinary intrusion into New Guinea, and I thought it might be helpful if I gave you some unofficial and informal background about him.

He is quite an extraordinary individual of American birth but brought up as a child in Central Europe and multilingual. There is no question about his intelligence or training in pediatrics and virology, and I found myself very interested by his enthusiasm for the pediatric and cultural study of the development of children in primitive communities. On the other hand, his personality is quite extraordinary, and is almost legendary amongst my colleagues in the U.S. [John] Enders (Boston) told me that Gajdusek was very bright but you never knew when he would leave off work for a week to study Hegel or a month to go off to work with Hopi Indians. Smadel at Washington said the only way to handle him was to kick him in the tail, hard. Somebody else told me he was fine but there just wasn't anything human about him.

Actually, I got on better than I expected with Gajdusek during his 15 months at the Institute. During the last 4 or 5 months, he did some first-rate work on autoimmune reaction; and we parted on excellent terms. My own summing up was that he had an intelligence quotient up in the 180s and the emotional immaturity of a 15-year-old. He is quite manically energetic when his enthusiasm is roused, and can inspire enthusiasm in his technical assistants. He is completely self-centred, thick-skinned, and inconsiderate, but equally won't let danger, physical difficulty, or other people's feelings interfere in the least with what he wants to do. He apparently has no interest in women but an almost obsessional interest in children, none whatever in clothes and cleanliness; and he can live cheerfully in a slum or a grass hut.

He is not a first-class scientist in any field, but I doubt whether anyone in the world has anything like his knowledge of children in primitive communities in very many parts of the world. I introduced Ian [Burnet] to him while he was in Melbourne, and I knew Gajdusek had ideas of going to see Ian at Lufa when he left; I understood that it was mainly to enquire about head-binding in infants, which he had made some observations on in New Britain. In his last letter to me, in reply to one of which Scragg had a copy, Gajdusek said that his main reason for going again to New Guinea was to find centres for subsequent child study projects which he expected to develop with support from one or another foundation (e.g., Ford) when he returned to the U.S.

Anderson will come up to Moresby as soon as Qantas is functioning again—he hopes immediately after Easter. He finds Gajdusek completely unacceptable as a collaborator, not so much from the current situation but from a deep personal antipathy which developed soon after Gajdusek's arrival here.

As I said, I hope this background may be helpful in dealing with someone who is quite unique in my experience, and I should guess in yours!

GUNTHER to DCG   16 April 1957
Dear Professor,
Since writing to you on 9th April I have received your letter of the 2nd, for which I sincerely thank you.
2. Further, since writing to you I have seen the two cases brought by Zigas. Clinically they are fascinating; I cannot blame you for wishing to partake in any research. However, I do wish you had gone about it differently. It is most unfortunate that you bypassed me and inserted yourself into a situation that you well know was reserved for Sir Macfarlane Burnet's supervision. You have now obviously alienated Sir Macfarlane's goodwill, and perhaps our increasingly close relationships with him and his Institute. You must be well aware that whilst we had to do something, the urgency was not such that we couldn't wait a few weeks.
3. Dr. Zigas has advised me of your intense interest and energy. I, of course, have heard of such attributes before.
4. I think most that I previously wrote was justified under the circumstances, and I feel certain you will agree when the discord settles. In the meantime, I am awaiting further advice from Sir Macfarlane.
5. I conclude by saying, once again, kuru is the most interesting medical epidemiological problem of my generation.

Yours faithfully,
J. T. Gunther

DCG to SMADEL   [20 April 1957]
Dear Joe:
Thank you for the copy of your letter to the NFIP (to Marion Morris). Perhaps I should have written to them myself. In any event, assure them that a series of papers on my Australian work is in press and that they can get all the reprints they need. I shall make certain that Hall Institute sends them manuscripts shortly.

Enclosed are excerpts [10] from my last letter to Sir Mac, which will further acquaint you with our work here on kuru. I am fully at work and things are going well. I have had a bit of a spat in the mails with Sir Mac, which is concerned mostly with Australian fears of having "foreign" (i.e., American) workers studying such an exciting problem in their Territory, and their claim that we are "stealing" a problem from them; but since Vin Zigas (a full-time Administration Medical Officer) and I are the only ones working on kuru, in spite of our pleas—in letter after letter—for colleagues and collaborators, I have felt justified in rather forcibly stating my point. In reply, Sir Mac has been most helpful and kind in his suggestions, in spite of an initial recoil at learning that I was interested and working on a problem which he considered "his territory," for some reason. We are still begging him and the Hall group to go through with plans to come up and join us, as my letters to him—which now number five, as information-laden as the enclosed excerpts indicate—have been pleading. All seems settled now; and as far as fieldwork is concerned, we are mighty busy. (Eighteen cases of Wilson's disease–like "parkinsonism" in children, and a total of over 60 cases now in our hands—with good histories on about a hundred additional documented cases—is a series not to be sneezed at.) I shall eventually show you the full correspondence on the spat—now resolved, but which would have doomed our study had I backed out.

Every bit of progress is being communicated to the Public Health administration in Moresby and to Melbourne as we make it. Our sending two cases to Moresby was expensive (chartered plane, etc.) but a farce, since not a damned thing was really done and our patients were more neglected than they are in our bush hospital. I hesitate to say that so directly to Moresby Administration or Sir Mac, but it was clearly the case; Zigas accompanied them, and what little was done, he did. Vin Zigas will be senior author of whatever we finally get out of this; and, thus far—and unless our Australian collaborators/competitors come through with some help—there will be no others.

Please, Joe, ask someone in the NIH library to dig out reprints on Wilson's hepatolenticular degeneration, and any current work and theories on etiology, epidemiology, copper metabolism, ceruloplasmin, etc. in such disorders, as well as anything on current concepts and therapy of parkinsonism-like syndromes, and send them to me at the same address you used before, by airmail. Obviously, I could use some such leads here in the bush. Also, the papers on the Guam (if I recall correctly) "epidemic proportions" of amyotrophic lateral sclerosis—if that was it—for we have a somewhat similar situation, although a different degenerative CNS disorder. Any other ideas and suggestions will be immensely helpful, including any drugs you think might be critically tested in parkinsonism, for we are really equipped to test them on one, two, or three dozen cases, with controls or placebos.

Airmail reaches Kainantu, and an arduous jeep-track reaches us in the bush at Moke if we are not off on patrol. So we are in contact with the world.
[11]

Sincerely,
Carleton

P.S. Any of the drugs (or others) mentioned in my letter to Sir Mac would be most welcome (i.e., Artane, Parsidol, etc.).

DCG to BURNET 20 April 1957
Dear Sir Mac,
I thank you for your clarifying letter of April 9; and I am most disappointed to learn that Anderson has interpreted my work here, which I had planned as simply some clinical pediatric observation for my own intellectual curiosity and as material assistance to his intensive microbiological epidemiology, as a field-pinching maneuver. Zigas and I had, in our first weeks of study (in anticipation of Anderson's arrival), concentrated only upon gathering and ferreting out cases and observing

them clinically—with emphatic attention to the child patients. The family-history–taking has developed into an art which involves the use of the Fore and Keiagana medical vocabulary and nosological classificatory system. Fortunately, all cross-checking has revealed an astonishing accuracy in their diagnoses, and especially in their spotting early kuru cases and "knowing kuru when they see it." Thus, we have been able to dig up fairly extensive family histories on many of our patients. These cannot be forced back more than a generation or two, but our concentration on child cases again has an advantage here. Old grandparents are apt to be alive—and thus, when found, serve to give us the history of the paternal and maternal families at least one (often two) generations back. In no case can we find individuals who can supply histories of events prior to their having reached childhood; but the old folk do well in accurate recall during their lifetime, and cross-checking of their stories usually provides support for all they report.

By this technique it has been possible to find kuru in the parents, siblings, children, or aunts, uncles, and cousins (by our classification) in most of our patients. Since the disease is in high incidence, this in itself would not be remarkable; but the occasional discovery of a family of from two to four siblings all of whom have died of kuru—and good verification of this story—forces genetic considerations upon us. In fact, the incidence of two to four kuru deaths in siblings in one family, with these cases developing years apart and often at very close to the same age in all children of that family, is the strongest bit of heredofamilial ataxia evidence we have. In addition, there is already some tendency evident for the so-called "law of anticipation" to hold: thus, the children of an affected parent, or of a parent in whose family the disease has occurred, seem to have the illness at an earlier age than it appeared in the parental family.

Finally, we have now, in our case-hunting, been patrolling far afield and begun to establish the limits of the disease. Nothing is as yet certain, for we have undertaken an immense job in this survey—which involves prolonged monotonous discussion, interviewing, and questioning in every one of the hundred-odd villages of the affected tribal and linguistic groups (questioning which has been exhausting not only me but my informants as well). But as my command of Fore and Keiagana begins to surpass my current smatterings, things are working out well. It now seems evident that spread from the Fore to Keiagana, Kanite, Auiana, and Kimi linguistic groups has been restricted to only those adjacent regions of probable or known intermarriage. The verification of these sharp lines of demarcation is a most important matter, for thus far we can spot no cultural epidemiological features of ecology, geology, crops, diet, etc. which could account for such a sharp demarcation other than intermarriage.

Finally, I am most aware of the intense current interest, in biochemical and neuropsychiatric circles, in the possible metabolic-biochemical basis for hepatolenticular degeneration, which—symptomatically, at least—so closely parallels kuru. If the primary defect in the disease does turn out to be the gene-determined synthetic deficit which prevents normal production of ceruloplasmin—the copper-binding globulin—then the environmental factor which might determine expression of the genetic deficit could be sought and perhaps even controlled. We are trying to pursue this line of thought by planning to send food specimens, etc. to the Department of Agriculture in Moresby, where Dr. Zigas has already arranged for copper-levels on blood specimens on two cases. Trace-metal determinations in clinical medicine are fraught with danger, and we shall have to run many a control before we can believe anything that turns up. Our two patients had quite normal blood-copper levels, however. We shall send more—along with controls this time. We have, incidentally, found no evidence for liver disease of any sort as yet.

We have now located over 60 active cases of kuru, and none of our patients have yet died. Of this 60, we have carefully and fully examined some 40 cases, on whom lumbar punctures have been done. Two classical cases of moderately advanced disease have gone to Moresby with Vin Zigas for one week and have now returned to Moke, where morale is again up among our patients. Moresby was most rushed with an overload of patients for the thoracic surgery team from Germany, and it was difficult to get things done. However, Dr. Price in Pathology gave great assistance and did several additional tests for us. No abnormalities of liver function were found in liver function tests. Unfortunately, EEGs, pneumoencephalograms, etc. were impossible. We shall continue to consider the possibility of sending a patient to Melbourne eventually, but can now sit quietly thinking over the matter and planning. I would still suggest your considering taking a case eventually, since there is no EEG (to my knowledge) at Fairfield, and since Dr. Wood's service is so well equipped for clinical research.

At the moment, we have 18 child patients, some 6 of them near death but maintained on supportive measures (as many neurologically deficient patients can be somewhat indefinitely). However, decubitus ulcers are defeating us in two cases at present, and we may soon have a chance at autopsy material. Whether we shall succeed in getting full-autopsy permission or not is a bit uncertain. We shall do our best, at the risk of disrupting the slowly established confidence which we have gained from our other patients.

The patrolling and extensive history-taking, the statistical, historical, and genealogical studies will take time. We shall take the liberty of sending you further specimens for clinical chemistry; and if at any time anything turns up that may offer a microbiological lead, we shall send this on to you and Anderson immediately. We have been expecting Anderson any day, but from your recent letter do not know whether to still expect him or not. Please tell him that we have been and are anxiously anticipating his arrival.

We are trying to get our hands on Artane (trihexyphenidyl), Parsidol (diethylamino-2-propyl-1-N-phenothiazine) (called Lysivane in England), and other drugs conceivably helpful in parkinsonism—such as the trial use of BAL, calcium versenate, and potassium sulfide (orally) as Cartwright et al. used in trying to control copper uptake and excretion in Wilson's hepatolenticular degeneration—for clinical trials. Thus far, antihistamines, low doses of ACTH, sulfonamides, chloramphenicol, high vitamins and iron therapy, and phenobarbital have been of no effect. Dr. Scragg is sending us Tridione, Mesantoin, and other anticonvulsants to use in trial; and we may get enough testosterone to try the wild lead suggested by expression of the disease almost entirely in prepubertal males and in young and adult females but very rarely in postadolescent males.

Thank you again, Sir Mac, for your suggestions and interest. Please show Anderson this letter and tell him that I am most sorry about the misunderstanding. Had I been told more about kuru in Melbourne, discussions on the spot would have obviated any possible misunderstandings. Rest assured that Vin Zigas and I shall keep you informed of our every move.

I hope to see Ian [Burnet] eventually, when I return to Goroka. Vin Zigas has learned that the one kuru case reportedly seen in the Lufa region now appears to have been a Fore native who wandered over that way. We are now trying to determine whether his disease developed outside of the Fore region—a most crucial matter, for it alone would remove suspicion of local environmental factors in our snooping throughout the region.

Please send my greetings to Pat and Lois, to Anderson and French, Ada, Gottschalk, Dineen, Spector, Wood, and Ian [MacKay] and all others in Melbourne who may inquire about me.

Sincerely,
D. C. Gajdusek

c.c. Dr. Gunther

[12]
DCG to GUNTHER   20 April 1957
Dear Dr. Gunther,

I am sorry to have a note from you which indicates that you are most annoyed with my having missed you in Moresby, and with your ignorance of my planned work in the Territory. I had indicated in my brief talks with you in Melbourne with Sir Mac that I was planning to return to the Territories to complete the work of last year, and also indicated my hopes of setting out for new locations for long-term child growth and development studies which embrace the entire field of child developmental and disease patterns. This plan you apparently concurred with, and I saw no need of further burdening you with correspondence when I applied for entrance to the Territory. Thank you for having taken care of the arrangements as you did.

On my arrival in Moresby I immediately asked to see you, and was informed that you would be busy with administrative duties and that I should work through the Department of Health, which I did fully. I asked specifically on Friday March 8 to see you, and was told then that you were at home ill; I had the entire day available, with no other plan or hope but that of seeing you. At the Public Health Office, the office of the Department of Native Affairs, and at the Administrative Offices themselves—all three of which I visited inquiring for you—I was told that:
1) you were most tied up with administrative functions and duties, and that it would be rather uncertain as to when I might see you;
2) you were ill and at home, and that I would be ill-advised to disturb you.

Thus, I discussed every aspect of my plans with all three departments of your government, and set forth for the Highlands on Tuesday March 12, after having failed again on Monday March 11 to make contact with you. Dr. Scragg gave me every possible assistance; and I told him clearly that I was on my way to Southeast Asia, Europe, and home to the USA after having completed a year and one-half of work at the Hall Institute. I asked him specifically for suggestions as to where in the Territory somewhat isolated and/or resistant-to-cultural-change areas were to be found where we might plan future long-term child growth and development studies, and he gave me kind advice and assistance which was most satisfactory. I specifically asked him for areas of skull-flattening, strange diseases, strange or unusual child-care practices, etc.; and he pointed out to me the Okapa region as one such. [13]

I had heard brief mention thereof in Melbourne, and having learned in Moresby more of kuru and the nature of the region and of the Hall Institute's planned virus studies there, I asked specifically to see the disease and the Okapa region. [14] Furthermore, I addressed a note [15] to Sir Mac, telling him of my interest and expressing a bit of surprise that I had left Melbourne after months of open discussion of New Guinea matters—including my own plans and hopes and interests and previous studies—without having been told about kuru more specifically. When I left Moresby, I specifically planned with all departments and District Commissioner Seale to visit *both* Lufa and Okapa before heading for Tari and other regions; and at Goroka it turned out that Lufa—which was my first intention—was impossible, with Ian Burnet just called out from there for patrol near Goroka. Okapa was then chosen for me as the best first choice, although all plans had been to spend the weeks waiting for Anderson at Lufa.

Immediately upon contact with Dr. Zigas, it was evident that kuru presented problems of pediatric and child growth interest; and I promptly addressed a long note to Sir Mac explaining that I was going into Okapa to look at the disease clinically, to concentrate on pediatric aspects and the study of its effects on children, and urging him to let me know when Anderson and the Hall Institute group might arrive. A few days later it was already evident that the disease was *not* quite what I had been led to expect in Moresby (i.e., most probably a post-infectious problem or a toxic pattern); and I thought it my duty to advise him promptly thereof and to nevertheless urge him and Anderson to take immediate great interest in it since, although the post-infectious possibility and the toxic possibility could not be completely excluded, an even more interesting possibility—that of a genetically determined heredofamilial degenerative disorder of the CNS (especially of the corpus striatum) in almost "epidemic" proportions—was now forcing itself upon our consideration. In reply to my third lengthy letter of preliminary information in anticipation of the Hall Institute study—my own work had been only clinical observation and concentration on pediatric clinical aspects until that point—Sir Mac addressed to me an annoyed note which indicated more concern then for Administrative protocol than scientific interest in elucidating a fascinating problem. I promptly replied with a full statement of the state of my work, of my hopes and anticipations for complete study of the problem by the Hall group—to whom we had addressed all clinical specimens, along with complete account of every aspect of our work—and an appeal for epidemiological study of a vast problem for which the arrival of several dozen qualified investigators would hardly crowd the field, *then as now*. It was most flattering to see suspicions that I might discover all there is to discover about such an illness—whereas parkinsonism, post-encephalitic parkinsonism, and Wilson's hepatolenticular degeneration are still great unsolved mysteries to us in Europe, Britain, and the U.S.

In reply to my letter (the fourth), Sir Mac has now indicated that he favours the approach Dr. Zigas and I are taking, and he gives us fine advice and encouragement. I am myself deeply disturbed that my preliminary observations—rushed to Melbourne to better equip Anderson in preparing his expedition here—were instead interpreted as a stealing of the field from them, a field which I fear a dozen years of study will not exhaust. Dr. Zigas and I have been anticipating Anderson's arrival ever since we started work, and still hope he may come. We have certainly not eliminated toxic or post-infectious factors, although they appear ever more unlikely. For having spotted this early and for having seen fit to warn my Australian colleagues, I found myself "in hot water" in the bush; and Dr. Scragg's suggestion that I retire to Lufa came at the inopportune time when I was saddled with a huge epidemic of upper-respiratory disease in infants with a high incidence of pulmonary complications and over 100 cases of similar illness in school children, as well as meningococcal and pneumococcal meningitis—all in addition to our kuru studies. I promptly replied that I thought moving then inadvisable, and that interruption of our kuru case studies and the case-hunting which we were doing in anticipation of the Hall group's arrival would, at that stage, jeopardize the entire study for them (as it most certainly would have done). I begged to be permitted to await Anderson at Okapa.

I am sorry, Dr. Gunther, at the extent of our misunderstandings, and hope that this historical exposition may serve as some explanation thereof. I am writing at the moment to Sir Mac with further observations and with a full account of our more recent findings and current studies, and I feel certain that he and Anderson will have little cause for chagrin therein. There is a great deal more to be done on kuru; the disease will remain, I fear, long after Zigas and I have finished our current studies. It is a matter which your Medical Service—through the astute observations and remarkable zeal of Vin Zigas—has done well to bring to the attention of medical research, for it will be a challenge to all of world medicine, and one not easily solved.

I shall send to the Department of Health and also to you directly, Dr. Gunther, detailed accounts of our work. If they are at times in the form of carbon-copies of letters to Sir Mac, please pardon this inconvenience; we are working hard in the bush with minimal assistance and facilities. I thank you for your frankness, and hope you will accept my apologies for the inadvertent insult my failure to see you in Moresby confronted you with. Expecting, furthermore, your pardon for the boorish American use of a terminal preposition in my penultimate sentence, I remain

Respectfully and sincerely yours,
D. Carleton Gajdusek

[handwritten]   Please, Dr. Gunther, show this correspondence to Dr. Scragg; I shall send Sir Mac a carbon.

GUNTHER to BURNET   24 April 1957
Dear Sir Mac,

Thank you for your very interesting letter re Gajdusek.

There is no doubt he is erudite. The day after I posted my letter to him expressing annoyance, I received a most charming letter from him, but without any explanation of what and why he was doing it.

With warm regards,

Yours sincerely,
J. T. Gunther

GUNTHER to DCG   3 May 1957
Dear Dr. Gajdusek,
  We seem to be sorting ourselves out. I want to see kuru research stabilised. It is my desire that Sir Macfarlane assumes the mantle of Director, and Anderson his Field Manager who will coordinate the Administration's efforts with those of others involved in the field.
  The humanitarian problem is paramount, of course. In all our thinking, the end result must be the relief of this disease and its eradication. If it is possible that this can be advanced before we have completed epidemiological studies or the aetiology wholly known, then this must be done. You, more clearly than I, will see the need for a coordination of effort. We will make our funds and supplies available for this coordination through one central authority.
                                                          Yours sincerely,
                                                            J. T. Gunther

DCG to WOOD   Kainantu
                                                                  5 May 1957
Dear Dr. Wood,
  Dr. Anderson has just arrived at Moke to look at our kuru cases, and after a brief visit has returned to Kainantu. I am with him, for I am going on to Goroka and Okapa and down to Lufa to look at a few stray cases of parkinsonism collected outside of the kuru-region to determine, if possible, whether or not they are kuru. Anderson has decided to write you a report shortly, but believes that at the moment any infectious disease epidemiology is a long shot hardly worth taking. Our primary concern—that of getting brain and other material for neurohistology and general pathology—remains unsolved, for we have not had a death in the 66 kuru cases we have now collected, and the few deaths from kuru in the past month have been in cases not discovered by us until they had died. We now have our hands on half-a-dozen near-terminal cases; but to sit by them in order to be there when and if they die disrupts most other study possibilities, and in addition it is almost impossible to convince the natives to stay with us as death approaches, for they want above all else to be on home-ground at death. Thus, they are willing to come to us and to let us try every possible procedure and therapeutic test until they think the patients are near-terminal. Then they must be at home—often days of rugged trails away. However, we are hanging on to a few near-terminal cases, and hoping to be able to secure autopsies without disrupting the confidence we have laboriously won from the people.
  I have been patrolling extensively in an attempt to find the exact demarcation of the kuru area. It appears that it is going to be most sharp, and that study of local factors—dyes, ectoparasites, mineral deficiencies or excesses, and other ethnocultural matters—will best be done at a line of such sharp kuru/non-kuru demarcation. In all, we have done lumbar punctures on 45 kuru cases and examined another 20. We usually keep a hospital census of from 15 to 25 cases—predominantly children, on whom my studies are concentrating. However, I could round up as many as 30 or 40 hospital cases at one time at Moke (as I have in the past) for special projects such as an extensive motion-picture of all phases of the illness, which we are contemplating.

I actually am writing to let you and Ian [MacKay] know that I received along with Anderson's visit the manuscript of [AICF] paper II (which looks fine, but which I have not yet been able to sit down and read) and the information about the change of journal. Admittedly, the papers are very lengthy and involved, and—especially since they should be published together—it is a bit much to expect of the *JAMA*. I think the matter is fine the way you have left it, and I am quite satisfied to have them appear in the *Archives* [*of Internal Medicine*]. I shall look over the papers soon, and send a note of comment shortly. Thank you for all your help, assistance, and cooperation. Thank Ian for me, if you will, for all his work, and again accept my gratitude for all your enthusiastic help and encouragement during the year-and-a-half I spent in Melbourne. Gray tells me of your intense interest in kuru and willingness to do anything that may help. I thank you greatly. It is a truly fascinating disease, and one which, I fear, we shall not solve—any more than Wilson's disease, parkinsonism, or even Von Economo's post-encephalitic parkinsonism have ever been "solved." Yet it is a most intriguing challenge, and we have only begun to work. I shall keep you and Sir Mac completely informed of every stage of our progress and of anything new that turns up.

I am intrigued that manioc has come into the area about coincident with many accounts of kuru history and origin. Since it is a prussic acid–containing tuber (which, if not properly handled, is toxic), I am wondering whether any type of prussic acid poisoning, under any extraneous circumstances, can lead to parkinsonism or basal gangliar disease. The diet study we are launching may help on this; thus far we can spot nothing unusual in Fore diet customs or practices.

Ceruloplasmin studies and the use of BAL in Wilson's disease are intriguing, and we shall follow through on these as far as we can. I shall make good use of the chemically cleaned glassware Gray has brought up, and send specimens for Cu and Mn studies to him and Mr. Holden.

Gray will talk to you about the possibilities of our getting any further liver-function or other metabolic studies done on field-collected specimens. At the moment, I am a bit interested in urine amino acids, etc.—in line with cysteiuria, phenylketonuria (in phenylpyruvic oligophrenia), various fructosurias, and ochronosis-type phenomena—for some very advanced cases have acid urines with many cystine crystals in them; this is, considering their state, probably nothing unusual. However, these other genetically linked metabolic leads should be followed through, as well as trace-metal possibilities.

Please show this note to Sir Mac and tell him that shortly I shall have a much more extensive report for him, with (I hope) some real concrete statistics on incidence, death-rate, geographic distribution, age- and sex-preference, etc. Our earlier guesses seem to be holding up. Somewhere around one percent of the inhabitants of the kuru-involved region have active kuru at the moment, whereas in a few centers from two to four percent of the population have the disease. It has accounted for up to one-half or more of the deaths in some centers in the past five years, much fewer in others.

My greetings to everyone in your unit, to Ian [MacKay] and Sara [Weiden] and Taft especially for all their help and cooperation, and to your ward staff for all their enthusiastic aid.

Gray will have much more to tell you. I am sorry that he arrived without our knowing in advance, for I had only 17 cases on hand to show him, whereas a few days earlier I had three dozen. But not expecting anyone, I had conceded to their insistent pleas for discharge. He saw all phases of the classical disease, however. Lois should have a few hundred Kodachrome slides of the region, and of kuru cases and other illnesses.

I have asked that they be sent to us here, heavily insured. However, please feel free to look at them all and show them about before she sends them off. I hope they are okay.

Please see that a copy of the *Nature* note gets off to me when it finally appears, for I have no manuscript of it with me.

My greetings to your daughters and to Mrs. Wood.

Sincerely and respectfully yours,
Carleton

ANDERSON to BURNET
Kainantu
5 May 1957

Dear Sir Mac,

I met Dr. Gunther at Moresby and Dr. Scragg at Lae on Friday; and Scragg and I came to Kainantu Saturday, visited Okapa, saw 16 cases of kuru, and discussed the situation with Gajdusek.

Scragg and I are quite happy to leave Gajdusek to complete his part of the work in one or two more months. Zigas is well in on the work.

Scragg will give necessary support and provide New Guinea consultants in nutrition, botany, anthropology, and entomology for Gajdusek and Zigas. I have agreed that, subject to your agreement, the Hall Institute will do whatever it can to assist. At present the only assistance requested will be trace-metal estimations—Cu,Mn,Co,Hg, and C—and perhaps liver function tests and qualitative amino acid estimations, plus a little serology.

I anticipate being in Moresby on Wednesday, 8 May, and perhaps coming to Australia soon thereafter.

Yours sincerely,
Gray

[16]

ANDERSON to ZIGAS
Goroka
7 May 1957

Dear Vin,

First I want to say how grateful I am for the hospitality Mrs. Zigas and yourself gave me. I certainly hope we can see you in Melbourne soon.

We arrived in Goroka as planned and saw 4 cases sent from Chimbu:
1. Kadaga   M 46   Chuave. Ill "long time." Had a spear injury. Does not resemble kuru.
2. Kameia   M 38   Kamoneku (Chimbu). Ill 15 months. ?Encephalitis lethargica. Not like kuru.

3. Pile   M 38   Oldar (Chimbu). Ill 3 months. ?Encephalitis lethargica. More like kuru than 1. or 2., but probably *not* kuru. This man may remain at Goroka for several weeks.
4. [name and specifics not given] Birth injury. ?Erb's. Possibly an early tremor of healthy arm, but very slight. Probably *not* kuru.

Dr. Scragg also saw these, says categorically they are *not* kuru.

The film was projected last night and came out well.

I am beginning to wonder whether kuru might not be due to lesions elsewhere than in the basal nuclei. Perhaps cerebellar, midbrain, or spinocerebellar tracts.

With best wishes to yourself and Carleton.
Gray Anderson

BURNET to SCRAGG   14 May 1957
Dear Dr. Scragg,

Dr. Anderson has given me an account of his visit to New Guinea, and I am so glad that things are now running smoothly in the kuru investigation.

He mentioned his discussions with you and Dr. Gunther on plans for research in the Territory, and said that he had told you I should be in London for a short visit soon (June 24–July 6). If there is anything I can do by discussion with people in London to assist in obtaining a suitable man as a director of research, I shall be glad to help.

I leave Sydney for London on June 14 at 3 p.m.; and if, as seems possible, you are in Sydney that morning and thought a discussion would be advisable, I shall be at the Wentworth Hotel. I come up from Melbourne on the afternoon or evening of June 13.

With thanks for your help to Dr. Anderson,

Yours sincerely,
F. M. Burnet

CLELAND: Report to Department of Territories, Canberra   Port Moresby
The Secretary   [n.d.]
Department of Territories
Re Dr. S. G. Anderson: Encephalitis Epidemic,
South Fore Area—Investigation

Dr. Anderson arrived in Lae on the 3rd May and proceeded immediately, with the Acting Director of Public Health [Scragg], to the Okapa area to consider the present position concerning kuru at Okapa.

Detailed discussions took place in which the following programme was decided upon, and this programme has already been put into operation:

1. Dr. Gajdusek has established himself in the area and it is apparent that, at the present stage of investigations, he is the best man possible for the duties involved. We are endeavouring, by casting a broad net, and under his supervision, to find some definite line of investigation that we can follow. On the finding of such a lead, then other specialists available in the Territory will be sent to the area, or we will ask your permission for visits of other specialists from Australia. Dr. Gajdusek has agreed that, in any publications regarding the results of the investigations, he will

place Dr. Zigas, our own preliminary investigator and the original finder of the condition, in the position of senior author. While we have no written guarantee of this, we must rest on Dr. Gajdusek's honesty as a scientist to give due recognition to Dr. Zigas and also to the Department of Public Health for the assistance and help they have given him and will continue to give in the future.

2. The role of the Walter and Eliza Hall Institute and of Dr. Anderson will be that he arrange full liaison in regard to the processing of any pathological or other material that we are unable to deal with in the Territory. This also envisages that, should some definite indication as to the cause come to view, then the Walter and Eliza Hall Institute would concentrate on the detailed investigations of this cause. If it appears to be associated with an encephalitis of virus origin, then it would be necessary for Dr. Anderson to undertake full field investigations along this line. Amongst the investigations to be done in Australia will be detailed histological examination of brains, examination of blood for trace-metal content, and examination of spinal fluids.

3. Visits are being arranged for the following local specialists to the area:

Dr. W. E. Smythe, linguistic expert and pathologist, with his detailed knowledge of the language, to determine the history of the condition and to ascertain more definite genealogical information as to whether the condition is hereditary or not.

Miss Hamilton, dietitian, will visit the area in about a month's time to determine whether there is any evidence of any peculiar food habits or methods of treating food which may result in chronic poisoning.

A visit from the chief of the Division of Botany, Mr. John Womersley, may be necessary following the visit of Miss Hamilton to determine the nature of any unknown plants consumed.

Dr. Campbell, a specialist-physician, will visit the area shortly to investigate the neurological aspects of the problem. If he is unable to make a definite decision as to its pathological basis, it may be necessary for us to obtain a specialist-neurologist from Australia.

An entomologist will determine the present rodent ectoparasites and mosquitoes and make a comparative study of two areas.

4. A Medical Assistant has been stationed in the area, and he will proceed with the establishment of a hospital to look after terminal patients and to endeavour to determine whether there is any prodromal condition which we may have missed. It is intended that the hospital be maintained in the area until such time as finality is reached or it is apparent that the condition is insoluble. He will also, under the direction of Dr. Zigas and Dr. Gajdusek, determine the actual areas involved by the condition. Such a delineation will necessitate a patrol over the Lamari River into the uncontrolled Kukukuku area, and a visit from Okapa to Lufa to determine the extent in the opposite direction. The Medical Assistant will also, on the departure of Dr. Gajdusek in approximately two months time, maintain a continuous follow-up and investigation of all new cases and keep Dr. Zigas in touch with conditions and developments in the area.

5. Dr. Zigas and Dr. Gajdusek have made the following determinations with regard to the condition. They have found 60 cases, and it appears to affect approximately 1% of the population of 15,000; one third of the cases are children from the age of four; one third are female adults between the ages of 20 and 40; and the remainder are split up into other age groups; overall, there are approxi-

mately ten women affected to one man; there appears to be some familial incidence, but it is not definite enough yet to make any genetic determination. They have probable evidence of one case recovering, but with a further period in the area they may have more evidence on this matter.

D. M. Cleland
Administrator

[17]

DCG to SMADEL [mid-May 1957]
Dear Joe,
I thank you for all your efforts and successes on my behalf, and I assure you that I am fully at work on kuru, perhaps the most important epidemiological problem in the world at the moment; in fact, it potentially holds many leads. It is certainly the most fascinating that could be visualized in any part of the world.

. . . I write at the moment to let you know that we have had a kuru death and a complete autopsy. I did it at 2 a.m., during a howling storm, in a native hut, by lantern light; and I sectioned the brain without a brain-knife. But the brain, in 10% formol-saline, is off to Melbourne for neuropathology—along with pieces of all organs I sampled, fixed both in Zenker's and 10% formol-saline. We are sending to you by air freight three small tubes containing fixed tissue specimens in 10% formalin in saline (10% formaldehyde); in addition, there are three small tubes with Zenker's-fixed tissue. Sorting the mixed tissue out, once fixed, is fraught with some uncertainty and I may have been wrong in identifying them. However, here is a list of what I think you are getting. Tissues assorted among the three bottles: liver, kidney, muscle and the fascia of anterior abdominal wall, lung, pancreas, aortic wall, section of cervical cord, ovary, a piece of cerebral cortex, and the spleen. There may be some adrenal and pancreas and heart tissue somewhere in these formaldehyde-fixed specimens; the pituitary in its entirety is off to Melbourne.

[18] Every bit of pathology which can be thought up would be much appreciated, although grossly we found no abnormalities. The enclosed carbon-copy of our clinical record—incidentally, we are keeping similar clinical records on some 60 patients—will give you some details about this child patient who died of typical kuru. The next autopsy I shall try to get to you in entirety, rather than to Melbourne. The first goes off to the Royal Melbourne Hospital.

Joe, there has been much politics and jealousy coupled with the study of kuru by Vin and myself. No other investigator has even looked at the illness, until S. G. Anderson paid us a half-hour political visit—designed mostly to smooth over controversies with the Administration, which Sir Mac had stirred up by noting that it was a bit unfortunate that Australian workers did not have this study in their hands. However, Sir Mac was told off in a letter by me in reply to his notes, and since then has been most cooperative and helpful and has offered every assistance possible. In fact, I hear on the Australian National Broadcasting network about the great team of experts at work on this scourge. All this while I am struggling to get Gram's stains working, Leishman's stains prepared, and our Pandy's solution standardized—without a hint of help from anyone thus far but Vin Zigas, who is stationed in

Kainantu and takes the risky mountain-track jeep trail here at every possible opportunity, spending every bit of time he can with me. I have a native hospital, built with the help of the patrol officer, the only other European here in this area of 30,000 native population, much of which is "uncontrolled." *Tukabu,* the ritual formalized murders, keep him busy when he is not dealing with cargo cults in the more sophisticated of our Fore and Kimi and Auiana neighbors. The natives have been fine friends, and have thus far cooperated well. We have 22 cases now in the hospital, over half of them children, and I shall regularly be bringing in some of the 40-odd others back for follow-up. They all (with a few notable exceptions) seem to be following much the same rapidly progressive course, and the uniformity of the syndrome is remarkable.

Heavy-metal toxicity is a thing we are pursuing, and a clinical trial of BAL is underway. I have not yet seen any response to antibiotics, phenobarbital, Benadryl or Pyribenzamine, cortisone acetate, ACTH, aspirin, or vitamin preparations. We have had a remarkable chance to evaluate these potential therapies, but our patients have degenerated while on high doses of each of these. If you can get any of the tranquilizers to me—ample dosage and literature for treatment of a half-dozen or dozen cases—or, more particularly, any of the new anti-parkinsonism drugs, I should be most anxious to try them. Here, I have no source of supply, although the Administration (which is flaunting its highly publicized intensive investigation of the scourge "by a great team of experts," 100% of which Vin and I do) is really supplying us with about whatever we need. I have the native-material house nearly finished and will be able to move our lumbar-punctures, autopsies, and urinalyses and reagent preparations off the *kiap*'s (i.e., the local patrol officer's) dining table shortly. In addition, I am sending you under separate cover a stack of carbon-copies of some of our case records—minus the temperature and treatment charts, which I do not have in duplicate. These you may study, but please hang on to them for me.

I have a letter from Geoff Edsall informing me that my manuscripts will be awaiting me in Kuala Lumpur. It will be at least a month or two before I can wind up the current phase of kuru research and head, via Dutch New Guinea and Borneo, for Malaya. Kuru is a most difficult thing to abandon; it is too good a problem. [19]

May I hear from you again? Any ideas and suggestions will be most appreciated. No news as yet from your neuropathology experts?

Sincerely,
Carleton

ANDERSON to ZIGAS and DCG 16 May 1957
Dear Vin and Carleton,

I had a comfortable trip back to Melbourne, and have discussed matters with Sir Mac Burnet.

Mr. Baseden of the soil chemistry laboratory of the Department of Agriculture, Port Moresby, says he is technically able to do all qualitative trace-metal estimates likely to be required, and permission for this is being sought from Mr. L. Dwyer through Dr. R. Scragg.

Would you please send all samples for trace elements to Baseden or Dr. Price, but not to us, and also make letter contact with Baseden? He will send us appropriate portions for qualitative spectrographic analysis, if I can arrange for this to be done in Australia.

Simmons will be writing to you soon. Blood-groups distribution is fairly normal, but there may be some unusual paper-chromatographic features.

You might bear in mind ritual painting with toxic ores or dyes.

Dr. D. Cowling at R.M.H. would be happy to examine blood smears—thin films preferably on slides, unfixed. If you wish to send oxalate blood, he will do haemoglobins.

All best wishes.

Yours sincerely,
Gray Anderson

ANDERSON to DCG and ZIGAS   17 May 1957

Dear Carleton and Vin,

A few additions to the enclosed letter.

1. You may have found 100 embossed aluminum-foil bands in the parcel. Would you please send them airmail to Dr. Price, Ela Beach, Port Moresby?

2. We are posting you bottles of sterile glycerine, prepared:

$$\left.\begin{array}{l}\text{45cc glycerine}\\ \text{0.9 g. NaCl}\\ \text{105 distilled water}\end{array}\right\} \text{135 cc of this solution, add} \left.\begin{array}{l}\frac{M}{15}\,Na_2HPO_4 \quad 5 \text{ cc}\\ \\ \frac{M}{15}\,KH_2PO_4 \quad 1.0 \text{ cc}\end{array}\right\} \begin{array}{l}\text{autoclave}\\ \text{pH 7.4}\end{array}$$

3. Dr. Wood advises that the liver function tests run at Moresby cannot be added to.

4. Dr. Ian MacKay is investigating the possibility that amino acids in urine could be determined. If they are going to do this, they would like serum too.

5. If you send down brain in formalin, Dr. Holden will run some of the brain for metal content. To this end, would you send some formalin in another container for control testing for the same metals.

6. There is a good article on manganese poisoning in Moroccan miners in *Brit. J. Indust. Medicine*, 1955, vol. 12, p. 21. The clinical picture is like kuru; it may begin during or several years after exposure to manganese. White cells—relative lymphocytosis; 17 ketosteroids in urine diminished; BMR increased; manganese eliminated in faeces (50 mg/kg) and hair, but not often in urine.

7. I will be making a formal report to the Administration, at the request of the Secretary of the Department of Territories.

8. Thinking the matter over, I believe that the relative absence of the disease in males may be a most important clue.

Yours ever,
Gray

DCG to BURNET and ANDERSON 19 May 1957
Dear Sir Mac and Gray,

We are working ahead as rapidly as possible, but the mere routines of examining our 24 now-hospitalized kuru patients and arranging for follow-up visits from the 40 others keeps us working from dawn to midnight. There is less to report than previously. I have now patrolled about half of the Fore population, and our total list of active kuru patients has not yet reached 70. Documented deaths in the past five years are probably above 200 (there is no time to analyze our data now—perhaps 300), and incidence figures remain much the same as my former estimates.

Vin Zigas has his hands full trying to cover medicine at Kainantu and spend as much time here as possible, and when he is away (as during the last ten days) it is risky for me to be away from our Hospital doing epidemiology or case-hunting—without which, incidentally, new cases do not turn up. A death would be whisked away back to the village and terminal patients whisked off were I not around to continually bribe, cajole, and insist. We have had several terminal cases run off, and duress in such matters is against my principles.

Anderson's visit was most welcome; but we were most disappointed that he could not remain here with us to work along on the project, which is still an immense project of study and investigation, I fear. That infectious disease epidemiology might be worth having, he and I and Vin appreciate; but we must admit with him that it is very unlikely to pay off, and the great effort and time it would cost is not likely to be well repaid, I fear.

Last week we had our first death in the Hospital of a kuru patient; and at 2 a.m. in rain and cold wind, a *docta boi* and I did a complete autopsy in our treatment/laboratory hut by lantern light, and then at first cockcrow got the body borne homeward with the mourning mother well rewarded with axes and salt and *laplap* (which she appeared to take as much interest in as in the death—which, it must be said, she had long, long ago resigned herself to). Thus, we got the "dastardly deed" done without awakening much local curiosity or attracting too much attention to our butchery. There were no gross pathological findings of interest. All thoracic and abdominal viscera were disgustingly normal, with no sign of any pathology. Liver showed no sign of any hepatolenticular degeneration–like pathology (hobnail, if I recall correctly), with no multilobular degenerative changes. Spleen, kidneys, intestines, adrenal all looked normal. Adrenals a bit small perhaps, and kidneys a bit hyperemic, I thought. Brain and meninges looked normal; but for the young age, the dura appeared thickened, I thought—although there was no exudate, and no adhesions to cerebral hemispheres were found.

The brain had to be sectioned with a carving knife and it was no easy matter, for although it was removed within one hour of death, it was mighty soft—and I fear that no pathologist will think much of my sections. The accompanying sheet summarizes the materials you are receiving by air (including the whole brain) and the treatment they have received. Zenker's-fixed materials which allow other staining (including Giemsa) are in 70% ethanol, having been washed and brought up through 50%; but no iodine removal of excess fixative has been attempted. I hope the isotonicity of our locally prepared saline is not off. We have not yet checked it against red cells—I shall do so today. With one-quarter volume of 40% formaldehyde added, it certainly is no longer isotonic as used.

Identifying fixed, pooled tissue slivers was no easy job, and getting them to a size to fit those few jars and flasks which we have which are leakproof has also been a problem. Thus, I cannot vouch for the correctness of our lists, yet everything listed should eventually be tracked down by histopathology.

In changing the brain to a new container, I looked at it enough to ascertain that Dave Cowling will have difficulty getting every piece into proper topological configuration, and that good full sagittal sections will hardly be possible. I trust he can find the lenticular nucleus and putamen and caudate nucleus, etc. We are now hoping to get further brains, and trying to get a brain-knife, better flasks, jars and containers, and our fixatives prepared in advance—thus there is a possibility that this will not be the only post you will receive, but please treat it as though it were.

I have located several intriguing cases which extend out of the Fore region into Auiana and Awa linguistic groups, and these we shall have to pursue shortly. I am now fulfilling the formalities of administrative red tape in applying for "permission to enter restricted areas," in order to cross the Lamari River to track down this most important, rather firm, limiting barrier of kuru. This will require spending some time in population nuclei across the Lamari from the Fore centers of the disease, and trying to spot what differences in ethnological pattern or ecology might account for the absence of kuru across the river—if it is really absent.

We were glad to get Gray's account of the four "Parkinson-ism" cases which he saw at Goroka, and that they are probably not kuru has eased the pressure of the immediate need of tracking down the Lufa cases, etc. However, we too have located a non-kuru congenital tremor akin to parkinsonism; but I believe, from first cursory examination, that it will fall, rather, into the congenital heredofamilial cerebellar tremor group. Also, a few odd cases of other heredofamilial diseases are on our hands, including an odd deafness of paroxysmal nature—with similar disease in mother and daughter. These are just now being studied, and more about them later. They form an interesting contrast to kuru. Bacterial meningitis has now completed the triad. Meningococcal pneumococcal had been on our hands; but yesterday we turned up a real *H. influenzae meningitis*, with bugs plentiful on smear, in a lad of about four years. Thus, it is forcibly emphasized that reports of encephalitic-like illness here in the Territory must always be suspected as perhaps bacterial unless confirmatory microscopic examination of the CSF is provided. The local medical men in the centers are well aware of this and do a good job of CSF-examination, I find, but tend to miss tapping infants who often present the most "atypical" pictures with their meningitis ("atypical" in that meningismus may be absent and fever may at times be low).

We have now had a series of adult patients on acetylsalicylic acid without benefit, another series of children on high doses of parenterally administered crude liver extract, and a series of five carefully chosen cases on high doses of cortisone acetate. None of these regimens have helped. I shall continue trying cortisone, and soon start on the BAL you supplied in response to our interest in copper and other trace-metal possibilities. Thank you for it, for it will be a really worthwhile therapeutic attempt. Nothing else works at all as yet. We shall also try testosterone and stilbestrol later. It is, however, essential to run these clinical trials long enough and on enough patients with each drug to be fairly certain before discarding it.

Mail which reaches me indicates that my Pneumocystis carinii paper in *Pediatrics* is out, as well as the note on my AICF test in *Nature*. Can you send me a set of reprints—half a dozen of each, perhaps? Keep me apprised of progress on the definitive AICF papers. I was pleased with the manuscript on paper II, and shall soon be writing to Ian [MacKay] about it.

The patent ductus patient (from everything I can determine here in the field, she is a patent ductus) has been relocated; and we are trying to get her off to Moresby, where Dr. Scragg has indicated they are awaiting her for cardiac surgery. I trust they have adequate diagnostic facilities there—but hoping for cardiac catheterization would be gilding the lily, I fear.

We sent off a shipment of new clotted blood specimens for the storage of serum specimens on kuru patients last week, some 16 of them. Did they arrive OK? Air and transport schedule was such that there is some hope that refrigerated clotted blood in venules might be of some use on the third day after collection, which accounted for our sending them as clots.

Carleton

SIMMONS to DCG   20 May 1957

Dear Carleton,

Sorry for the silence, even though I have been really working hard. Enclosed is a copy of the Cape York paper, which I think shaped up quite well; I will be interested to hear your comments on it. [20]

I enclose the results of blood-group testing and serum tests on your two lots of Okapa natives.

Dr. Curtain of the Baker Institute will post off to you this week tracings of the serum tests showing the unusual β-globulin component. I gather the β is generally raised in all this group, but most unusual in almost six of them. He found no abnormal haemoglobin, and has checked his work with a sample from a sickle-cell anaemia patient—the first found here. Dr. Curtain will write you fully about his findings.

Dr. Anderson rang me about doing more tests on the samples you sent down, and that will be OK. Dr. Curtain rang me today and I told him we now had a "Spinco" paper chromat 7 apparatus set up, and Dr. John O'Dea who is using it also has a Spinco ultracentrifuge. Curtain will contact O'Dea about ultracentrifuge tests on some of the unusual sera.

You see we have not been entirely idle in relation to the New Guinea problem. Apparently the rise in β-globulin is spectacular in some of the cases—I have of course given messages to Lois for you on these matters. The map she drew and had photographed is fine, I think.

The New Britain paper is half finished, and I will proceed with it when I hear more from you. Last week I gave our Director four papers for approval—three for *AJPA* and one for *MJA*. Please make any suggestions for alterations or corrections in the Cape York paper as soon as possible.

Regards,
Roy Simmons

P.S. Lois has fifteen bloods from you today; she will send them to me tomorrow.

COMMENTS ON RESULTS—19 NATIVES FROM MOKE, EASTERN HIGHLANDS, N.G.

The results follow the pattern of those reported for natives of Chimbu, Nondugl, Mt. Hagen in the Central Highlands of New Guinea.
Med. J. Australia, Sept. 8, 1956, 365–371.
Other summary, see *AJPA*, 1956, 14: 275–286.

There is nothing in the blood groups of the 19 natives with a neurological degenerative process to suggest that blood groups would afford genetic data on the condition, or that the blood-group frequencies for natives in the Eastern Highlands would differ greatly from those in the Central Highlands.

There are two other recent papers from the works of the M.S.W. Red Cross V.T.S. and the Department of Public Health, Port Moresby.
 1) Blood groups of a third series of New Guinea natives from Port Moresby. *Oceania*, Sept. 1956, Vol. 27, 56–63.
 2) A medical and anthropological study of the Chimbu natives in the Central Highlands of New Guinea. *Oceania*, Dec. 1956, Vol. 27, 143–157.

The second lot of 7 follow the patterns of the first 19.

RIVERS to DCG   21 May 1957

Director
National Foundation for Infantile Paralysis

Dear Dr. Gajdusek:
  Dr. Joseph Smadel has taken a great deal of interest in you, to the extent that he would like to get you out of the jungles and back to the United States.
  The red tape of the United States Government is considerable, as you know, and there is a certain amount of it in the National Foundation for Infantile Paralysis. I have taken the bull by the horns and have sent to Sir Macfarlane Burnet in Melbourne, Australia, a National Foundation check for $1,000 made out to you.
  I have a little money in the Medical Director's fund with which I can take a flier on certain projects or for certain outstanding young men. I am told that you are such a person, and I hope that this flier on you will not be considered one of my mistakes.
  The check that Sir Macfarlane holds is to be used without restrictions, with the exception that I hope very earnestly for your return to Washington and report to Dr. Smadel at least for a conversation some time within a reasonable period.
  I have read some of your letters to Doctor Smadel and Sir Macfarlane. I wish I were young enough to do what you are doing. Kuru or epidemic Parkinson's disease intrigues me, and I know Sir Macfarlane would be greatly interested also. However, you can only do so much, and Sir Macfarlane and others must finish the job.
  I hope that you can terminate your activities in that part of the world soon and be on your way to a talk with Doctor Smadel.

With very best wishes and a great understanding of men with inquisitive minds, I am

Sincerely yours,
Thomas M. Rivers, M.D.
Medical Director

cc: Dr. Joseph Smadel
Sir Macfarlane Burnet

RIVERS to BURNET 21 May 1957
Dear Sir Macfarlane,
Enclosed you will find a copy of a letter that I have written to Dr. Carleton Gajdusek, and a National Foundation for Infantile Paralysis check for $1000.00 drawn to Dr. Carleton Gajdusek.

I believe the enclosed copy of the letter explains the situation. Perhaps I might say, in addition, that Dr. Gajdusek was a National Foundation fellow and relinquished his fellowship to do what he is now doing. It is my understanding that his cash is running low, and the enclosed check is to take care of his immediate needs and get him back to the United States. Dr. Joseph Smadel is very interested in this man, and would like to put him on the staff of the National Institutes of Health.

I am doing all of this because I have been informed that Dr. Gajdusek is a rather unusual person, and because Dr. Smadel is very much interested in him.

You will know how to handle the matter when he shows up for the check. I wish to thank you for helping me in this situation.

With very best wishes I am,

Sincerely yours,
Thomas M. Rivers, M.D.
Medical Director

CURTAIN to DCG 23 May 1957
Dear Dr. Gajdusek,
Mr. Simmons of the Commonwealth Serum Laboratories passed on some of your very interesting specimens to me with the suggestion that I do electrophoresis of serum and haemoglobin. The electrophoresis was carried out on Whatmann 3 MM paper in pH 8.6, 1 = 0.05 veronal buffer. The haemoglobins were all identical with adult normal, but the serum showed some striking abnormalities. The most consistent was an enormous elevation of the beta-globulin (present in all but three of the sera investigated). A few of the sera showed a somewhat high alpha-globulin. I am enclosing tracings of the original scans of the patterns, and, where possible, patterns of duplicate runs. I am dubious about the eight normals which I obtained from Lois Larkin, as I understand these come from a different part of the territory. I have included a composite pattern showing the limits. The figures on the tracings are component percentages.

I carried out the Sia test on your sera and found that two were positive, Asomeia and Niaia—the latter very strongly. This heavy precipitation on dilution with distilled water is supposed to be a necessary, though not sufficient, test for macroglobulins. I have contacted John O'Dea at C.S.L. and have made tentative arrangements to have the Niaia serum investigated in the ultracentrifuge.

We expect to take delivery by next week of a new moving-boundary electrophoresis apparatus with a very sensitive interferometric optical system. I shall run through these sera, and samples not yet touched, as soon as I have set up.

I have discussed your material with a few of the people around Melbourne with some experience in the electrophoresis of serum, and there is general agreement that the patterns are quite remarkable. I, personally, have only seen such patterns from cases of lymphosarcoma and, rarely, myeloma. Rises in gamma-globulin are common enough in a variety of conditions. The only marked rise in beta- that I have read of is in dystrophica myotonica, Zinneman and Rotstein, *J. Lab. and Clin. Med.*, 47, 907 (1956).

I am most interested in abnormal serum proteins and would appreciate further specimens—particularly normals from the districts where these cases occurred.

Yours sincerely,
C. Curtain, Ph.D.

WOOD to DCG   23 May 1957
Dear Carleton,

Please excuse the delay in answering your letter, but things became rather complicated with letters arriving from all directions. So I thought I would leave it to Burnet and Anderson to cope with the problems. Now Gray Anderson is back, I do want to say how much I have appreciated all you have done in New Guinea; and he is also appreciative. He really has worked very hard indeed since his return to get the virus tests etc. under way, and we have offered to help where possible.

It would certainly appear that the disease is not an infectious one, and more likely to be the result of their mode of life rather than a pure genetic influence. The absence of the disease in older males may give the clue.

Good luck to you,

Yours ever very sincerely,
Ian Wood

BURNET to DCG   24 May 1957
Dear Carleton,

Thank you for your letter of May 19th and for the specimens which arrived safely. The brain has gone to Dr. Hicks, and Dr. Graeme Robertson will collaborate in the histological work. Lois has the other specimens, and I think they are all being routed in the directions required.

I was rather impressed with your statement that the organs of the dead kuru case all appeared perfectly normal macroscopically. That in itself seemed to me quite interesting, and pointing rather strongly to the concentration of the lesion in the CNS in one form or another. I feel you are on the right lines in determining whether there really is a sharp boundary between kuru and non-kuru areas; and that, if there is such a boundary, is the natural place for intensive epidemiological study. If any positive findings emerge from the histological work, I shall let you know immediately; but it will probably be quite some time before a negative report will be worth making.

Our major interest in the past week or so has been the Singapore influenza. Dr. French has just completed its characterization as an A-strain with an extremely marked serological shift from all previous A-types. We gather from press messages that Hilleman in Washington also agrees that it has no serological relationship to former types, but we have not yet heard whether he agrees that it is an A-type. From information I got today from Fenner, the epidemic appears to have started in Northern China and was widespread in Peking and Shanghai in February and March. It appears to have reached Singapore via Hong Kong.

Today Dr. Scragg has been in for a discussion on various matters, and is interested in the possibility of Singapore flu reaching the Territory and what measures should be taken against it. My feeling was that it would be very well justified to immunise at least all their Papuan employees and particularly key personnel. Whether anything more than that should be done will depend on how circumstances develop.

I am leaving for Europe in three weeks' time, and under the circumstances I gather it is unlikely that I shall see you at Geneva.

With best wishes to Zigas and yourself,

Yours sincerely,
F. M. Burnet

[21]

BURNET to RIVERS 27 May 1957
Dear Dr. Rivers,

I have received your letter and check for Dr. Gajdusek, and I shall wait on his instructions as to how to deal with the latter.

It has been an interesting experience having Gajdusek about for eighteen months or so, and the recent kuru period the most interesting of all!

As you may have gathered from him, Gajdusek really had no business in that field at all. The journey to New Guinea was primarily to be a more-or-less social call on my son, who is a patrol officer for the area next to Okapa. But when he heard of the kuru situation, which we were arranging to investigate in about a month's time, nothing could stop Gajdusek and in a week's time he had taken over the whole show and was working 20 hours a day as the self-appointed representative of the Hall Institute. Both the Administration and I were extremely annoyed for a week or two. However, I have a sort of exasperated affection for Gajdusek and a great admiration of his drive, courage, and capacity for hard work. Also there is probably no one else anywhere with the combination of linguistic ability, anthropological interest, and medical training who could have tackled this problem so well. So everything is now in order; we are acting as a base for any laboratory work, including histology, and Gajdusek is being given full Administration support.

(In parenthesis, I might add that I think that we could have built up an Australian–New Guinea team that would have got the same investigation done equally well, if perhaps at a more leisurely pace.)

Kuru is one of the most fascinating disease problems I have ever come across, and I can assure you that, whether Gajdusek stays on the job or returns to Smadel, the investigation will be continued until at least the aetiology of the condition is clarified.

While with me, Gajdusek did an excellent job of work on some auto-antibody reactions which emerged accidentally from an attempt to detect hepatitis antigen in human livers. This has opened up quite a field that our clinical people are now cultivating.

I thought it might be of value to you to have this informal account, because for a time I thought the situation might be difficult to straighten out and that there might even be minor international complications. I know the Administration is very sensitive about its health obligations, and I think they were worried at the possibility that Gajdusek's action could be interpreted as the taking over by an American of work that they had neglected. This of course is very far from the case. However, Gajdusek has behaved admirably in cooperation with the clinical man, Dr. Zigas, who was seconded by the Administration to work with us, and is now working with the full authority of the Administration.

I look forward to seeing you at Geneva.

With kind regards,

Yours sincerely,
F. M. Burnet

DCG to SMADEL 28 May 1957

Dear Joe:

Kuru is becoming ever more interesting and baffling. I am discontinuing cortisone acetate on five moderately-severe-to-severe cases in children, having reached toxic levels over a course of two weeks without a sign of improvement in the moon-faced subjects. Another six, five of them children, are on high doses of BAL, the therapy having been conducted along with as-controlled urine collection routine as the windy, stormy mountaintop we inhabit would permit. We managed to collect specimens in clean enamel kidney dishes washed with catchment water out-of-doors to avoid dust and smoke from all the houses (including our hospital ward), and with a minimum of debris entering the specimens. Some were collected in chemically clean glassware, and all have been transferred to either vacuumatic bleeding-venules or other similarly clean containers. The serial specimens collected during the first 18 to 24 hours of high BAL doses after an initial specimen should settle the matter of real copper intoxication, mercury intoxication, and other possible heavy-metal poisoning, which must be considered in this disease. I am sending one set of these serial urine specimens to Moresby for copper levels by the agricultural chemistry lab. However, I am rushing this note to you to ask if you could dig up mercury, manganese, and similar heavy-metal–level testing (since heavy metal intoxication is the major possibility we must now rule in or out, probably out)—including all possible metals which might be the offending agent, by even the remotest possibility. Can you find someone interested enough in this really fascinating matter to give us a hand?

I shall send you (for storage, at least), along with the pathological specimens and the brain, a set of serial urines on such BAL clinical trials. They are precious specimens and have cost us heavily in time and effort to obtain under these primitive conditions, where even the suspicion of sorcery worked on body parts or excreta is a great hindrance. We would like any and every study which neuropathologists and biochemists can think of. Copper is indicated because of the close neurological parallel between Wilson's disease symptomatology and our disease. Mercury, I read, can in certain forms of chronic poisoning, give rise to neurological tremors. Manganese is said to do the same in some animals, I am told. Thus, we are grasping at every straw.

New cases have now developed under our eyes; all our 60 to 70-odd study cases are slowly (some rapidly) deteriorating; and we have raised the number of cases we have ferreted out of the Fore's 8,000 to 10,000 population to 75 active cases. In the past six months there have probably been 100 deaths from kuru in the estimated 15,000 population affected by kuru; and another 100 active cases, at least, are present (we have our hands on 75). These 100 can be expected to be dead in six to twelve months, with some exceptions (probably not over 20% still alive by then). Thus, some 1% of the population, or more, is now dying per year of kuru, and from our epidemiology it appears that at least as many have died yearly over the past five—perhaps ten—years. Finally, some centers have up to 6–10% of the population now sick with active progressing disease; and many centers can attribute fully one-half of all deaths to kuru in any of the past five years. Could any more astounding and remarkable picture be found anywhere?

The picture is so uniform and its progression so classical that it is hard to believe. We have taken 8mm cine films, and three reels are off to you; I hope to airmail out with this letter another three 25-foot rolls of 8mm cine Kodachrome. These are the property of Dr. Smythe, who is visiting to give us a hand temporarily. It is a chance for us to get films of many of our best patients, along with the peripheral activities and ethnology of this remarkable population. Please do all you can to get at least one good copy of the films run off for file awaiting my return for editing. Already it will be possible for you to spot the salient features of kuru, if the film is at all well taken. We have filmed many patients and demonstrated some of their instability, ataxia, grinning, emotionalism, and tremors. These films, coupled with our earlier ones (which we shall dig up and send to you), will offer a chance of seeing the very same patients at various stages of their illness. I think they will be far more important than the rabies film I brought back from Iran.[4] Finally, I would very much like to get more usable, more manipulable 16mm film, but have not yet found the means.

We may send some tape recordings of the foolish speech, giggling, and chatter of some patients, with examples of hesitancy, blurring, and slurring—along with samples of control speech and records of local traditional music. I can collect these with our recorder (which equipment is far from perfect), if you can have as high-fidelity copies made for us as the tapes themselves warrant.

[22]

Again, many thanks, Joe.

Sincerely,
Carleton

---

[4]*Rabies in Man*, a 16mm film distributed by the U.S. Armed Forces Institute of Pathology.

DCG to SMADEL [late May 1957]
Dear Joe,

I received your notes on the treatment of Wilson's disease by Dr. P. Bailey, and the two notes on brain fixation by Greenfield, and Bailey on brain post-mortems. We have, as you may have already learned from my recent note, had one post-mortem on kuru now; and the brain is off to Melbourne. I would just as soon send the second brain off to you. This first brain was taken one hour after death and fixed after sectioning (rather sloppy, I fear) in 10% formol Ringer's and transferred in 10% formol-saline on the third day. It may give us some answers. My last note told you of extensive pathological materials off to you; I hope the small pieces are adequate for some studies. We have had several (eight cases) blood-copper determinations already done—at a well-equipped agricultural research lab—and they are about normal, although higher than in a control series we sent with them. They were not ceruloplasmin studies but rather total whole-blood or total serum-copper levels. Other copper studies are now planned; and BAL treatment will now be started, after seeing that even cortisone has been without effects.

New cases are always turning up. There have been other deaths, but we could not get posts. I have my hands on a few others and will try my best to get posts, but these are not people you can push. They are proud and have their own ideas, which are most intelligent; and although they have conceded that I can cure their meningitis and pneumonia, they have decided that this magic is too strong for me and that my prolonging life by treating and controlling decubitus ulcers is no blessing at all. They want to die at home; and once fully incapacitated, they want to die as quickly as possible. With such apparently hopeless neurological disease, you cannot blame them. Since we have done lumbar punctures on over fifty (repeatedly on some) and taken 30 ml blood specimens, collected urine and blood for CBCs, and done sedimentation rates frequently, and loaded them with painful shots (everything from crude liver and cortisone to parenteral antibiotics) without any effect—and they know it—I have only more respect for their "hands off" attitude. But, to humor me and repay my many miles of mountain-climbing to track them down, they haul litters over miles of cliff-faced and precipitous jungle slopes to bring the patients in for another shot at our therapeutic trials and experimental poking. I admire and respect them thoroughly.

Thank you for all the help; keep sending literature, new ideas, suggestions, and anything else you can think of. Could eating prussic acid in uncooked manioc cause anything like this? They now have manioc as an introduced crop from the neighboring tribes for the past decade, just when kuru got bad. Any drugs, special equipment, etc. which may be of use would be of help. The usual drugs, antibiotics, and standard supplies I have plenty of. Can you handle photographs? Could you get me cine equipment (16mm) and film on loan for a fine kuru film? It would be really impressive.

Sincerely,
Carleton

DCG to SMADEL [late May 1957]
Dear Joe,

Developments are rapid. We had a second kuru death [and post-mortem], a classical case in a middle-aged woman who had long been under our observation.

Just after receiving your brain-handling instructions, we altered our procedure and did not slice the brain but fixed it whole in 15% formol-saline, opening the ventricles. Brain, pituitary, cauda equina (we cut too low for much of the lumbar cord, in our rush), cervical cord, bone, and almost every organ of the body is now in ample slices or whole in 10% formol-saline (except for the CNS in 15% formol-saline). We shall change later to 10% formol-saline, wrap in gauze and cotton, and try to get all this off to you. We even have a sample of the formol-saline for trace-metal controls, in case you can use the tissue for this purpose. I shall keep the controls here in our jungle highland lab, pending word from you regarding the several problems. We have excellent air service by bush pilots who connect with Qantas Airways in Lae, and from there there is biweekly service to KLM Airlines in Hollandia, and thence via Europe around the world to you; perhaps better, twice or three times weekly from Lae to Sydney, and then via Pan American Airlines to you. I can thus send this material and anticipate no trouble.

This is fixed noninfectious tissue. However, could you clear with the local PAA and have them clear with Qantas that I am authorized to send—as I usually do anyway, without such preliminary preparations, and, incidentally, get away with— shipments to you by "Convert to U.S. Government Bill of Lading"? What I am really driving at is for PAA to get Qantas here (i.e., in Lae and Moresby) to know that they can accept such shipments and should push them rapidly forward. I suspect we shall get them off without trouble; but Lae is a long way off, and Vin may go all the way in to be sure things work okay. Kuru brains are not a commodity on the open market, nor will they ever be; we are lucky to get any. The material is really precious; and we need every assurance possible that it will (1) get through and (2) reach those interested and prepared to study it. Sorry for all these technicalities.

We have 70 kuru patients being intensively followed; and we are now preparing a preliminary note on this exciting disease, because we have finally awakened to the fact that we have sent off enough reports and data for a dozen reports to be written by those who have never looked at the disease. Our epidemiology is by no means complete, and we are still in the dark as to etiology. Patients are walking in and deteriorating so rapidly and in so stereotyped a fashion that it is hardly believable.

[23]

Naturally, everyone would like to get their hands on kuru brains; we were lucky to get two and may get further ones, but our ex-cannibals (and not "ex-") do not like the idea of opening the head, although other dismemberment does not seem to perturb them. However, death away from their remote villages does! We both (Vin and I) feel that you at the NIH could do more than anyone else on the pathologic material, and thus are sending the major portion of this—our second post—on to you. The first brain, at Melbourne, is from the same child whose portions of viscera we sent off to you. The spleen in that lad was really small. The spleen in the woman was unusually small and shrunken; and nothing else abnormal was noted except for (1) a very fatty heart with thin myocardium and much fat and (2) a peculiar subpial arachnoid fluid ?exudate collection which appeared blood-stained and may well have been the brain-removal trauma, but which was over one cerebellar hemisphere and extended down along the cerebellar stalk into the angle between the cerebellum and midbrain. I think this is an artifact we produced, but it did look funny; and since we are dealing with a tremor-ataxia disease, perhaps the most dramatic such in existence, it behooves one to look hard at the cerebellar surface and subarachnoid space.

[24] More [trace-metal studies] are being done, and the accompanying letter to Dr. Price will keep you informed about what we are trying to get done this way. However, just how accurate and how extensive such trace-metal studies can be, I cannot evaluate from here; and this brain or the next may be the last we ever get. So please ask your neuropathologists and chemists to think about what might be done. I still think kuru is one of the most important problems in existence in medical research, for it has all the earmarks of one that may give a real lead into chronic degenerative neurological diseases. Trace-metal toxicity is only our current craze—it must be fully investigated—but we still have family histories of the disease in well over half the cases and have even seen multiple family cases. It rarely lasts over a year in the village, from onset to death. Fully developed kuru is invariably fatal, all the natives say, and it certainly seems to be; early kuru ataxia and tremors occasionally regress and disappear, and we have seen this occur. Thus far, this has only been in hysterical temperamental women, who may have been well but hysterically mimicking early kuru. Since kuru is a major, perhaps *the* major death threat, magic threat, and cause for vendetta killings in the tribes today, such hysterical mimicry would not be remarkable. Certainly, most of our seventy current cases seem doomed; and our child cases have advanced before our eyes, in spite of every type of treatment. BAL will now be started as soon as we get a gang of them back to normal from the moon facies and other toxic effects which an intense and ineffectual course of cortisone acetate produced in them as their kuru progressed.

I shall shortly send you my black-and-white negatives on kuru, in the hope that you can get some printing and enlarging for a file that will eventually be used in the publications. I shall catalogue and describe each negative frame in a note to you. Please let me know, Joe, if I am overburdening your wide interests in medical problems by inflicting kuru on you this way. I really think it deserves every bit of emphasis I am giving it. Should you be able to see our ward filled with "epidemic Wilson's-parkinsonism" (which is what this is clinically, but probably not pathologically or etiologically), I think you would fully agree. A native-material house is rapidly nearing completion, with a 24x24-foot room, a table lining all four walls for our specimens, the slides, papers, records, etc., and ample pacing room. Thus, we and our visitors will be able to eat on a table without post-morten specimens at one end, and be able to do lumbar punctures at night on a table other than the autopsy/tea/lab bench/typewriting/emergency surgery table that must now be cleared for meals three times a day. Our kuru cases have been moved into a majestic new hospital that we have built by the cooperative efforts of some one thousand Fore-Kimi-Keiagana natives, and it is quite a cozy place in the stormy downpour which beats daily on our mountaintop.

Sincerely,
Carleton

DCG and ZIGAS to PRICE [late May 1957]
Dear Dr. Price,

Vin Zigas and I are ever more grateful to you for all of your help. We have read over the results of copper-levels on our patients and on the controls, and also the controls from Ela Beach, which you added to the study. It appears that if copper metabolism is involved in kuru it will be a most subtle matter requiring extensive controls and, eventually, probably ceruloplasmin studies, etc. However, we think every line of possible approach should be exploited, and we ask you to continue requesting copper studies for us. In addition, is there any chance that manganese and other trace-metal studies might be done, or is copper the only one set up in Moresby at the moment?

Gray Anderson explained that Mr. Holden at Hall Institute would only be prepared for copper—perhaps also manganese—but hardly any others. He offered to check for us with CSIRO laboratories in Melbourne to see what they might do. Thus, should any of these new specimens which we are now transmitting to you not be exhausted by your copper and other studies, will you please freeze them down in chemically clean glassware adequate for trace-metal studies, in order that they may be used in the future should we be able to line up other trace-metal studies? Again I am sorry to place so many burdens and problems upon you. We have received so much real help from you already, we are tempted to repeatedly call upon you for more.

[25]

The shipment now being sent to you is summarized on the accompanying sheet. It includes blood specimens in sterile, chemically clean vacuumatic ampules from adult kuru patients, all with typical advanced disease, and from some normal Fore control subjects from the same villages. These were collected for copper-level studies, along with other trace-metal studies if they can possibly be done. Furthermore, we should like luetic serology, liver function tests, and any renal function tests (NPN or BUN, etc.) possible—along with total protein and any protein partitions which the volume of the specimens will permit and which your time and facilities can provide. Such huge orders and requests are embarrassing for us to make of you; as we have previously indicated, we both know full well what an immense job such laboratory studies can be. However, these are things we cannot do here at Okapa Patrol Post, and they need doing. Kuru remains the same unsolved problem, and we are trying everything for "leads."

Along with the blood specimens is a bottle of bile aspirated by syringe from the gall bladder at post-mortem. Any chemical and trace-metal studies possible on this would be of value. In addition, we are sending four other tubes which each contain slivers of autopsy-collected organs: spleen, liver, kidney, and psoas muscle respectively. These have been placed into bottles which were chemically cleaned for trace-metal studies in Melbourne and thus should satisfy the chemists if they can manage to use such material. Since no fixative or solutions have been added, these specimens have been kept frozen in our ice-chamber and in a Kainantu deep-freeze as much of the time as possible, for they were not collected under totally sterile conditions. We hope they reach you in usable form. We are not likely to get much further such material!!

Having requested such an absurd amount of work, we might as well go on. Could you make up further Zenker's solution and send it to us along with acetic acid? We do expect at least one more autopsy—perhaps two—shortly, and we have used fixative in vast quantities, taking rather extensive collections of every organ in the body in our last two posts.

Could you ask the chemists, Mr. Baseden and Mr. Southern, to supply us with chemically cleaned glassware for trace-metal studies, adequate for collecting further pathological specimens, further blood specimens, and, in particular, for supplying urine specimens, etc. which we shall collect in connection with BAL (British anti-lewisite) therapeutic trials which we shall now begin to conduct. Incidentally, cortisone carried to fully toxic levels failed to alter the course or severity of the disease.

Since BAL can stimulate the excretion of a wide range of heavy metals deposited in the body, the BAL-trial specimens will be particularly valuable for copper studies as well as others; and we thus request in advance that aliquots be set aside for these other studies.

Should you or your colleagues, Dr. Price, have any suggestions to make to us, please let us have them. We are seeking advice and discussion from every side, for while we have this opportunity to study kuru under fairly good conditions, it would be foolish to have overlooked observations which could easily have been made or experiments which could easily now be conducted. It will not be for much longer that we can keep some 70-odd kuru patients under our thumbs, and the local Fore will not welcome being guinea pigs for much longer—unless we can provide evidence that we have something to offer them, which is unlikely.

Please tell Mr. Baseden and Mr. Southern that we have in mind writing to them personally with our gratitude and thanks for all their endeavour; but thus far it is all we can do getting these notes off on secondary specimens we send out to you, to Melbourne, and elsewhere.

Again accept our thanks for your enthusiastic cooperation.

Sincerely,
D. C. Gajdusek
V. Zigas

SMADEL to DCG   29 May 1957
Dear Carleton:

Your long letter (date not given, presumably about 22 May) describing recent developments in kuru and the autopsy arrived yesterday. Today, the tin box containing the fixed material reached here in good condition. I shall consult with Dr. Bailey and others regarding who might study most profitably this valuable pathological material. I will write you a letter shortly with answers to some of your questions. I am glad to know that your two papers have just come out in *Nature* and *Pediatrics.* I'll look them up.

Enclosed are carbon-copies of letters from Tom Rivers to you and to Burnet. Please write dear Tom a good letter, thanking him for his fine efforts in getting you $1000. Best wishes.

Sincerely yours,
Joe

DCG to SMADEL   [30 May 1957]
Dear Joe,

Enclosed is a summary of a large shipment of urine specimens which is fully explained on the accompanying sheets. [26] They summarize our tropical bush attempts at 24-hour urine collections under hellish conditions for controlled experiment. However, negative results would be rather conclusive, and our collections are as accurate and carefully made as can ever be done under such conditions. Any help with any and every trace-metal, urine amino acids, and any other studies you can do would be much appreciated.

[27] Neuropathology, whether completely negative or with interesting positive findings, may considerably clarify the kuru matter. All other pathology will be of value, as well. Yaws is present, but we have a number of positive CSF Wassermanns without any other sign of CSF treponematosis; on New Guinea patients, with their gamma-globulin derangements, this may mean nothing. But can silver impregnations for spirochete also be made? Some 60-odd kuru spinal fluids have thus far been negative. We shall send you some CSF for trace-metal studies later.

Thank you for all the help and interest.

Sincerely,
Carleton

P.S. [28]

Please let me know about the cine films when you finally see them. This is a really good problem, Joe, and I hope you someday get to see it. A wild—really wild—suggestion: could anyone in the USA interested in CNS diseases (especially the degenerative ones like parkinsonism, Wilson's disease, etc.) be interested enough to undertake the wild venture of bringing a set of these kuru cases to the NIH for real study??? Nothing short of this will ever settle the issue, and our child-patients could be managed well. The trauma to the natives of bringing two to Moresby—where less was done than we do here in the bush—could not be exceeded by New York or Washington; these fellows adapt quickly to airplanes, jeeps, jets, etc.! Australia has not really offered us anything for investigating a few cases in their centers—other than Fairfield Hospital, which is *not* equipped—and funds are hardly available. I just pass this wild suggestion on; I could easily get a set of mildly ill but rapidly progressing child-cases (with interpreters and guardians) off to you if and when funds and administrative authority were cleared, and more might be settled that way than by thousands and thousands spent in "field ventures." This *is* a new disease, Joe, and an important one. I have been flatly rejected on plenty of these "wild" schemes, but when I see how millions of dollars are spent on far more hare-brained schemes, I consider them worth making. Enough for now.

DCG to BURNET and ANDERSON   4 June 1957
Dear Sir Mac and Gray,

I have been so busy that it has been impossible to keep you as well informed as I should have liked to have done. Slowly but surely we have dug up new cases and persuaded the old known cases to come into Pintogori [our hospital site] for complete study. Furthermore, I have extended the intensity and, I hope, perspicacity of our clinical studies. Complete blood counts with good staining of films are finally worked out; these are being done and have been done routinely on over 30 of our patients, a job requiring days of work. Nothing remarkable in the RBC, WBC, differentials, or hemoglobins has yet been found. I shall send stained films to Melbourne shortly for whatever rechecking Dave Cowling or others with good hematological background can arrange.

We have used up every ampule of BAL that Anderson brought with him—treating ten patients, mostly children, on a single five-to-six–day course of high doses of BAL. [29] Urine collections were made on all but two of the ten patients, these two being incontinent and the specimens being always missed. Catheterization was out because (1) we have only one catheter and (2) the local gossip ensuing from the use of the catheter would have been such as to jeopardize our rapport with the natives. Urine specimens, collected in every available container which was or "might be" chemically clean, were sent on for copper levels and mercury levels (I hope) to Moresby, with instructions that aliquots thereof be sent to you in Melbourne to supplement the specimens we sent to you. If anyone at CSIRO or elsewhere can run spectrographic analyses for any excess of heavy metal or trace metal, we should be most appreciative. Mercury, copper, and manganese come to mind first; some forms of chronic mercury poisoning do result in basal gangliar–type tremors, I read.

Nothing unusual in local diet yet gives us any leads. I have caught many natives eating a particularly strong form of raw ginger, but am assured that everywhere in the territory this occurs without kuru ensuing. Cortisone, carried to toxic levels, readjusted and now maintained on high moon facies–producing doses, has not helped any of the five child patients a particle. BAL cannot yet be evaluated. No change occurred during its administration. I am now running a group on Dilantin, which finally arrived, and shall soon start with Tridione. One man, an adult male patient, is now on good doses of testosterone. I would like to know something about maximal safe testosterone doses, and can find nothing here in the literature available. Can you help me? The hunch is that since adult males are particularly spared, perhaps the male sex-hormone has something to do with it; and we can more easily boost a male who has the disease to normal male testosterone levels than a female.

Thus, clinical study at the moment is intense, since patients are on near-toxic levels of cortisone, delta-1-hydrocortisone, and agranulocytosis-inducing Tridione and Dilantin (which is also renowned for stomatitis and other reactions). Local Fore resistance to hospitalization is increasing; they know damn well that we do nothing for the disease but prolong its misery by supportive measures, and they are anxious to return to their technique of starvation and neglect in darkness, which ends in a speedy exodus once the illness is truly incapacitating. We have had several more deaths in patients in our "series" of 75-odd severe cases. From one adult whom we had followed closely, we were able to get a complete autopsy. Tissues are in fixative and being kept pending firm fixation and arrival of adequate containers, but will then be sent on for histopathology. The case showed no gross lesions at all; the spleen was noticeably small and shrunken, and sectioning all organs grossly revealed nothing unusual. We are still trying to get further brains, and will do our best. I have asked Moresby for more BAL, for we have not a speck left for further trials or—more important—for follow-up courses should any effects at all be seen from the first course on our ten BAL-treated cases.

Epidemiology is continuing. I have tracked kuru into the Usurufa linguistic area, but again it is found only in the Fore border area where Fore-Usurufa intermarriage occurs. [30]

[31]

Thank you, Sir Mac, for your kind letter of 24th of May. It was a great relief to Vin and me to learn that the first brain had arrived safely and was already in Dr.

Hicks's hands. A note from Simmons informs me that he has done an immense lot of blood-group genetic work on our specimens; and I believe that it will all fall very well in place once we can sort out the records and results, for the genetic stock of these people may turn out to be a key question. I shall shortly send new batches of blood for further red cell, hemoglobin, and serological study, along with new CSFs. Since the vacuumatic bleeding venules we use are apparently satisfactory for trace-metal work (as the Moresby copper determinations using them seem to demonstrate)—and, in addition, I can always supply a few controls which are empty—I wish to request any other work on trace metals which anyone in Melbourne can manage to do. The serum-blood specimens should—if handled properly for trace-metal work once the venules are opened—be adequate for trace mercury, zinc, copper, manganese, etc. In addition, they may be OK for amino acid studies, to be correlated on any urine amino acid chromatography you can do. Can any of the "BAL collection series" of urine specimens which you received be used for urine amino acid chromatography?

[32]

I am very much intrigued by the N-N-dimethyl-paraphenylenediamine dye test estimates of ceruloplasmin which I have heard of but the reference for which I have not yet seen (*Science* 125:117, 1957). Can someone look it up for us and determine whether there is anything therein which we might undertake?

Certainly trace metals must be pursued to the fullest, since they do offer a distinct possibility. However, I am becoming more and more skeptical that we shall find anything. New dietary leads may turn up. Thus far, raw manioc and raw ginger are the only two I can find that are at all remarkable; and old Territorials assure me that this is a pan-Territorial trait, not particularly Fore.

The full epidemiological survey from which we shall establish definitive statistics as to incidence and death rates from kuru is being slowly pushed forward, and the final delimitation of the zone of occurrence of kuru is being attempted. I shall shortly be devoting much attention to patrols designed to settle these unfinished matters. Thereafter, I think we shall have reached the phase of diminishing returns—and the Fore the limit of their patience and toleration—and at that point I shall sit down and start analysis of our records and charts prior to leaving the highlands. I shall bring everything to such a state that Vin can follow all of our patients easily by a visit every 3 months or so to the region and by the use of his doctor *bois* to roam the hills, in such order that a return in 12–18 months could be brief and yet definitive for determining what had happened to the natural history of kuru in the interval.

[33]

Again I thank you both for all your help and assistance. Tell Gray that I shall see that all Institute needles and instruments are shipped back when I pull out. In the meanwhile, a new supply of chemically clean, trace-metal–free bottles and needles would be a most helpful addition to our supplies. I have nothing to put autopsy materials into for such determinations. In fact, if you can design trace-metal–free—or at least "controlled"—formalin (10% formol-saline) and/or ethanol for proper preservation of tissues which may still be used for trace-metal studies, I should very much like these. We cannot get away from metal tin–contained formaldehyde and other fixatives here in the bush. Trace-metal–free bottles, empty and sterile, would be helpful for further blood specimens and CSFs.

Sorry, once again, for so many requests. If I ever get to Lufa, I shall try to see the tremor cases Ian [Burnet] has dug up. He is now on patrol.

Sincerely,
Carleton

P.S. Your letter announcing the arrival of the check from the Foundation [NFIP] reached me after this was written. Thank you, Sir Mac, for all of the administrative details which you have been saddled with and cared for for me. I am deeply grateful. Please send the check on to me at this address, where I can easily negotiate it through the mails.

May you have a fine trip through Europe. I trust the meetings will be interesting, and I hope that you may find it possible to return through the States once again. If so, I may possibly be there already.

Lois [Larkin] has repeatedly voiced an interest in seeing New Guinea, and I suggested that she use her time off to come up for a visit here if it could conveniently fall into the period when we still have accommodations and facilities available—that is, now and for the next month or two. I hope that this might coincide with a natural slump in your lab work coincident with your leaving for Europe. Certainly, I should like to see her again—a decided understatement.

ABBOTT to DCG 4 June 1957
Dear Dr. Gajdusek,

Thank you for your letter dated May 19th, 1957, indicating the progress you and Vin Zigas are making in the study of kuru.

You may rest assured that the necessary patrolling expenses you ask for will be forthcoming.

I have shown your letter to Dr. Gunther, who wishes to join in our thanks for it.

Will you please say what is required in the way of jars, flasks, etc. for pickling specimens, so that I can try to get them to you expeditiously.

T. K. Abbott

DCG to SMADEL 7 June 1957
Dear Joe:

Enclosed you will find a manuscript of a paper we are preparing as a preliminary note for a "general" journal, such as *Nature,* since it is now obvious that our great mass of data and records will require a set of papers which will be months—perhaps years—in preparation. Kuru remains fully in our hands; we have had no one else make the arduous journey in to see it; but we have, as you can see, interested a number of others in peripheral problems. It remains every bit as good a problem as originally, and our findings have provided a great mass of negatives. We are about where we were with epidemic hemorrhagic fever all the time—perhaps even more in the dark—and all our patients continue to get worse; we are now poised before a rapid series of deaths in six to twelve of our patients. Whether we shall get any posts or not, I cannot judge; but the formol-saline–fixed brain and cord and pituitary, and all viscera and Zenker's-fixed portions of tissues on our second complete autopsy, are all off to you along with this letter. Keep an eye out for them; and please direct them to as good a histopathologist and neuropathologist as you can, for you may never again get a kuru brain in the U.S., unless we are very lucky.

I am out of containers and venules as of today, for we are sending two dozen further kuru blood specimens to Melbourne, from where the sera will eventually go on to you in the U.S. The exciting alpha- and beta-globulin matters are being pursued; but, as you may know, globulins in all New Guinea natives are haywire, with no relation whatsoever to our temperate-zone American "normal" standards. However, the beta rise is so spectacular—approaching a macroglobulinemia—that it must be further pursued.

[34]

The enclosed correspondence to Sir Mac will let you know a bit more of what we are up to. He himself has not seen kuru. Anderson, who was interested in it, came in for a few hours while in New Guinea, but is not really working on it. However, others in Moresby and Melbourne are giving us a mighty hand, and Anderson is lining up much of what we request with the labs that can handle it. However, nothing is as urgently needed as good study of the brains and other tissues. Please let us know what turns up. Enclosed, also, you will find the carbon-copy of the history of the deceased patient whose tissues you are getting. Please file it away for me. There is much added data on her, which I do not have in duplicate.

Sincerely,
Carleton

ANDERSON to ZIGAS 7 June 1957

Dear Vin,

We received the brain about a week ago, and it is being sectioned by Dr. Graeme Robertson of the Royal Melbourne Hospital. The other organs have been looked at by Dr. J. D. Hicks of the Royal Melbourne Hospital; and he found no lesions, with the possible exception of a mild brown atrophy of the heart muscle.

We received the urine samples which either you or Carleton sent, and Mr. Holden will ash these and send a portion to Dr. Rees for spectrography. We may be able to have amino-acid patterns estimated, but this is not finalised.

With kind regards to Carleton and yourself.

Yours sincerely,
S. G. Anderson

ANDERSON to ZIGAS 14 June 1957

Dear Vin,

The celloidin sections of the brain are not yet available; but from the paraffin section so far tested Dr. Robertson has not found much to comment on, with the possible exception of diminution of the Purkinje cells of the cerebellum. This, however, will need to be confirmed. Dr. Robertson has prepared the enclosed sheet of instructions, and would like another brain or brains prepared in this way without being sliced. He feels this may throw some extra light on the condition.

With kind regards.

Your sincerely,
S. G. Anderson

[35]

ANDERSON to DCG 19 June 1957

Dear Carleton,

I thought you might be interested in three separate quotes from a book *De Re Medica* put out by Lilly, page 470. They are:

> In adult testicular failure (primary postpubertal hypogonadism), an average initial dosage of testosterone propionate by injection is 25 mg. three times weekly. The initial requirement of methyltestosterone by sublingual or buccal administration is 5 to 30 mg. daily. When methyltestosterone tablets are swallowed, the necessary daily dose is 15 to 100 mg. or more. After a satisfactory response has been obtained, dosage should be reduced, and the minimum necessary for maintenance should be determined by trial.
>
> ....
>
> Overdosage is to be avoided, particularly in women. Total quantities of 500 mg. or more of testosterone propionate per month (or equivalent quantities of methyltestosterone) are likely to produce masculinizing effects in women, such as enlargement of the clitoris, deepening of the voice, baldness, growth of beard, coarsening of body hair, and acne vulgaris. The first three of these changes, once accomplished, may show little or no regression when treatment is discontinued. *It is strongly advised that women receive no more than 75 mg. of testosterone propionate or its equivalent per week, or 300 mg. per month, except when necessary in the treatment of serious conditions such as mammary cancer.* Especial caution is desirable when the patient already has an unusually coarse growth of facial hair or other submasculine characteristics.
>
> ....
>
> The large doses required in breast cancer may induce hypercalcemia (p. 345), sometimes of dangerous degree. This complication should be watched for. If it appears to threaten the patient's life, it may be treated by intravenous infusion of 2.5 percent sodium citrate solution.

Yours sincerely,
S. G. Anderson

ANDERSON to DCG 20 June 1957

Dear Carleton,

I have various notes from you and will reply to portions of them. Firstly, you have asked for clean sterile bottles for further blood collections; but I have also received a copy of a letter which Alex Price has sent you stating that he will prepare chemically clean bottles for you. Therefore I will not send you any at the moment unless you make a further request for them. In this case they could be prepared and sent to you.

If it is not possible to get BAL from Moresby, we could obtain further supplies from the R.M.H. at your request.

The people at the Women's Hospital who are examining urines for amino acids strongly prefer samples which are sterile. It might be worthwhile sending separate samples for this purpose; and if there is any suggestion of abnormality in the first few, I think, they would be willing to examine further samples.

I enclose a typescript copy of the article in *Science*. Mr. Holden is attempting to procure reagents, and we may try a test here on material from mental cases in Melbourne. Holden is somewhat sceptical of the value of the test but if anything comes of it we will let you know.

I have passed on your message to Sue Ungar, the new librarian, who, I am sure, will do what you have asked.

Mr. Holden is preparing some chemically clean formalin, and is sending it up in concentrated form to Dr. Price with the request that he might dilute it with glass-distilled water and forward it to you.

I have just received three CSFs and the blood specimen. I will send the CSF specimens down to R.M.H. for examination, will keep the serum from the blood, and will send the blood clot and cells to Simmons of C.S.L.

Yours sincerely,
S. G. Anderson

SIMMONS to DCG   25 June 1957
Dear Carleton,

Did you receive the Cape York paper sent to you by airmail on 20/5/57—i.e., five weeks ago?

I told Washburn the paper was following two others sent, and have now been asked "where it is and have I sent it to USA?". I am holding reply until I hear from you, giving me confirmation that you approve of it.

Have kept in close touch with Curtain and have a copy of his first report. Have since given him further samples, and have more cells and sera to pass on to him tomorrow.

Lois Larkin is away on holidays. Which of Lot 4 are kuru and which are normals? She did not know! [36]

I go to Brisbane on 29 June 1957 for one week to give some lectures there.

Regards,
Roy Simmons

SMADEL to DCG   25 June 1957
Dear Carleton:

In case I have not previously done so, I hereby acknowledge receipt of a large batch of urines, another autopsy, several letters (including the case history of the patient from whom the second autopsy material was obtained), and, finally, the draft of your paper on kuru. It is a very nice paper, not only good but interesting. I hope you have already sent it off to the journal. Why *Nature*? Why not *Lancet*?

Mr. Wipf, Administrative Officer, National Institute of Neurological Diseases and Blindness, has brought together material for photographing the kuru patients. The list is attached. It will go off within the next day or so. Look, chum, I practically had to swear to Wipf that I would be responsible for all this stuff. Either you come home with it, or leave your head in New Guinea. We have received six rolls of color film from you to date. These have been sent out for development and for preparation of two color dupes. When all of it is returned, I will file under appropriate conditions the original film. Then a number of us here will run off one of the copies to view it, as it may be helpful in getting ideas about kuru. The second dupe will be sent to you promptly by airmail.

I am leaving on the 4th of July for the Fourth International Poliomyelitis Congress in Geneva and will be back about August 1st. The way things shape up now, I may be one of a number of Americans who go to the Ninth Pacific Science Congress at Bangkok, which goes on for several weeks in late November and early December. Just what days during this period I shall be in Bangkok is still unsettled.

If my general calculations are correct, you will probably be out of New Guinea about September or October and will spend a month or so in Kuala Lumpur. Frankly, I don't see how you are going to do all this traveling on the $1,000 Tom Rivers got for you, but you manage to get around on relatively small amounts. It is certain, however, that this particular check is the last one, so you better get back to the United States before you run out of money. I really don't see much point in your spending two months in Kuala Lumpur; but if you do and still have any money left, you might go to Bangkok for the Congress.

Dr. [C. G.] Baker got some information for you on chronic cyanide poisoning. Copies of this are enclosed.

Sincerely yours,
Joe

DCG to SMADEL 29 June 1957
Dear Joe:

I have just returned from twenty days of patrolling over some rugged, little-known country, tracking down kuru. I managed to delineate the western boundary of the kuru region, and to find where kuru came into the Kimi neighbors of the Fore and where it did not. It extends only along the Fore-Kimi boundary, and in some cases can be traced into the Kimi kuru-free regions following the marriage of a Fore girl into the Kimi, the disease occurring in her years after she left the Fore kuru-region and started residence in the previously kuru-free Kimi region. This border area is the most profitable region for kuru study, for kuru is an isolated event in a rare family here while in the central Fore it is the major cause of all deaths. However, in spite of more and more clinching evidence for a genetic pattern, I found a few "red herrings" such as a Kimi girl from a supposedly kuru-free Kimi community getting kuru after coming to the Fore kuru region. For one such case, there are dozens of the genetic-suggesting others, but the one "red herring" is enough for thought.

We now have over a hundred active kuru cases, and our cases are starting to die off rapidly. We get autopsies only with difficulty, but now have two more brains on hand—both first fixed in 15% formaldehyde and then transferred to 10%, in which they are being shipped to you. Case histories are being sent to you on these two deaths, both in patients we have followed and treated extensively: one in a small girl and another in a middle-aged woman. Both were classical kuru. Thus, these will be the second and third kuru brains sent to you. Neuropathologists in Melbourne kept telling me that perfused brains were essential, and that without perfusing through the carotid arteries and circle of Willis, the specimens would be inadequate. If this is so, please let me know, send me full instructions and any necessary canulae, etc. We will have plenty more brains coming, I think, and we can definitely follow through much more complex procedures. I am now getting ready to try to get a brain in trace-metal–free formaldehyde and trace-metal–free containers. Can you supply any proper plastic containers or bags for such a brain? We can get distilled water and trace-metal–free HCHO flown in to us from Moresby and Australia.

The girl whose brain is coming died while I was out on patrol, and a native medical assistant followed our instructions and obtained it and dropped it into previously prepared fixative. The second brain (that of the woman) I obtained with Vin Zigas, and we did a complete post—finding nothing unusual, except again the same slight exudate-like material over the cerebellum, which may be from brain-removal trauma. In any event, the full brain and meninges are fixed and sent to you; and samples of liver, kidney, pituitary, adrenal, stomach, intestine, abdominal wall muscle, and spleen and uterus are also included. All are HCHO-fixed, with the exception of a small bit of Zenker's-fixed tissue, so labeled.

Once we leave here, kuru pathological material will be no more; and thus I beg that you consult everyone who might have a bright idea. I fully suspect no pathology to be found—as is the case in many a toxic or degenerative neurological disorder—and thus the most subtle and original techniques are indicated. One brain sent to Melbourne, sectioned here, has revealed nothing thus far except possibly a reduced number of Purkinje cells in the cerebellum. That our patients are dying of a rapidly fatal progressive neurological degeneration is certain, and some one hundred are now in our hands. We have tracked down histories of about five hundred kuru deaths in recent years, and have only just begun. Furthermore, on my last patrol, we tracked down a high incidence of kuru nearly to the Papuan border, extending into very little-known country. We are about to launch searching patrols into the uncontrolled Kukukuku regions which border on high-incidence Fore regions, to try to establish the furthest southern extent. That no yet-discernible ethnic or ecological feature—other than marriage with Fore—seems to characterize the kuru limit, seems more and more certain. We have, I suspect, the highest incidence of a genetic neurological disturbance ever seen in a large population; it is a mighty striking disease, as you must have already seen if the films are at all intelligible.

Please keep any notes I have jotted on the film packs with the respective films. We have no other record. However, I think that Vin and I can reconstruct a good narration from recall when we get the films for editing from you. We already have pictures of the progressive stages of illness in several individual patients; and when we can dig up our first film and get it to you, you will have early films of many patients who have subsequently died and were filmed in their progression. Two rolls of cine film are going out with this letter; one has several new cases in children and follow-up pictures on several old cases. Another has kuru cases (follow-ups) and a good deal of panorama of the kuru area and adjacent country in which the boundary between kuru region and non-kuru region can be shown later. [37]

Thank you, Joe, for all of your encouragement and support. We are really getting things done. Vin and I have the entire matter on our hands, but we are getting support in equipment and special tests from Moresby and Melbourne as we request it. The draft of the preliminary report I have sent to you. Many additions and improvements can now be made, but I am awaiting your comments. Lois Larkin, Sir Mac's technician who did all the AICF work with me, will be coauthor of our third paper on this; she is up here "on her vacation" with us, and with her help we are really going to get our Kuru Center moving full speed during the next few weeks before launching off on patrols again. BAL and cortisone did not help dramatically, but to decide whether they help a bit is most difficult. Patients survive months when well-supported in our kuru ward, and go off to die in the village only days or a few weeks after our temporary discharge. To differentiate between the specific effect of drugs, and that of pure nursing support, is a most subtle matter.

I am ever more convinced that kuru is one of the most exciting and fascinating medical problems unsolved today, and one which offers to medicine many an insight, especially in this modern age of fear of gene-induced dangers to populations—for where in the world is another population so threatened by a lethal set of genes?? Obviously, I am still thinking genetically—but remain open to toxic, infectious, and other possibilities, if they can be found. No luck yet.

We are isolated, and every word from you is a great encouragement. Here we cannot possibly handle these many technical matters which I presume to thrust upon you. I think kuru is interesting and important enough, Joe, to warrant this intrusion. I am sending a copy of this to Tom Rivers, along with a copy of the draft of the "preliminary report," with a note of thanks to him for the check. I am living frugally here; but, without salary, I should have been stranded when I pull out were it not for his help. Thanks.

[38]

Sincerely,
Carleton

DCG to SIMMONS　　　　　　　　　　　　　　　　　　　30 June 1957
Dear Roy,

I am very sorry to be thus delayed in thanking you for the final manuscript, and for the extensive blood-group and other data on our kuru patients. Kuru cases on our hands now number over 100; and they are all, with few exceptions, running a rapidly progressive course toward death. Ten of our cases have died thus far. All looks more and more like a chronic progressive heredofamilial degenerative disorder of the CNS; and we can find no sign yet of toxic factor or any other environmental variable at work other than intermarriage with the Fore, who are the linguistic-cultural group entirely affected. Kuru is the cause of over half of all their deaths, and currently affects over 1% of the entire 15,000 kuru-affected population, reaching an incidence rate of 5–10% in some population groups in the center of the region. That it is genetically determined seems more and more certain, and clear cases of Fore introduction of kuru into neighboring Kimi, Keiagana, Usurufa, and other linguistic groups are now on our records: these we find at the boundary of kuru/non-kuru regions in these other linguistic groups.

Thus, Roy, in spite of my own involvement, I think that objectively kuru can be said to be the most exciting and potentially most important medical-genetic problem in the world today. Its full elucidation, its epidemiological study (for where else in the world is an entire population so genetically threatened?), and the study of every aspect of its spread in the Fore and neighboring groups becomes of immense value. For that reason, I write to urge you to consider whether you might not be interested in the immense job of looking at normal-population blood samples from kuru-involved and kuru non-involved populations in this region. It is, incidentally, one of the least studied and investigated New Guinea regions; and we have, as you know, several quite distinct linguistic groups. Specifically, I should like to request continued study on those kuru patients' blood whose specimens we get to Melbourne, and at least a blood-group study on a set of normal Fore, non-kuru subjects. In addition, we are now planning an extensive patrol into the uncontrolled Kukukuku region; and these Kukukuku border directly on high-incidence Fore kuru-country and are, apparently, kuru-free. There is on this side no intermarriage. Thus, the yet-unstudied, racially extremely distinct, Kukukukus become of particular interest to us.

We are without any adequate further blood-collecting venules and tubes. Could you send me a supply of your special blood-specimen collecting tubes (with directions) in time for our Kukukuku patrol, which is still one month away? Other collections can be made more leisurely; but Kukukuku patrols are rare and difficult, and I hate to miss this fine opportunity. It would contribute considerably both to straight blood-group anthropology of New Guinea and to kuru-study simultaneously.

If you agree with us and can find the space for storage and the time in your most busy life for the work, consider also the possibility of doing such work on a sample of Fore, Kimi, and Keiagana blood specimens.

In the event that this is asking for more than you can possibly fit into your current program, simply let me know. Naturally, I am already most indebted to you for all the work you are doing on the kuru patients' specimens which you have been receiving. Tell Cyril Curtain that I shall try to get a letter out to him by next mail. I shall request the Hall Institute to send to him—via you—serum on a control-group of Fore, for control checks on his high beta-globulin patterns, a most important matter.

I hope to begin digesting our data shortly, and I shall try to write to you at greater length before too long.

The New Britain data I am still stalling on, but I might be able to get it off to you in the next mail.

Sincerely,
D. C. Gajdusek

DCG to CURTAIN [early July 1957]
Dear Dr. Curtain:

Please excuse my laxity in writing. As you have no doubt heard, I am rather swamped with a research project on kuru and the entire problem is one of staggering dimensions to Vin Zigas and myself—who have undertaken it together and are fully dependent upon you, those at Commonwealth Serum Laboratories, Hall Institute, Port Moresby, and U.S. laboratories who are providing us with extensive laboratory support for our investigations. Kuru is a really exciting disease, and at the moment we have over 100 patients dying from this parkinsonism-like illness—some one-quarter of whom are children, on whom we are concentrating our studies. In addition, our epidemiological studies have located over 500 deaths from kuru in the affected population during the past five years, and this work has not yet been more than half completed. It is fair to say that nowhere else in the world is such a collection of a rapidly progressive degenerative disorder of the central nervous system available for study. Certainly, we have not yet found any good explanation for kuru, nor any environmental factors responsible for its pathogenesis. Much suggests that we are dealing with a heredofamilial disorder of genetic determinants.

Your finding of distorted globulin patterns is the most interesting bit of laboratory evidence we have yet had presented to us. Most of the studies done have yielded highly negative results. Obviously, we are extremely interested in learning more about it and in controlling the observation as amply as possible. I am deeply grateful to you for the records and data you have sent. In attempting to correlate it with our extensive clinical and laboratory records, I am unable to identify some of the names on the graphs and paper-chromatography slips you have sent. Since it has been possible to identify all of the names on the summary sheets which Roy Simmons has sent of his blood-group genetic work, will you please check your labels and names with him—since his specimens were, I believe, the source of all your specimens—and correct those names you have to coincide with his? Otherwise it may never be possible to identify the specimens you have already examined, and to ascertain what illness they had or what stage of kuru. Some specimens Simmons has had are not from kuru, others are from atypical cases, and others are from a wide variety of stages of the disease on various medication—and diets—or on no therapy. All this can be sorted out if you can accurately identify your specimens as per those of Roy Simmons. Thus, for example, graphs labeled Niaia, Inau, Lakuo, and Faruou, and a paper strip labeled Faruou, cannot be identified. On a few others, we believe we have placed them properly.

Since receiving your letter, many further kuru-serum specimens have been sent to Hall Institute, and Roy Simmons has received cells and serum on most of these. I hope you will get (through him) further clots for further hemoglobins and, especially, adequate serum specimens for further electrophoretic and other studies on the serum proteins. An abnormality in the globulins would be a most intriguing finding in kuru. The "controls" Lois Larkin gave to you are really not adequate, coming from very different people culturally and racially. No sera from this region are available outside of those we have sent down. Some we have sent are most definitely not kuru. Specifically, Simmons's specimens labeled Ivuti (Ilagi), Kako, and Uvae are not kuru. In the very near future I shall be making large control blood collections on like ages, like sex, but not-kuru normal people who are from the same regions and genetic stock as our cases. Other "control" series shall be from the same locality but from kuru-free lineages, etc. I certainly shall request that you receive clots and ample serum specimens both from further kuru patients and from these "control" series. Furthermore, serial samples from various stages of this rapidly progressive, fatal, rather uniform-in-pattern disease have been collected and are being collected; and it would be most interesting to have results on the same individual in early stages, moderately severe disease, and terminal stages. These specimens are already available at Melbourne, and many further will be coming. I have requested Lois to supply you—if you are willing to continue the investigations—with proper specimens and controls.

[39] Thus, I shall welcome further reports, and especially any suggestions, advice, or observations you care to offer. We are always anxiously awaiting news of results on the many specimens we send out; and Vin Zigas and myself are anxious to provide you with any specially collected specimens which you believe may help in further unraveling this complex problem of kuru. Your finding is the most interesting to date, and we are most anxious to hear from you again. Thank you for your help and interest.

Sincerely yours,
D. C. Gajdusek

P.S. Lois is here in Okapa "on vacation" working on kuru with us, and when she gets back will have everything quite straight for providing you with any type of specimen you require. However, among those already in Roy's hands are many which may serve your purposes well; and S. G. Anderson has his hands on newer specimens we have sent down, and can provide you with what you need in the interim.

Good luck,
Carleton

[40]

DCG to SIMMONS   4 July 1957
Dear Roy,

I am sorry that in all my rushed notes I have never properly discussed the Cape York paper with you. [41]

Lois Larkin is here with me at Okapa, giving us a fine hand "on her vacation." We have just tabulated all four combined tables you have sent us on kuru patients, and all checks well with our records. The names, as you record them, are all correct or close enough to identify easily. However, by the time Cyril Curtain transliterates them onto his report and record, several are no longer recognizable or identifiable. Please ask Cyril to check his specimen-naming and designations against yours—since you are the source of his specimens—in order that none of the results gets so removed from our file-designation that they cannot be located and used.

In summary, it appears that you have done blood-group and -type studies on 44 kuru patients. Of these, only 4 are atypical kuru. These are: Oni, Fopota, Iokoio, and Mara. The first is probably not kuru, as things have developed; but there is still some doubt. The next two (Fopota and Iokoio—or Yokoio) are "recovered kuru," which at present I think may be a different disease—i.e., hysterical mimicry of early kuru in hysterical-temperament women who go on to recover, a rather rare occurrence. The fourth was probably never kuru; but she "thought she had it," and we did too for a while. They cannot be called "normal," however, for these 4 represent the group of differential diagnostic problems the exact nature of which is not yet clarified.

In addition to the 44 (including the above 4 "exceptions" cases), you have already done three specimens from non-kuru cases in Fore natives; these are: Kako, Ilagi (Ivuti), and Uvae.

Since the fourth group has reached you, many further kuru and some non-kuru specimens have been sent to Melbourne. Dr. S. G. Anderson may have already sent you cells and sera on these. If not, he can if you are ready for them and give him a ring. In any event, Lois will get specimens off to you when she gets back.

Returning to the Cape York paper, I did not understand that it had not yet gone off to the journal; here I misread you. Even in your most recent letter I again misread you, not understanding that the paper was not already long since off to the journal. Lois now points this out to me. I am sorry for my sloppy reading and misinterpretations. I am in full accord with your sending the paper as is.

Sincerely,
D. C. Gajdusek

DCG and ZIGAS to ABBOTT 4 July 1957
Acting Director of Health
Territory of Papua and New Guinea

Dear Dr. Abbott:

I thank you for your kind letter of June 4th, and I beg that you excuse my laxity in keeping you informed about kuru research. I have been so rushed that reports have simply not been forthcoming. We have had a total of 12 kuru deaths among the 110 kuru patients we have now brought our total "Active Kuru Cases" to, and of these we have had four autopsies—three complete and one with brain only obtained. We are awaiting reports on histopathology, but thus far preliminary communications inform us of nothing startling except perhaps a slight diminution of cells in the cerebellum. Grossly, we found nothing remarkable; and thus Wilson's hepatolenticular degeneration is fully ruled out.

I had arranged for thorough blood-group genetics to be done on kuru patients' red cells and serum (including abnormal hematological antibody studies); and these have been completed on some 44 by Roy Simmons, with whom I have collaborated in the past. These results may eventually be of value when it is possible to compare them with normal non-kuru control groups and with Fore, Keiagana, Kimi studies—specimens for all of which studies we have already been collecting. If things do work out to substantiate the current suspicion of a strong genetic predisposition as perhaps important, if not determinant, in this disease, then such studies will be of immense value. They are already partially completed.

Other collaborators have managed to detect rather startling protein abnormalities in some of our patients, and we are pursuing this matter at length. Of course, it again must be controlled by normal Fore and other Eastern Highlands population studies, which we are trying to carry out. Dr. A. V. G. Price has been of immense help to us, and we are sending him many specimens—blood, urine, and autopsy—for trace-metal studies and other clinical investigations. He has supplied us with everything we have requested, made valuable suggestions, and provided many important determinations. Thus far, we have no evidence of copper toxicity from the data he has sent us, but other trace-metal studies are under way.

Antibiotics, sulfonamides, phenobarbital, acetylsalicylic acid, Dilantin and Tridione, cortisone, delta-hydrocortisone, testosterone, and BAL have all been tried without any dramatic effects. We are still trying BAL before we can be certain that it is ineffective; and since a single course requires three or four injections on day 1, and twice-daily injections for four to five succeeding days, the supply of ten packages (12 ampules of 2.0ml each) which S. G. Anderson brought up, on my request, was soon exhausted and the new supply of three boxes is only enough to give a second course (which is now due) to four of our patients—children, at that. Could we get a further supply of BAL (oily injection of dimercaprol)? We have ample Tridione to complete the current clinical trial on 11 well-chosen cases. We did not have ample Dilantin-sodium, however, to make the trial of it a fully satisfactory clinical investigation; and if more could be supplied, we should be most grateful.

Testosterone and stilbestrol investigations are only now starting. We will need a further supply of testosterone propionate or methyltestosterone for this work, if it can be provided. The marked sex inequality in the distribution of our cases provides the basis of therapeutic trials in this direction.

I have just recently returned from 18 days of patrolling in the Baro Ianai (Yani River) Basin area, where I located kuru far south to the most southerly extent of the Fore, in high incidence (4% of the most southerly Fore communities were currently suffering with active kuru). However, what was more important, I found the limits of kuru for fully half the circumference of the kuru-affected region; and in this region of kuru/non-kuru boundary, epidemiological study proves much more profitable and rewarding. Here it is possible to find isolated kuru cases in almost kuru-free communities, and to contrast lineages and households with kuru with those without kuru, and to compare kuru-present with kuru-free communities of the same cultural stock. Such studies are not possible in our Kuru Research Center at Moke–Okapa, for kuru incidence is so high that there are no "control" households or lineages or hamlets.

In the low–kuru-incidence region, I could trace many cases to Fore women brought into the community by marriage. Some were from Fore families with much kuru in parents, siblings, and other close relatives; a Fore girl married into the Kimi developed kuru in the virtually kuru-free Kimi community only after years of residence there. This would seem to cinch the genetic matter, were it not for a few—decidedly in the minority—cases (still to be further investigated) of Kimi girls from low–kuru-incidence communities developing kuru when marrying into the Fore kuru region. Thus, matters are by no means clear as yet. However, it is essential to delineate the kuru boundary; for it is here that investigations may prove most valuable, and study of this boundary and any shifts in it may in future years provide valuable data on kuru and its behavior. I certainly hope that before then we may locate some other factors of importance in pathogenesis or some specific etiological agent for kuru, but thus far we are faced with no leads which have led to anything but a further incrimination of the discouraging genetic-familial hypothesis.

Lucy Hamilton is with us, giving us a fine hand in tracking down dietary practices. Toxic factors keep suggesting themselves, but none has yet proved a very tenable possibility. It will require something mighty unusual to account for the peculiar 15:1 female-to-male sex-ratio among adult patients, and 3 or 4 to 1 among child patients. The child and early adolescent cases (on whom our studies are concentrated because of my preference for working with children and better familiarity in clinical investigation in pediatrics than in adult medicine) comprise some 20–25% of the total number of active cases in the region. In general, there are four adult cases to every child or early adolescent case.

We are anxiously awaiting the [Australian] film unit, but unfortunately it was unable to arrive when we had—as we have just had—maximum in-patient census. We are now forced to yield to our patients' insistent pleas and release many of them to their villages. While the film unit is here, we shall do our best to round them up again. We are also planning a patrol to visit a few patients in the fully native setting. Most of those nearby can no longer be so classified, for we have dressed and bandaged ulcerations, urged supportive measures, and in general interfered with the native practice of neglect and avoidance once the disease has reached incapacitating stages—practices which, in the long run, may be the most humane, for they ensure a rapid exodus once incapacitation is such that there is no further possibility of entering into any part of normal native social life. Asomeia, one of the two patients we sent to Port Moresby, is now in terminal condition—in very pitiful condition. She cannot live much longer. Manto, the second patient who visited Moresby, is likewise deteriorating rapidly, but is not yet quite terminal.

Our new kuru ward provides ample space for hospital cases, and we have finally provided ourselves with good laboratory and treatment room facilities; thus, hospital housing is fine. The new native-materials house for our kuru research personnel—all of whom are still living parasitically on the *kiap*, Patrol Officer Jack Baker, who is giving us wonderful assistance—is nearing completion and will provide ample roofing and table space for the work. Funds for patrolling and odd kuru research expenses, Vin Zigas tells me, are fully exhausted; and if there is any possibility of an added grant to cover the costs of trade items, supplies, etc. (which are essential if we are to keep finding cases, and studying them and caring for them), it would be most welcome.

[42]

In my latest letter from Bob Traub (Commanding Officer of U.S. Army Medical Unit at the Institute for Medical Research, Kuala Lumpur, Malaya), he mentions how very helpful you were to him in Borneo and urges me to look you up. From all Vin Zigas has told me of your interest in kuru, and from all the Traubs have told me of your interest in investigative work, I am certainly most pleased to have this opportunity to correspond with you; and when kuru research is more settled, I shall certainly hope to meet and discuss matters with you and Dr. Gunther in Moresby. At the moment, I am trying to complete as accurate and thorough an epidemiology as possible; for should we be stalemated by kuru (as we shall be if all the suggestions that it is a degenerative heredofamilial disorder of the CNS— rather than a toxic or post-infectious phenomenon—no evidence for which we can yet find [—turn out to be right]), then the most important contribution we can make to kuru study and to the field of human genetics is a thorough incidence, distribution, and case-density pattern survey as a baseline for follow-up studies in future years. It will then be the future behavior of kuru in this and surrounding communities which will give us the added precious information. At the moment, however, without a fully validated epidemiological record of the illness, future patterns could never be interpreted, for there is nothing to refer them back to. Thus, our rather ambitious current goal is to document *every kuru death* in the memory of the 15,000 affected population, check these by cross-interrogation and the extensive family histories of our current cases, and to plot this data out on a population map and tribal-linguistic map of the region (which we are now also compiling, with the extensive aid of P.O. Baker, who has given much of his administrative time to helping us on our kuru study). This ambitious project is over half-completed already. When the kuru perimeter is fully substantiated and patrolled—and this too is over half-finished—the current phase of kuru study will be over. By that time we should have simultaneously completed adequate therapeutic trials of all medications which we thought offered some promise, and also have carried out as thorough a clinical study as bush conditions will permit. At that phase, I shall try to travel to Moresby to see you and others interested in kuru; and I hope that Vin Zigas, on whose shoulders so much of the work and support in this study lies, can accompany me.

[43]

We are pleased to learn that Kako—the Fore girl with patent ductus arteriosus, whom we found in our studies and sent to Moresby for ligation—has been operated upon successfully. Our error, in sending the full case history not with the patient but in our report-letter on kuru to Dr. Scragg, may have caused some unfortunate delay in the appropriate people getting her case history; we are sorry. Thus far, we have not seen the patient again, but hope she will soon be back in her village, where her kinsmen are inquiring about her.

Finally, Dr. Abbott, please excuse my addressing you improperly, if you are not now Acting Director of Health. I do not know whether you or Dr. Scagg has the post at present, and I have wanted to thank you for your letters and assistance. Will you please show our report of current work to Dr. Scagg and Dr. Gunther, both of whom have asked to be kept informed of its progress.

Sincerely yours,
D. Carleton Gajdusek
V. Zigas

[44]

DCG to SMADEL 8 July 1957

Dear Joe:

Enclosed is a copy of a recent letter to Dr. Abbott, currently Acting Director of Health, letting him in on our work. It should serve to inform you of what we are up to also.

I have some two or three letters out to you at the moment, and am anxiously awaiting even a few words of comment on the brains we have sent and the manuscript. From current notes from Melbourne, they are now down to 4% formaldehyde for fixation, whereas we have used either 10% or 15% formaldehyde for the four brains we have already obtained (three of which are off to you). Furthermore, Melbourne—only Graeme Robertson is looking at the brain, and he has had only one thus far, our first—requests a thorough perfusion of the brain before fixation. This procedure I shall try, but we have no appropriate canulas, nor is the cold wind and rushed excitement of "bush autopsies" conducive to careful and accurate perfusion. We shall, however, do what we can. I would very much like added word on:

1) value or lack of value of perfusion, and any suggestions for doing it well in the bush—we have a good roof, syringes, and needles and might even get intravenous saline through if we had canulas to tie in;

2) best possible fixation methods—we used 15% formaldehyde on the brains which you got, and transferred them to 10% thereafter (one to two days later), and sent them as soon as they were somewhat hardened; and

3) any detailed suggestions for more elegant preparation of brains and other pathological specimens, since first report from Melbourne claims nothing much on neuropathological study, other than (?)diminished cells in cerebellum.

Our patients are dying of advanced neurological disease; and many with death imminent are on our hands, and brains and full autopsies may be possible. Thus, we might be able to supply whatever your neurohistologists and general pathologists desire, if you only tell us. Cases now total over 110; we have some thirteen deaths since we started studying, and I suspect all the others will go the same way within one to two months.

Enclosed is a copy of most of our record on Yabaiotu, a woman whose brain was shipped off to you last week. Histories on the other two specimens are already in your hands, I believe, as well as that on Kinao (our first autopsy, viscera of which you received; the brain went off to Melbourne). Your records should thus be complete. Please ask those who look at the material to keep in touch with us, for here in the bush any and every lead may influence the course and direction of our work.

The markedly abnormal beta-globulins found in many of our sera need further study. Lois Larkin is going to mail the entire extensive kuru-serum file off to you as soon as she returns to Melbourne. Thus, you will have all the serial specimens on the patients, as well as many control sera for eventual serological study or—perhaps more important—further check on the protein abnormalities Cyril Curtain has been finding on the specimens he got from us in Melbourne.

I have written to Tom Rivers and thanked him for his great assistance. If you think that the current letter contains anything he should see, let him see it. There is not much new since my last long letter to you and to him, just plenty of work. Tridione, testosterone, and second courses of BAL are the current therapeutic regimens being tried. We never did get any of the anti-parkinsonism drugs which I hear about from the U.S.; and if you can track us down a supply of them with literature, I should be most interested.

Sir Mac has a paper follow-up on my auto-immune complement fixation test, and many are jumping on the bandwagon. That it is a really hot test, in certain macroglobulinemias and most disseminated lupus erythematoses, is certain; and I beg you to ask anyone who might have multiple myelomas, macroglobulinemias, disseminated lupus cases, and this type of case on his hands, to save (at $-20°$ C.) serial serum specimens for me. I shall rush through them immediately upon my return. The Melbourne crowd has actively kept up with my test, and it is proving most intriguing in certain cases of DLE and other diseases. The two definitive papers still held up by *Archives of Internal Medicine* have me angry; there will be a whole series of "follow-up" papers out by others before they get off their editorial asses, unless they hurry up. I hope they write soon.

Sincerely,
Carleton

BERNDT to SCRAGG[5]   9 July 1957
Dear Dr. Scragg,

Thank you kindly for your letter of June 21st, 1957 (ref. No. 32/35/M1032–RS/O'D), in reference to the presence of a disease known as "Kuru" in the Eastern Highlands of New Guinea. I am sorry I was unable to reply to this immediately, but have been inundated with work just lately.

Yes, I am interested in your comments on "Kuru." My wife and I had two periods of fieldwork in that region, within the linguistic blocs of Kamano, Usurufa, Jate, and Fore—working first from Kogu and secondly from Moke-Busarasa. During this time we both saw and heard of a number of cases of this, which is known in

---

[5]This letter was forwarded to Okapa by Dr. Scragg on 16 July 1957, with an offer to obtain copies of the journals mentioned by Berndt.

pidgin English as *guria* (possibly a derivation of *chorea*?). *Kuru*, if we remember rightly—without looking up our notebooks—is both the Fore and the Usurufa word for this condition, known by other local words among the Kamano and Jate (e.g., *guzigli* or *guzili* among the Jate). I have commented on the similarity of this to the *teti* shaking which accompanied manifestations of "cargo" movements in that region (see *Oceania*, 1952, Vol. XXIII, No. 1: e.g., p. 57); the *teti* or *taborabainu* is mentioned also in another paper (*Oceania*, 1954, Vol. XXIV, No. 3, especially on pp. 207–8; and in No. 4, pp. 272–3). Again, I have discussed this form of *guria* incidentally in a volume I am completing: it was (and I assume still is) regarded as a form of sorcery, and is relatively common throughout the Jate and Fore region, with cases also among the Usurufa. In the cases we observed in the Kogu and Moke-Busarasa districts, the subjects could walk with the aid of sticks or with help from others; there were (e.g.) involuntary twitchings, a feeling of "abnormal" coldness, dilation of the eyes (which appeared glazed and watering), and (especially) lack of control over limbs. Despite muscular difficulty in uttering words, they did not appear to be mentally affected, although we did record material inferring this. The attacks are apparently recurrent, the symptoms becoming intensified. We have recorded a great number of cases in what can be regarded as all stages of development of this form of "sorcery," which has several locally defined subdivisions. The detailed cases make interesting reading. I tried to interest medical authorities both in New Guinea and in Sydney at the time, but without much response. My impression was that this form of *guria* was psychosomatic, and differed from St. Vitus's dance. If a detailed medical study were to be made, it would be interesting to dovetail it with the anthropological data. Personally I would be pleased to have a look at any medical data forthcoming from Dr. Smythe's survey, and would be pleased to comment on any aspect which enters my own sphere of competence.

I would say that one hears more about *guria* in the Fore linguistic area, and much *guria* sorcery is said to have come from there, but I would not say it was "centred" in the Moke district: to my knowledge it is fairly well distributed through the Jate and Fore groups, as well as to a lesser extent through the Usurufa and southern Kamano. My rough tentative calculation of Fore-speaking people is 10,000 at the very outside; but of course, as you know, the term *Fore* is used fairly widely, and actually includes a number of differing language-groups. (For instance, it is used in administrative reports from Goroka to refer to people in the far south of Bena sub-district). I would agree that our own observations would tend to confirm that possibly 1% of the population of the particular area with which I am acquainted were affected by this disease. During the course of our anthropological research work, we obtained very detailed genealogical data: again, I would agree that it has been present long before European contact; but from genealogies I would infer that *up to 30 years* is a very conservative estimate.

I would agree that this disease appears to affect women more than men, and affects those of all ages. Our Fore figures aren't immediately available, but in the Usurufa district of Kogu, e.g., in genealogical material involving a time-depth of approximately 50 years: 268 Kogu males are involved, and of these 2 are said to have died of *guria* sorcery; of 226 Kogu females accounted for, 10 are said to have died of *guria*. We have no information on what the possible "cause" may be, beyond that which is regarded as magical or supernatural. A number of examples imply that death took place when maggots appeared on the surface of the body: death, it is said, "always" takes place, and no known cures are mentioned; but there are a couple of cases, though, of recovery from *guria*.

Perhaps I should comment here on one particular case. When we returned to Kogu for the second time, my wife found one of the local women whom she knew moderately well (Tu'efa) to be suffering from this—as she had not, apparently, been 6 months before. She accounted for it by sorcery performed on her during a visit to Ke'jagana (Jate). Although she was having a severe attack, and could not move about without help from others, this did not prevent her from formulating and conceptualizing her experiences, except that she had considerable physical difficulty in articulating words. My wife went to some trouble to explore this point, since one feature of the advanced stages of the disease is said to be a loss of "mental," as well as of physical, control. About three weeks after our return, Tu'efa appeared to be completely recovered, although both she and others declared that she could expect to have further attacks later on. You may be interested to know, however, that shortly before we left Kogu for Moke-Busarasa, several women were warning my wife against going further south, owing to the danger of this type of "sorcery." Tu'efa, who was present, added "you saw what they did to me," and proceeded to illustrate very realistically the behaviour of a victim. The movements of her limbs (the way she fell to the floor, and so on) appeared almost identical with those she had displayed when actually suffering the previous attack. My wife even thought for a moment, in fact, that her actions had "touched off" another attack; but presently she stood up, quite self-possessed, and repeated her warning "That's what we do, so don't go there."

We are interested to know that Dr. Smythe is working in that region; we do of course know of Dr. Smythe by repute, but have not met him. We assume he is also carrying out linguistic research: it may interest him to know that we have collected a great deal of linguistic material from the Kamano, Usurufa, Jate, and Fore; and my wife has a preliminary paper in *Oceania* on some translation problems in that region. My wife concentrated on the Usurufa and Kamano languages, and I on Jate; and we both attempted a little Fore, but we never became so proficient in this, although we have a fair amount of textual and (written) song material in it.

We hope you will keep in touch, and we shall be pleased to elaborate on any of the above points or answer specific questions.

With all best wishes, I am,

Yours sincerely,
Ronald M. Berndt
Department of Anthropology
University of Western Australia

DCG to SMADEL   10 July 1957
Dear Joe:

Only two days after writing the enclosed letter (and still no way to get mail out to Kainantu), and our cases have jumped to over 120 active cases in our Study Series and we have had two further deaths *with complete autopsies*. Both were done immediately after death, and brain obtained less than one hour post mortem. Thus, we have two further brains on our hands already—one for you and one for Melbourne—and both went immediately into 4% formaldehyde and will be changed daily to fresh formaldehyde until hardened.

Please rush details of more elegant fixation, or at least straighten out the conflicting percentages and directions.

Enclosed is a long manuscript of our second kuru report, soon to be sent off to an Australian journal—which we must do for political reasons. It was written only yesterday but is already out of date, for I have now brought our study cases to over 120 and we have these two further autopsies.

In addition, we have just received our first *early* case of acute meningoencephalitis: nothing like what any of our kuru patients have ever had, as far as we can ascertain, but we have a boy of six with 248 cells, almost all mononuclear, in the CSF and mighty sick. We are sending feces, throat washings, CSF, and blood off to Anderson in Melbourne today—and hope that with the two days' delay in getting there, they may still net us something. We have nothing here in which to inoculate, other than ourselves—and we have not reached that stage of desperation as yet.

In spite of the interest in this case—our one previous case mentioned in the enclosed paper was recovering when first seen, but still had CSF pleocytosis—we cannot link this sort of illness with kuru, as hard as we may try. And we have only two cases of benign aseptic meningitis, undiagnosed, in four months of study—and over 120 cases of kuru!

Please ask your neurologists and neurophysiologists and internists to read over this paper—and the previous one—and send me any comments, suggestions, corrections, or disagreements they can. It is difficult writing and working here in bush isolation, and I sadly feel the lack of colleagues and critical discussion. I shall be most grateful and indebted to you for all and every editorial comment you make, and all that you can secure from neurologists at NIH. This applies to the enclosed and to the previous paper. Both are, of course, still confidential outside of the professional audience who may be willing to proofread and criticize for us.

A jeep is at last coming in to take out Vin Zigas with mail and specimens, and I must close and mail this off without further proofreading. Please excuse my unedited reflections, as usual.

Sincerely,
Carleton

DCG to ANDERSON 12 July 1957
Dear Gray,

These specimens—coming immediately by air express—are 79 specimens on the [district boarding] school children. The first 30 are acutely ill in the first days of a virus epidemic of epidemic parotitis with marked upper respiratory illness. White counts are normal but predominantly lymphs. There is no parotid tenderness, and only about half of them are having parotid swelling. I cannot call the illness classical mumps. The cough and respiratory symptoms predominate. We have gone on and bled those who have not yet become ill, and most will probably be ill in the next few days. Thus, I firmly suspect that these bloods will be pre-bleeds of the best kind.

If you can manage to do any virus work, fine. If not, please save all serum specimens at $-20°$ for eventual pair-serum serology. If you want to try for isolations, the first thirty specimens (i.e., Nos. 1 to 30 inclusive) are all acutely ill in the first to third day of symptoms with typical disease.

In addition, all of these specimens should have clots sent to Simmons for Fore background genetic studies and for abnormal hemoglobins, for we have full lineage and genealogy on each child! Furthermore, many have a very good chance of eventually developing kuru—coming from kuru families.

We have two cases of acute meningoencephalitis, and the blood and CSF on the first went down to you. Blood and CSF on Toneaso, a girl with acute aseptic meningoencephalitis, is in this shipment along with two lots of CSF. The CSF is for chemistries (which we need to help diagnostically) and serology, and the second is for isolation attempts (if you deem them still worth trying). Serum is for both serology and for isolation (use clot as well, if you wish—we got virus from clots on old South America bloods). More later—but courier is waiting for these shipments.

Please write.

Sincerely,
D. Carleton Gajdusek

[P.S.] Blood collected just now, today.

CURTAIN to DCG   15 July 1957
Dear Carleton,

Thank you very much for your reply. I am sorry about the names, but the confusion must have arisen from misinterpretation of Roy Simmons's rather flowing hand. It is very difficult to distinguish his i, m, n, u, w, and e. I have all the original tubes in the deep-freeze, and shall drive over to C.S.L. this week and check his labels on these with the master-list.

We have electrophoresed a further 23 samples on paper, some new and some repeats. The pattern of the elevated β-globulin remains consistent. I now have my moving-boundary apparatus running, and shall run as many as possible of the sera on this. I succeeded in salting out the globulin from one of the more generous samples, "Amakiora." In the moving-boundary instrument, this appeared to be very sharp—indicating a high degree of electrical homogeneity and a fairly big molecular weight. Immunologic double diffusion in agar showed that it was closely related to a β-globulin obtained from a lymphosarcoma and somewhat distantly related to one of my myeloma proteins.

Does autopsy reveal any hyperplasia of the R.E. system? I am interested in applying Coon's fluorescent-antibody technique to localisation of the paraproteins produced by hyperplasias, although I don't know how one could organise it technically with your material.

As to the haemoglobins, I did these on all the clots corresponding to the serum specimens I reported on in my last letter. In all cases they were identical to normal adult. One word of caution: I am inexperienced with abnormal Hbs, having worked on only one family with Hb. Here I had some trouble with adsorption of samples on the paper, although the results were clear enough. Ultimately, I shall organise the alternative light source and optics of my M-B machine (to work in red light) and repeat all Hbs.

I shall send pictures, scans, paper strips, and corrections to names next week after I have seen Roy Simmons.

Thanking you again for the opportunity of helping a little with your exciting investigation, from my point of view a fascinating mass-dysproteinaemia.

Best wishes,

Yours sincerely,
C. Curtain

REES to SOUTHERN
Department of Agriculture, Stock, and Fisheries
Port Moresby

Melbourne
17 July 1957

Dear Mr. Southern,

The spectrographic survey of the ashed blood and tissue samples from sufferers from Kuru disease has now been completed, and the report is attached hereto. I am also forwarding a copy of the report on the samples submitted by Dr. Anderson of the Walter and Eliza Hall Institute. I presume you have the key for identifying the samples.

The analysis is of course qualitative at this stage, although any heavy metal present in significant quantity should have been detected. The compositions appear to be fairly normal to us; but if you wish a quantitative estimation of any specific element in any particular samples, I would be glad if you would let me know at your earliest convenience, since the equipment will be reorganised for some rather different investigations of our own.

I trust that this information is of some assistance to you, although it is evidently negative insofar as metal toxicity is concerned.

Yours sincerely,
A. L. G. Rees
Assistant Chief
Division of Industrial Chemistry
Commonwealth Scientific and Industrial Research Organization

ANDERSON to SOUTHERN   19 July 1957
Dear Mr. Southern:

I have just received the two reports from Dr. Rees, which I am sure you also have received.

The samples we submitted were numbered, and I enclose the key.

As you will see, of the nine kuru bloods, all contained copper at the level detected by Dr. Rees and three contained manganese. Of the three control bloods from Okapa, none contained either copper or manganese according to their test. I have spoken with Dr. Rees on the phone and he will proceed to quantitative estimation of copper, and possibly manganese, on these bloods.

With best wishes.

Yours sincerely,
S. G. Anderson

SIMMONS to DCG 19 July 1957

Dear Carleton,

By air freight today, express: three one-gallon flasks—each with set of 100 glucose-citrate bottles, needles (surgical), sterile swabs, instruments, lists for names, return labels, etc. The total 300 small bottles require about 8 drops of added blood, and storage at ice temperature. These would only be suitable for blood groups. See instruction sheet.

I chased Australia for venules and found practically none, so—as I think we should do chromatography on the 200 sera, and abnormal haemoglobins on various lots—I have improvised with a gross and a half (216) of 1-ounce McCartney bottles each containing an 18G 2″ needle for bleeding. The complete outfit is sterile. I bought the needles from Surgical Manufacturing Co. at a cost of £18/18/0, and have arranged for all the air freight charges to be sent to me. I have money, originally from the Wenner-Gren Foundation, which I can use for this purpose.

The aluminum caps must be tightened before dispatch, and a rubber grip is enclosed in the parcel for this purpose. Curtain is prepared to cooperate further and we could ask John Owen of Melbourne University to cooperate on the abnormal haemoglobins. Curtain has his moving-boundary apparatus set up for sera. It took Miss Casey and me six hours to set up the improvised McCartney outfits.

I suggest you use the McCartney bottles giving us cells and sera for all tests, and that the number you collect from any group be enough to give us a representative sampling. My limit will be 200–300 samples, as the amount of work with eight blood-group systems is huge and I just don't have about four free weeks to spare.

My appeal is to sample adequately, but with discretion.

To further help out: the return flasks or cartons may be sent to C.S.L. air freight collect, and again I will pay for it.

Seeing that you didn't read my earlier letters properly, I suggest you read this only twice. The paper (Cape York) has been sent to Washburn [editor of *AJPA*].

I gave Curtain a full list of the first 44 names. I have results for about another 24 to send you, and possibly a dozen not yet grouped. (I spent a week in Brisbane on a lecture tour). Where names appeared to be duplicated, I did not repeat the grouping tests—I have missed a few which were truly not duplicates, but you gave no advice.

Regards,
Roy Simmons

P.S. The 18G needles should permit you to bleed patients allowing the blood to drop into the McCartney bottles.

SCRAGG to DCG 24 July 1957

Dear Dr. Gajdusek,

The following letter has been received from Dr. Anderson and is passed on for your information:

> I have just had a letter from Dr. Rees of CSIRO enclosing a list of spectrographic results, a copy of which he has also sent to Mr. Southern of your Agriculture Department. The substance of it is that Mr. Holden of this Institute ashed 12 bloods, 3 from

controls in Okapa and 9 kuru cases. Of the 9 kurus, 3 showed manganese in qualitative spectrographic examination, but there was none in the controls. The 9 kuru cases all showed copper, but again there was none in the controls.

These results, which suggest an abnormal amount of copper, and perhaps manganese, in this qualitative test must be confirmed by quantitative tests; and Dr. Rees of CSIRO will proceed to do this.

The results on blood ashed in Moresby and sent to Dr. Rees did not quite agree. In the bloods from Moresby, 4 controls had copper present and 7 kuru cases also showed copper. It is obvious we must wait for further results from Dr. Rees, but the position is at least interesting and it may be worthwhile swinging the investigation in Okapa to investigating sources of copper and manganese.

Yours sincerely,
R. F. R. Scragg

DCG to SMADEL 25 July 1957

Dear Joe:

Enclosed is a letter and a slightly revised manuscript (no essential changes; just a few grammatical corrections, clarifications, and figures brought up to date) for the *New England Journal of Medicine*—which you chose wisely and, from my point of view, very well when *Science* suggested a more medical journal. I do not know the exact address, and thus beg you to rush the letter and revision off to the journal as promptly as possible.

By now you have no doubt received the follow-up paper just going off to the *Medical Journal of Australia* in order to pacify Administration here. However, we are counting on the *NEJM* article—prepared and submitted two months earlier—appearing first; and thus I write to let them know we are in full accord with their having the paper, and to let them know that a series of papers is soon to follow, and thus to urge as prompt attention to our report as possible. We know it is a good first report of a new disease, and think it deserves prompt handling.

Venules have arrived and they are fine, although the 6–8 ml size often leave the labs screaming for more serum. We are most grateful. We also have extensive new studies under way about which I shall write shortly. Just off to you are two new brains—whole brains—from two typical cases on which we have observed progress from early to terminal disease. Both are from children: Aranaka, a boy of about seven, and Mulinapa, a girl of seven to eight. Both died of typical childhood kuru, rapidly advancing. Also, Zenker's- and 5% HCHO-fixed tissues on Mulinapa are sent (from a wide range of viscera, muscle, fascia, etc.), and the same range of tissues (but only HCHO-fixed) on Aranaka (in this case, in the brain tin along with the brain).

We need, more than ever, some idea of what—if anything—pathology is showing, since we soon will call an end temporarily to fieldwork and any possible leads may determine the course of our further work these last months.

In addition to all this, there are Zenker's-fixed (now in 70% ethanol, as usual) and 5% HCHO-fixed liver, kidney, and spleen, on a case of rapid liver-coma death with slight ascites in a chronic liver-failure case with rather acute symptoms and rapid exodus. This pattern in well-nourished individuals, in a well-nourished community in which we have yet to see clinically diagnosable infectious hepatitis, is not uncommon in both adults and children; and we would like some pathological study, for its etiology may be of much importance to our work and to medicine here.

Liver—as is obvious from the chunk—was shrunken, nodular, and firm; spleen slightly enlarged, hyperemic; kidney a bit small.

[45]

More study of our patients and the help of some clinical visitors has thrown us more into doubt than ever as to just where the neurological lesion is. Emotional instability is certain; a tendency to excessive hilarity, etc. is certain. The "mask-like" facies is rather a fixed facies—but not "mask-like": i.e., it is full of expression but quiet, with rare blinking and little motion until stimulated. Then it responds quickly and usually somewhat excessively, with euphoric grins and smiles, or even shrieks.

The entire postural tremor situation is most complex. It is not a true cerebellar, nor a true parkinsonism tremor. It is, rather, a tremor and other types of involuntary movements which are very irregular and difficult to describe. It is definitely an antigravity postural disorder. Thus, sleeping, or when curled up and firmly supported in any position—i.e., in another person's arms, pressed tightly in another's grasp, etc.—the tremor disappears. Any sudden relaxation of this passive support, or sudden shift of even one portion of the body's posture (as of the head, neck, one arm), produces a violent tremor plus choreiform response in the entire body, apparently aimed at reestablishing a difficult-to-maintain antigravity equilibrium. Toes are constantly gripping and searching when a patient tries to stand unaided—or, if more advanced, even when supported. Sudden loss of antigravity postural support, given passively by examiner to head or upper extremities, suddenly sets off repetitive, irregular tremors or a choreiform pattern of movement. Rigidity is minimal, if at all present. It appears late. Instead, there is an increased tone to the muscles that are associated with attempts at maintaining posture and preventing the antigravity tremors which fight the slightest instability of standing, sitting, lying, head posture, etc. and which initiate as a startle response. If well and firmly supported passively, even in late cases, this "intermittent rigidity" subsides to complete relaxation. This more detailed description may help some neurologists further locate the lesion for us. Please tell them to pass on their ideas. Nystagmus is absent; Babinski has not yet been positive; but ankle and patellar clonus are not rare in very advanced cases. When present, patellar clonus and ankle clonus are sustained and definite; only a certain few advanced cases show these.

Joe, please urge the *NEJM* to get our paper out before the "follow-up" studies start to appear. They have a chance at the "first report," and I think it is one worth taking!! We are referring in the second and third papers to our previous "preliminary" report, which we now hope will be in the *NEJM*.

Since the paper is now off to a medical journal, we might as well show a picture or two of the illness at the start. Of the black-and-white negatives sent to you, a certain group of frames were designated for enlargement for possible publication. If you could select the most informative among the following, look them over and send two or three to the *NEJM,* along with copies to me to label and describe and decide definitively on; it might greatly improve the paper. I am referring, of course, to the black-and-white still negatives we sent to you:

Roll 1: Frames 10, 11, 12—all three kuru children have now died.
Frames 34, 35, 36, 37—Aranaka (whose brain is now en route to you).
Frames 20, 21, 22—Kinao, who has died of kuru.

Roll 2: Any of these on this roll—all show kuru patients—which show kuru well, especially any showing strabismus or the laugh.

Although we have many other films, Joe, these two are the only ones now readily available in your hands, and from them you should be able to pick a few shots worthy of illustrating our "first report." If you will send these to the *NEJM*, and copies to us, we shall be most grateful.

[46]

If I can finish our work, the Kukukuku patrol, and the epidemiological delineation of kuru in time, I shall try to join you in Bangkok. Can you get me an invitation to present any of our kuru material? Or should I save it? We will have prepared by then a fairly definitive article for an American journal; it may be of such length as to appear in two to three sections. Furthermore, the pediatric study of our child cases—over two dozen of which we have studied most intensively—will also be prepared later this year.

Chronic manganese toxicity, as described in E. J. Underwood's *Tracer Elements, Human and Animal Nutrition* (1956), has come to our attention; and although not what we are seeing, it can be in Moroccan miners so suggestive of kuru that we are most intrigued. Yet we cannot find a clue as to the toxin source in our women and children patients. Please tell the NIH neurology group to rush to us any suggestions, further ideas, and advice. We need it, for we are ever being more and more shoved into the genetic corner—the least promising, I fear, but if so, the most demanding of thorough epidemiological study in these days of radiation and mutation behavior speculation.

I hope you had a fine trip in Europe. Tell me of it.

Sincerely,
Carleton

ANDERSON to ZIGAS and DCG 26 July 1957
Dear Vin and Carleton,

I have recently received a large number of bloods from your outbreak of mumps. The bloods generally did not separate well and did not produce much serum. The serum is at $-20°$ and the clots have been given to Mr. Simmons of C.S.L.

Asomeia brain contained no virus on CAM or in mice; but I have now put it up in yolk sac and in the amniotic cavity, with the possibility that there may be mumps virus in the brain.

The Women's Hospital have reported examination on 3 urines from Okapa. Nata I apparently normal, Aranaka I and II and Yani I all contained an unusual chromatographic spot as yet unidentified but probably amino acid. They think this is very likely due to bacterial contamination, so would like urine taken in the following manner with preservatives from 6 or 12 cases and from several controls:

Make up 20% HCl by adding 150 cc of concentrated hydrochloric acid to 750 cc of water. Place 50 cc of 20% HCl in a bottle, and add to it the whole of a 24-hour specimen of urine. Measure volume of this 24-hour specimen, and send this information to me together with 100 cc of the 24-hour specimen.

You may have heard from Dr. Scragg that there is a very tentative suggestion—at a qualitative level only—that in Melbourne ashed bloods from cases of kuru show copper and perhaps manganese whereas three controls did not do so. I think it would be well worthwhile to send me further specimens for similar analysis. Quantitative analysis is also under way.

I am sending you under separate cover about 40 chemically clean McCartney bottles marked with my initials in diamond pencil; each bottle contains a clean bleeding needle. Would you put as much blood as possible into each bottle, from 20 cases of kuru and from 10 women and children controls and 10 adult male controls. Would you please let me know the names, ages, and sex of each sample; and state whether each sample is a case of kuru or a control.

Mr. Holden has suggested that hair may be a good material in which to search for heavy-metal poisoning. I will leave it to you what hair samples you get from controls and kuru cases.

With kind regards,

Yours sincerely,
Gray

C. G. BAKER to DCG   26 July 1957

Dear Dr. Gajdusek:

Since Dr. Smadel has not yet returned from Europe, I am writing to give you the latest developments. This letter will acknowledge receipt of written material accompanying letters from you, dated June 29 and July 10. To summarize other materials received from you to date: (a) four brains; (b) two cervical cords; (c) four sets of tissue obtained in autopsies; (d) 36 urine samples; (e) 11 rolls of 8mm film; and (f) three strips of 35mm black-and-white stills. To summarize material sent to you from NIH: (a) 1 16mm motion-picture camera and case; (b) tripod; (c) exposure-meter; (d) 8 100-ft. rolls of 16mm color film; (e) 4 100-ft. rolls of 16mm black-and-white film; (f) 36 one-dozen packages of Keidel-type Vacutainers; and (g) polyethylene bags.

You will be happy to learn that preliminary studies on the first brain received from you have revealed degenerative changes in the basal ganglia and, more strikingly, in the cerebellum. The early impression is that there is no acute inflammatory process (including virus infection) although previous infection cannot be ruled out. The group of neuropathologists working on the tissues are starting extensive studies with special histochemical stains, but as yet they have not had time to report on these special studies. Also, it is too early to say whether changes found in the first brain will be noted in the brains that arrived subsequently. The fixation of the brains seems perfectly adequate, and the recommendations from the neuropathologists are that you continue to follow the directions supplied to you earlier from NINDB. One of the brains was considerably flattened, however, and they suggest that you take care that the brains are *suspended* in fixative for at least two weeks prior to their shipment. They do not believe perfusion is necessary. They suggest that the best way to ship the brains is to surround the adequately fixed brain with formalin-soaked cotton placed inside a plastic bag which is tightly closed to prevent loss of moisture. The bag, with its contents, is then surrounded by additional cotton and placed inside a sturdy container of wood or metal. If trace-metal studies on the brain tissue are not contemplated, the group felt it would be preferable to fill the outer container with formalin solution as was done with the earlier specimens you sent. The addition of fluid around the bag would give extra assurance that the tissues would not dry out.

We had some difficulty arranging for the copper determinations, but work is now in progress on the samples received from you.

Some of the neuropathologists have read your manuscript with great interest and agree that kuru is a newly discovered disease of importance. At this time they do not have any comments on the paper, but I wish to point out that it has not been seen by very many of the group as yet.

We have continued to arrange for the development of the films as they arrive, and will continue to store under appropriate conditions the original negatives and a copy of each print. As regards the prints for the 35mm black-and-white stills, we have arranged for printing as outlined in your directions accompanying the letter dated June 29. We will, therefore, keep on file here in Bethesda the postcard-size enlargements of each frame and each filmstrip. We have also made enlargements for publication, as you requested for certain shots taken in black-and-white. We are sending copies of the enlarged pictures according to your directions.

We are also maintaining files of the material sent us, both the original material as well as duplicate files in NINDB. It would be helpful to us if you can send us material on bond paper, preferably with original typing, since we can then make the duplicate files without having to retype the material. We are, therefore, sending, under separate cover, a supply of bond paper for your use. I realize that it will not always be possible for you to send us original-typewritten material; however, it would facilitate reproduction of material if you can, whenever convenient to you.

We have not, as yet, had an opportunity to see the movies taken by you, but expect to have a "premiere performance" of the movie "Kuru" upon Dr. Smadel's return.

We have been notified from the *New England Journal of Medicine* of the receipt of your manuscript. We have not yet heard of the action taken on it.

Dr. Smadel will return to NIH on August 1. I will call to his attention the developments that have transpired in his absence. I am sure he will write you after he has returned from Europe.

Best wishes from Bethesda for your continuing good work under difficult conditions.

Sincerely yours,
Carl G. Baker, M.D.
Assistant to the Associate Director

[47, 48]

RIVERS to DCG   5 August 1957

Dear Doctor Gajdusek:

Thanks for your letter of June 29th with a copy of a letter to Dr. Joseph E. Smadel and a copy of a paper on kuru which you propose to turn in for publication.

I am glad to hear that things are going all right with you, even though your task is somewhat difficult at times.

The paper which you propose to publish is good, and I would like to know where you intend to send it for publication. I have asked Doctor [H. W.] Kumm to have you order reprints, at our expense, in case the paper is accepted for publication.

It certainly looks as though you are dealing with a genetic disease. However, I think you ought to seek very carefully for some environmental factor peculiar to the Fore people. For example, there might be something peculiar in their diet or something taken at ceremonial rites. This is all I have to suggest.

Best of luck.

Sincerely yours,
Thomas M. Rivers, M.D.

[49]

DCG to SMADEL [6 August 1957]

Dear Joe,

I suspect that you are back from Europe, and trust it was a valuable and enjoyable meeting. I have not changed my work appreciably in the interval. I am still trying to tie up our kuru work and leave it in such order that a return in one or two years can promise a brief but accurate epidemiological follow-up.

We now have 150 cases of kuru in our study, and some two dozen or more of our patients have died. Two brains well-fixed in 5% formaldehyde, from two of our carefully followed patients, were recently posted via air freight to you. Did you get this double shipment along with tissues (both HCHO- and Zenker's-fixed), in addition to tissues on an adult kuru the brain from whom went to Melbourne? In addition I included liver, kidney, and spleen from a peculiar kind of cirrhosis we are meeting here. Any pathological comment will be much appreciated.

Naturally I am waiting on pins and needles for any pathological reports from the NIH—but I fully appreciate that neuropathology takes ages!! The immense kuru-serum file will soon be sent to you from Melbourne, along with aboriginal and Melanesian blood collections. In the kuru patients, immensely elevated beta-globulins have been found and this is being further studied. I shall send you a code whereby you can select out classical kuru sera for check on this phenomenon. It must still be fully controlled by checking the normal Fore population. However, it is a full macroglobulinemia in some cases, and even if present in normal Fore is of immense interest. Cyril Curtain at Baker Institute in Melbourne is following this for us. I hope to do more back in the USA.

Another several kuru patients died this week, one here at our hospital (and a good post was possible). Brain is hardening in HCHO now. Thus, we are getting plenty of pathological material. I only hope that it is the *right* material, and that neuropathologists and toxicologists will not start saying "I wish I had tissue prepared . . ." etc. when I finally close down our Kuru Research Center.

Along with this letter is a large shipment of new urine specimens. They are in chemically clean, trace-metal–free bottles but have no preservative added, and since not catheterized specs, they are contaminated, undoubtedly. They have been refrigerated since collection, however, and I hope they will stay so in transit. Thus, they are a most valuable collection. Five sets are serial specimens on five kuru patients—all children—treated with immense doses of BAL. Some were in their first BAL course (Haida'abo and Mande), others in their second (Yani), and others in their third (Nata and Atona). The first specimen in all cases is a pre-BAL specimen, and the full time schedules of the protocols are available and will be sent to you. Thus, these specimens are for ashing and trace-metal studies—especially for copper, mercury, and manganese. Manganese looms particularly interesting at the moment, since lengthy reports of manganese poisoning in Moroccan miners present, in somewhat different order and array, virtually every symptom of our kuru patients. We certainly cannot detect any source of manganese or other toxin as yet in our patients, and heredofamilial-genetic factors are holding up well; however, this chronic manganese poisoning in Moroccan miners is the only disease I have yet come across with many of the features of kuru, and it must be seriously considered in spite of our apparent lack of toxic exposure. We are having soils, house- and cooking-fire ashes, foods, etc. analyzed for trace metals. Please, Joe, do all you can to have these BAL-treatment urine series carefully studied for all heavy-metal excretion. We have done our best here in the bush to collect them, measure volumes, and keep them trace-metal free. In addition, we are interested in amino acid patterns on these patients' urines. If anyone is prepared to do some urine chemistry—in particular, paper chromatography—the specimens could also be used for this. I think only a few drops would be needed, and the first (pre-BAL) specs on each of the five patients would be the ones to use. In addition, I am sending a few other kuru patients' urines (one specimen each) for heavy-metal and amino acid studies. These patients are not on BAL or on other courses of therapy such as Dilantin, Tridione, cortisone, Benadryl, or testosterone—all of which seem to have no effect on the disease.

[50]

We can really use the 16mm camera and film you have sent, and will do our best to get good clinical films off to you. I shall try hard, but it will be my first crack at 16mm photography.

Thank you for your comments on the first draft and for sending it from *Science* to the *New England Journal of Medicine*. Please, should you hear from *NEJM*, assure them that kuru is no figment of my imagination but a really exciting disease of great importance, for I am sure that is the case. It will be most inconvenient for them to publish long after follow-up papers have appeared referring to their note as the "first report"; thus I am anxious to have word from them.

Detailed epidemiology now fills nearly one thousand cards; and I am getting our clinical files—complete charts on all 150 patients, plus the immense epidemiological card-file on kuru which represents hundreds of miles of walking and climbing in rugged New Guinea mountains—into good order. They are the total and full record of kuru, and it is these files which I hope to get off to you when I am finally ready to leave. From them we can study all that has been done, all that our laboratory tests have shown, and make all the analyses we wish of kuru. Here in the field I am only trying to keep our records as clear, full, and accurate as possible.

In order to complete the one uncharted sector of the periphery of the kuru region on our epidemiological maps, I have requested a flight over the unexplored territory south and east of here—the Kukukuku region in general—into which I shall be patrolling in less than a month. Vin and I should make this flight soon—in a day or so if weather permits—to spot and plot out any population centers (villages and hamlets) on the east bank of the Lamari across from the Fore kuru centers on the west bank of that great river whose banks are almost as steep as those of the Colorado at the Grand Canyon. Thus, I shall have a jeep come in for me tomorrow and head out for Kainantu, for the first time in months. We get the plane at Kainantu. Should all go well, and should we not inadvertently pay a call from the sky on the Lamari-shore communities, we shall see in a few minutes what we shall spend days and days climbing and hiking through a few weeks from now. The goal is to place a definitive boundary on kuru and to complete the current epidemiological documentation of the disease, which is now nearly completed.

Thus, Joe, I am still at work but can already see the end of profitable fieldwork on this attack. I hope that kuru has interested you as it has me, for it is certainly a medical oddity and one which bids to give us important information. Should it be genetic, all that it can offer is immediately apparent. Where else in the world is such an illness cause of over half the deaths in an entire race? If it turns out to be toxic or infectious—and, for the life of me, I can get no leads on this as yet—it is one of the most mystifying and baffling such ailments known! If genetic, the Fore are doomed, I fear! If either (i.e., genetic or some as-yet undetected toxic or infectious process), medical science has much to learn from kuru.

We have had the director of Psychiatry of the Royal Melbourne Hospital [Alex Sinclair] here visiting for a few days, and he is as baffled as we are. He fully agrees that such a neurological degeneration cannot be a psychosis or an hysterical phenomenon, although there are many things in early cases which lead one to think that way.

I am completing this disjointed, hasty note from Kainantu.

Specimens, as described, are now being packed for shipment to you. [51] I send these to you—since immense files of trace-metal specimens have gone off to Moresby and CSIRO and to Melbourne, and some excitement is being raised already by reports of manganese levels in some (of which I am highly sceptical, since there is too much intermediate handling of specimens in other laboratories). These are going directly from patients to you, with the hope that someone at the Neurological Institute will have interest enough in kuru to get them into the hands of those who can do trace-metal surveys.

The camera, film, tripod, and all equipment is now here in Kainantu and I shall sign the receipt voucher and return it to NIH immediately. It looks mighty impressive, and I only hope that our results with it warrant the expense. Kuru is such a fascinating thing to see that our material cannot be at fault; the mountaintop scenery of our ex-cannibal community is of such grandeur that the background cannot help but impress you. All depends on our ability to master the technical side of the equipment, and we shall do our best.

I left our Kuru Center with 36 kuru patients currently in the hospital on various unsuccessful therapeutic regimens, and I am now again begging for a supply (large enough for adequate clinical trial) of any of the tranquilizers. I read of Serpasil being used in tremors of various types including Huntington's and Sydenham's and parkinsonism; we have not been able to get our hands on a single dose. We would need dosage literature along with any drug supply!! BAL we have now used rather extensively without help. Cortisone and delta-1-hydrocortisone have been hopeless. Testosterone we have only now received in supply adequate to treat two patients on high-dosage schedule for a month—thus far without impressive results. With material on hand, the disease progressing relentlessly to speedy complete helplessness and death before our eyes, the Fore nation in turmoil because of it, and with ritual murders and savage killings in reprisal for kuru sorcery comprising the major administrative problem in the region at the moment, we certainly feel we should be doing more for our patients—even if these trials are based on the most remote chances of benefit. Administratively, we are licked. Sorcery seems as good an explanation for it as any we can offer them. We now see that somewhat over 1% of the total 15,000 population of the kuru-affected region are currently sick with kuru which can be expected to kill them in the next 6–12 months or sooner. The disease is continuing to appear at the same rate, and we cannot yet spot any antecedent infectious disease (although we have now found cases of undiagnosed acute aseptic meningoencephalitis, on whom we are collecting specimens for virus isolation which are off to Anderson in Melbourne—with nothing yet isolated—and serial serum specimens for later serological study). These cases of acute encephalitis or acute aseptic meningoencephalitis have no detectable relation to kuru thus far, and are unlike any kuru-antecedent symptoms we can elicit or observe. One case with 280 mononuclear cells in the CSF developed a howling friction rub, albuminuria, peripheral edema, and everything in the world to fill the bill for clinical "encephalomyocarditis"—although I said in my last review that it caused no myocarditis in man; and we are still wondering whether he will survive. Thus, infectious disease is proving as interesting as ever here, and we are collecting reams of serum and running full blood-group genetic studies with Simmons of Commonwealth Serum Labs on all kuru-affected groups; but our central problem, kuru, has us licked as thoroughly as EHF had us licked in Korea, I fear.

That this is the best site in the world for study of chronic progressive degenerative disease of the CNS at the moment, I am fully convinced. Although I never had any intention of studying this type of disease, the world is beginning to be aware of the necessity for so doing and I have stumbled into the problem. If we can't "crack" kuru—with hundreds of cases available for full study in any 3-6–month period, I see little hope for parkinsonism, Huntington's chorea, multiple sclerosis, etc., etc.

Administratively, our colleagues are trying to get from Vin and myself advice as to whether to build an immense wall around the Fore and let them die in peace, or whether to encourage "out-breeding" and spread kuru throughout the world. Some decision is obviously needed soon, and I certainly cannot make any.

I do not expect you to study over these long pages, with all you must have on your mind at the moment. However, these idle musings may help those few in NIH who have become interested in kuru to know what we are up to and to encourage them to send us ideas, suggestions, and requests. So please, Joe, pass this on to any who are involved. [52] We have dug up our first film attempt, which is important in that it shows many of our early cases in the early stages of the disease depicted as much further advanced on these later reels. Many are now dead.

[53]

A copy of this letter will be sent to Dr. Rivers and the NFIP, and I shall try to keep him informed of what we are up to. Needless to say, his grant is all that makes this "unsalaried year of investigation" possible; and I shall be back in the U.S. a pauper.

Our genetic study of Cape York aborigines is off to the *American Journal of Physical Anthropology*. The New Guinea and New Britain genetic studies are soon to go off—some of last year's "side work"—and I hope to soon get off my report of skull-flattening of infants in New Britain and its lack of effect on growth. However, the matter of immense importance to me—the series of definitive papers on our auto-immune complement fixation test (on which several other groups, including Sir Mac himself, are basing studies now in press) which is in the hands of the *AMA*

Archives of Internal Medicine and which fully documents what I briefly discussed in my recent paper in *Nature*—is still unsettled. I certainly hope that journal gets to work on them! They are a set of papers of which I am proud, and I do not need their editorial comments to tell me whether they are good or not: they are!

Sincerely and gratefully yours,
Carleton

[54]

DCG to BURNET, ANDERSON, and WOOD   6 August 1957
Dear Sir Mac, Gray, and Dr. Wood:

Please excuse, again, this multiple address. Reports on kuru cannot easily be sent out as regularly as previously, for the mere paper work of keeping up with our 150-odd clinical case records, the immense epidemiological project we have undertaken, and the shipping out of specimens in all directions is exhausting every spare moment. In addition, kuru correspondence has mounted to an enterprise which could keep a staff of secretaries busy. Thus, Ford and Gordon Smith in Sydney, the Berndts in Perth, Simmons, Cyril Curtain, Smadel, Dr. Price, and a host of others in Moresby, Lae, Australia, and the USA are in correspondence with us—often offering good suggestions—and we cannot keep up.

Lois will no doubt have given you a full account of kuru as we were tracing it down during her vacation-trip visit to Okapa. The two enclosed letters will give you some idea of what we are doing since she left. We have taken your advice, Gray, and sent an early report off to the Australian journal [*MJA*]. Should they be willing, we shall try to publish with them a fairly complete photographic documentation of kuru which we have fortunately succeeded in obtaining. Should they not, other journals will, I believe, for we have good material to offer.

I shall try to get the new trace-metal samples off to you very shortly, and I shall not use patients recently treated by intense courses of BAL. We have exhausted now our third supply of BAL—two from the P.H.D. and the supply you have brought us—and only 8 patients have been treated: only 3 with three courses, 1 with two courses, and the other 4 with only one course. Thus, to fully treat patients as the literature recommends in Wilson's disease, we will need a really large supply of BAL. I shall ask the P.H.D. in Moresby for more. Furthermore, we have 2 patients on testosterone—children at that—and our current supply will just about see them through a full high-dosage course. I will send a copy of this note to Dr. Scragg, and explain to him that a fair clinical trial of any drug probably ought to include 10 patients and a "full-dosage" course of whatever drug we are trying on all those selected for the trial. Cortisone and delta-1-hydrocortisone have arrived in ample supply to complete good therapeutic trial thereof, and they do not offer any help to our patients. Dilantin and Tridione have also been supplied in completely adequate amounts, and they too do not help. Similarly, Benadryl has been retried—only to make our ataxic patients drowsy and worse. We now need more testosterone (either methyl or propionate)—thank you, by the way, Gray, for the dosage literature—and more BAL. Finally, we have been trying to get a supply of any of the new tranquilizers ever since we started. They are expensive, little-used in native medicine, and not too readily available; that we appreciate. But we have found recent reference to some success with them in CNS tremors, chorea, etc.; and we feel they should be tried.

We have definitely had a few cases of benign aseptic meningoencephalitis with CSF pleocytosis and some dramatic encephalitic symptoms and other involvement. One, with 280 mononuclears in the CSF (which cleared in one week), had a 2+ albuminuria persistent for weeks, peripheral edema, no cells in the urine, and subsequently developed the loudest pericardial friction rub I have ever heard. We could hear the rub two–three feet from the patient's chest—without a stethoscope—and we all observed it in awe for the first hour or so of its sudden development. On cortisone, which I could not but attempt using, it subsided in one day. What on earth this boy had I will never know. He looks now as though he will survive. His name is Amuwaiompa—something very close to that, at any rate—and serial samples of his serum, acute-phase blood and CSF (for isolation attempts, as well as for serology) are in your hands, we believe. These cases bear no detectable relationship to kuru, however, and although we have seen several, we have no assurance of turning up another if we suddenly plan an intensive virological attack; thus everything possible at the moment has been collected and sent out to you. This syndrome is nothing like what any kuru patients or kuru relatives have ever seen, and no cases report such an antecedent or prodromal set of symptoms. [55]

The professional [Australian] film unit has been on the way here for over a month, but has not yet arrived and now may not do so. We thrice collected proper patients and got ready; but the film team met with unfortunate difficulty with their jeep, a flooded Umi River, etc., and the whole plan never did work out. Thus we cannot hope for very much from them. Our best material will not be around, I fear, when they finally get here, for we must soon empty our wards for further patrolling time, to finish our kuru-epidemiological delineation.

We sent the vast collection of blood—over 100 specimens from the "epidemic parotitis"—because it was obviously not typical mumps. We would like mumps serology, if someone will do it, on a few of the typical cases to settle this point. I suspect it will turn out not to have been mumps. Chickenpox is now running through the [Okapa Primary] school (measles and pertussis have done so; and epidemic virus pharyngitis or nasopharyngitis, epidemic bronchopneumonia, and epidemic PUOs—twice—have run through this group of 100 tightly packed Fore youngsters herded together in one large dormitory). A real museum for an epidemiologist, and I regret we have not had the time to follow it all more closely. However, you have bloods on them all, and second-specimen serum will now be coming on those who had the peculiar URI with a high incidence of non-mumps–like preauricular swelling. Study of one or two of the serum pairs should settle this point. The acute specimens, I designated in our first shipment. I fully appreciate that you may not have had the time nor facilities for studying them. The cases are all fully documented "tribally and genetically," and thus may prove good for Simmons for blood-group and hemoglobin studies on the clots. Cyril Curtain may find some of the serum valuable for "control" serum on the Fore children in his study of the globulins. Finally, I should like aliquots of them left for any serological studies we may later dream up.

I hope, Sir Mac, that you had a fine time in Europe, that you found the meetings worthwhile—although I am sceptical by now of such large international assemblies—and that the trip was enjoyable. My greetings also to Dr. Gottschalk. I am sorry I did not get to meet him as I had hoped to in Tübingen. I still hope to get there this year.

Please tell Ian [MacKay] that I am most anxiously awaiting any word he has of our series of papers in the *AMA Archives of Internal Medicine.* If none arrives shortly, I would suggest inquiring.

Sincerely,
D. C. Gajdusek

[56]

DCG to PRICE 6 August 1957
Director, Public Health Laboratory
Port Moresby

Dear Dr. Price,

Vin Zigas and I have received your extensive reports of liver function studies, and of BUN, Klein's, and Coombs' tests, as well as blood-grouping on our patients. We are most grateful to you. In the meanwhile, we have burdened you with new and further specimens. A small batch of additional specimens is off today.

I have written to Professor Ford at some length, thanking him for bringing the manganese poisoning in Moroccan miners to our attention, and telling him that a manuscript on kuru is in press in the *Medical Journal of Australia,* which should, we hope, answer in a preliminary way many questions we are receiving about kuru. A copy—our only available spare copy—is off to the P.H.D., Moresby, and should be in the Director's Kuru file. If you should have a chance to see it, we would be most interested to learn of any comments you would be willing to make on our preliminary report of the disease. That copy does not have the map and photographs which we hope the *Journal* will include with the article.

We now are following over 150 cases of kuru, and our estimate of one percent of the population in the kuru region being currently affected (and doomed to die within one year) is conservative, I now believe.

Manganese poisoning sounds much like kuru in many respects, but the story of kuru with the Fore is a much different thing from chronic toxicity in miners breathing ore-dust day after day. We cannot locate a single possible source of toxic inhalation, ingestion, or skin absorption peculiar to the Fore, to our kuru patients, or to women and children among the Fore. We are treating the "preliminary reports" of manganese in some kuru bloods with skepticism and caution but with hope. We need the exact levels and names and dates of specimens used before we can make any evaluation. We ourselves know which specimens were collected in smoky atmospheres (because we could do nothing else under the circumstances), and which are perhaps more adequately collected. However, we hope that ample additional specimens are in your hands by now for the work to continue. Further courses of BAL therapy have been conducted, and BAL is certainly not yet visibly helping our patients. We are using immense doses, to boot. However, our third supply of BAL is now exhausted, and we are hoping for more. We have only been able to treat some 8 patients with all that has been supplied, and we have none left for follow-up second and third courses on most of these; such additional courses of

BAL are recommended in most chronic metal-poisonings and in Wilson's disease. We hope more BAL comes our way.

We are awaiting the specially prepared formol-saline, and already have the plastic container on hand. The special glass containers for brains are also here. If you could ship us, in addition to the formol-saline, some trace-metal–free distilled water, we could use it to rinse the plastic and glass containers several times before placing the brains in the special formol-saline into them, thereby further improving the accuracy of specimen-collecting. We have several further postmortem examinations—our total now at eight or nine. We are anxiously awaiting reports from our neurological examinations.

I recently sent a set of cooking-fire ashes, garden-soil specimens, and *kaukau* (the staple food) prepared in two usual methods, for trace-metal studies. These are from Moke, a village near our Kuru Center in which there has been much kuru in the past, recent kuru deaths, and at present two newly developed kuru cases. We shall try to get ashes, water, garden soil, etc. to you from the direct immediate environment of the newly developed kuru cases. These, together with the specimens which went off last week, should settle the matter of whether smoke, staple food, or garden soil are particularly unusual in the kuru region. Perhaps, in addition to the trace-metal studies, a complete soil analysis for all soil factors usually studied would be in order. Could you ask the Department of Agriculture if they would be willing to do this? Should the specimens as supplied be inadequate, we shall try to get them as they desire them; just ask them to send us directions.

Obviously, we have not shifted our thinking to accept anything but the heredofamilial, probable genetic pathogenesis of kuru which seems constantly to be plaguing us. However, this is a rather hopeless thing to settle for, and dooms the Fore beyond all hope. Thus, all of our efforts are directed at finding some toxic (metal or organic), infectious, or post-infectious factor operating in kuru pathogenesis. Only such an etiological mechanism offers much hope of eradication or cure; and we are most unwilling to settle for a genetic basis to the illness, even though our visitors all too readily agree with these conclusions which we have thus far been forced to draw. Every effort is still justified, we believe, to attempt to establish some environmental factor operating in kuru pathogenesis in spite of the extensive genetic evidence offered by our clinical and epidemiological studies thus far. We hope that you and others will agree.

Again we thank you for your extensive support of our studies. Without it, much less could have been done on kuru. We feel that kuru study is only beginning, but this intensive period of field study will inevitably be interrupted in a few months for an equally intensive period of retrenchment of ideas and approaches. Prior to that, Vin Zigas and I hope to visit Moresby for a Kuru Conference with the P.H.D., which we certainly hope you will be willing to join.

I am sorry that the soil, ashes, fungus, and insect specimens for metal studies went off to you with so little warning or explanation. We had a chance to get them out of Okapa (and such chances are erratic), and no time to get more into your hands than the short summary sheet I included with the specimens.

We are further interested in possible toxicological studies on such strange barks, plants, insects, and fungi as are eaten by the women and children and, in general, not by the adult men—i.e., by those who are the kuru victims as opposed to those who are usually kuru-free. Obviously, this is an immense undertaking, and the first step is a full entomological and botanical identification of the specimens. Miss Lucy Hamilton is working on the collecting at the moment, and many specimens are already off to the botanist for identification. We shall be most cautious in choosing materials for toxicological investigation, for the task of surveying the immense Fore food-list we now have is hopeless. We are trying to identify which items in these lists are either peculiar to the Fore and other kuru-affected populations—very few, if any, thus far fall into this category—and which are consumed largely by women and children, who are the usual kuru victims. We can think of no other approach, for obviously no one is going to do an intensive toxicology on the hundred-odd items of Fore diet and consumption, on the very remote chance that one has some organic or mineral toxin which acts by chronic accumulation in the exposed individuals. Perhaps taxonomy correlated with good toxicology—especially pharmacological, botanical, or organic—may give some leads. Therefore, can you suggest to whom I should supply our current food and contact lists, in addition to Professor Ford and Dr. Gordon Smith?

Enough for the moment.

Sincerely,
D. Carleton Gajdusek

DCG to WILSON [ca. 6 August 1957]
Public Health Laboratory
Department of Bacteriology
University of Melbourne

Dear Dr. Wilson,

I am sorry to write so long after having received your letter on May 20. It is only now that our immense rush of clinical and epidemiological studies has come to making use of the most important assistance which Dr. Margaret Holmes has offered to us. We are sending to you by air-express a shipment of Stuart-outfit cultures from our kuru patients; these should arrive along with this letter, shortly before or shortly after it. Many are from kuru patients whom we have just brought in from their native villages, and thus they should have their native-village flora in their nasopharynx. Others have been around our kuru wards for some time, and may have "hospital flora," such as it might be in a native-material smoke-filled Fore Kuru Research Hospital! None are on antibiotics of any kind at present, and none have recently had any antibiotics. No fever; some of those in the Kuru Hospital may have picked up bugs from antibiotic-treated patients around the hospital, and thus have antibiotic-resistant strains not representative of highland New Guinea villages.

Included with the kuru patients' cultures are a few "controls" who do not have kuru. These are the following names: Anua, Tiu, Masasa, Taka, Evesa, Wanevi.

All other specimens are from classical cases of advancing, progressing kuru. We should be very interested in a few antistreptolysin titers on kuru patients. The immense kuru-serum file, which includes serial specimens on most of the 150-odd patients we are following now, is at the Hall Institute; and Miss Lois Larkin there is fully familiar with which are from classical kuru patients and which are from patients with other diseases in which we have had passing interest. Thus, she and

Dr. S. G. Anderson can supply you with appropriate specimens from this serum file. In a few cases, volumes may be limited since specimens are earmarked for intensive virological and other serological studies and for further chemical investigations. However, adequate volumes of most should be on hand, and storage ($-20°$ C.) should have been adequate, I believe.

We shall soon collect another series of cultures from our kuru patients, attempting to restrict collection to patients in their village settings (as most of the current specimens are) rather than to those who have been about our hospital for a while.

We shall be most grateful to you for any report of findings you obtain and any suggestions you may offer as to the future study of kuru. Please thank Dr. Margaret Holmes for us for her willingness to help us. I know full well what work is involved in such studies.

Sincerely,
D. Carleton Gajdusek

IMUS to SMADEL
Memorandum.   Subject: Kuru.   8 August 1957.

Before Dr. Bailey left, he asked me to follow up the work on Kuru with Dr. Gajdusek. We have worked through Dr. Baker on this, and a summary report[6] of all activities is attached for your information.

We have put together six reels of film into a twenty-minute movie. This shows the type of jungle in New Guinea, the natives, and about half of the film shows various patients afflicted with this neurological disorder. We shall be glad to project this film for you at your convenience.

Pathological studies of four brains show most of the damage in the cerebellar region. Dr. Shy is convinced that Kuru is not Wilson's disease.

This report has been presented to Dr. Masland. He believes that both the speed with which the disease develops and the wide range of age-groups affected offer evidence against a diagnosis of hereditary ataxia.

He suggests that it might be of toxic or deficiency origin. For example, it is known that various tapioca plants differ in the amounts of cyanide content, that preparation of native food may not eliminate this poison, and that demyelination can be produced by chronic cyanide poisoning.

Because no indications of an inflammatory response have been found, an infectious disease seems to be ruled out.

Blood tests for the presence of parasites should be undertaken, if not already done. Chromatography or electrophoretic studies of the blood might provide a clue.

The most important problem is the need for an epidemiological study, with experts in anthropology, genetics, diet and personal habits, and water supply collaborating as a team.

The reports on the pathology found in four brains to date will be sent to you soon.

---

[6]An administrative report summarizing transactions with DCG up through early August was prepared for Smadel; see Appendix 2.

Dr. Masland will be in Bethesda on Wednesday, August 21, 1957, if you wish to discuss Kuru with him.

H. A. Imus
Assistant to the Director
National Institute of Neurological Diseases and Blindness

SCRAGG to ZIGAS and DCG   9 August 1957

Please find enclosed the reprint of the article in relation to ergotamine intoxication, as well as a *British Medical Journal* with a similar article.

Dr. Anderson attached a note to the reprint in which he states that he sighted other articles referred to and does not consider them relevant.

Dr. Gunther is still reading your article for the *Medical Journal of Australia,* and I will forward comment on it in a few days.

In one of your previous letters, you mentioned that you would be coming to Port Moresby in the near future. I anticipate that I will be absent from the Territory from the 25th August to approximately the 30th September, and I will advise you if I return earlier. Whilst away, I will be in Melbourne and there may be some matters you would like me to bring up with the Walter and Eliza Hall Institute.

R. F. R. Scragg

SCRAGG to DCG   15 August 1957
Dear Dr. Gajdusek,

I have read with great interest your report on kuru; and I feel that it gives us a baseline on which future investigations can be built, and also a picture of the disease so that experts in other parts of the world can discuss the syndrome and possibly give valued advice.

In your letter, I note that you are still doubtful as to the total period which you will have to stay at Okapa. When I last visited you in early May, you considered that two further months would be adequate, and stated that you required no financial assistance. In view of the extra months that you will be staying there, I would welcome advice as to whether these conditions still hold.

It is usual for us to give full assistance to research workers working in this Territory, but circumstances in your case were not those usually occurring.

As the kuru project has developed, it is obvious that your assistance will be necessary possibly for years to come.

Accordingly, without attaching any strings to the assistance, we would like to give some financial aid to you—or, if necessary, reimburse the organisations now supporting you in your work. If you are willing to consider such assistance, please advise and we will then work out the necessary details.

You may be interested to know that I will be visiting Melbourne on the 28th August, leaving Port Moresby on the 25th, and will be seeing Mac Burnet and Gray. If there is anything you would particularly like me to discuss with them, a letter or radiogram should be sent to arrive here by the 23rd.

Thank you for keeping me informed as to your progress in the investigations, and I wish to place on record the appreciation of this Department and the Administration of the excellent work you are doing on our behalf in this matter.

Yours sincerely,
R. F. R. Scragg

GARLAND to SMADEL *The New England Journal of Medicine*
15 August 1957

Dear Dr. Smadel:

Some of my associates and I are greatly impressed by Dr. Gajdusek's description of "kuru" and the picturesque background of his studies.

There was unexpected opposition to its acceptance at last week's editorial-board meeting, however, on the basis of a lack of any proved morbid anatomy, pending the results of microscopic examinations—which will, apparently, go to another journal when they are available.

We have at present, therefore, a clinical description of a fatal disease of the nervous system occurring in a remote area of the world, with no positive pathological findings on which to base a definite diagnosis. I am personally eager to accept the paper, which describes the clinical features so admirably, but believe that the subject should be presented with caution, since authentically new diseases, I am told, are infrequent and difficult to come by. Much of the weight that such a paper will carry depends on the credibility of the evidence and the reliability of the author, and I think that we must depend on you for such assurances if the paper is accepted against the opposition that I have encountered.

I realize that you can't do much more than to repeat your previous statements, but I shall welcome any further information and also the photographs that you mentioned in your letter of August 6. If we decide to go ahead with the paper, I shall use it as a special article and publish it as soon as is now practical—probably early in November.

Any further decisions will be made by Ted Ingalls, who will be in charge of the *Journal* for the next two months, during my absence. He agrees with me in regard to Dr. Gajdusek's paper and will see this letter before it goes to you.

Sincerely yours,
Joseph Garland, M.D.
Editor

DCG to ANDERSON [mid-August 1957]

Dear Gray,

The HCl prepared in Moresby by Dr. Price is here, but he is still awaiting trace-metal–free distilled water from somewhere to prepare the trace-metal–free HCHO for us.

Enclosed is a copy of a letter I have recently dashed off to Joe Smadel, which will acquaint you with a few further activities in which we are engaged. I send it particularly for the paragraphs about the further cases of acute meningoencephalitis we have on hand. In the same airmail is going off to you CSF—two bottles of each—and blood specimens on two girls with acute meningoencephalitis from a kuru-region village in which there is plenty of kuru. That we must still insist that none of our patients admit to any such antecedent illness, even years before, is certain. Furthermore, kuru patients are never malnourished or obviously recovering from a devastating acute illness as are these acute meningoencephalitis patients. Thus, we may only suspect that kuru, if—and this is a mighty uncertain *if*—it has anything to do with an antecedent acute infectious disease of the type we have now discovered, occurs years after the acute illness, when the patients have fully recovered and no longer remember the preceding illness. This seems unlikely, but might be the case.

I cannot honestly claim any shift in my thinking. Kuru remains in clinical type a chronic, rapidly progressive degenerative disease of the central nervous system—probably of heredofamilial predisposition. However, we have been pushing both possible toxic and infectious etiology rather strongly in the last months, and I apprise you of this fact. Enclosed with the shipment containing these two patients' bloods (which are not early in the disease, but as early as it will ever be seen hereabouts, I fear) for virus isolation attempts (the CSF is labeled "postencephalitis," though it should be labeled "acute encephalitis") are a large set of blood specimens on kuru patients, along with several "control bloods." Further controls will eventually be collected along with further kuru specimens. The kuru bloods are all on patients who have not had BAL therapy. They should be fine, and the collection of them has been as cautious and dust- and smoke-free as we will ever get. We prepared the arm only with ether, which evaporates completely before venepuncture. These are all in the trace-metal–free bottles you recently supplied. [57]

Obviously, we are anxiously awaiting any news on the microscopic pathology of the viscera and/or on the neuropathological-neurohistological studies on the two brains we have sent down. It appears as though we may get our hands on another before we close down our Kuru Ward for patrolling and completion of the field epidemiology.

Sincerely,
D. C. Gajdusek

KLATZO to SMADEL
Preliminary Findings in Brain of Patient *Yabaiotu*, 50-year-old female

GROSS: Marked engorgement of veins over the hemispheres. Coronal section reveals 3x4 cm-in-diameter area of hemorrhage in the white matter of the left occipital lobe. Vascular channels throughout the brain dilated and on few occasions surrounded by petechial extravasations into the parenchyma.

MICROSCOPIC: Cervical Segment of the Spinal Cord: Symmetrical demyelination in posterior spinocerebellar tracts, less intense in ventral spinocerebellar. Some anterior horn cells show chromatolysis or vacuolation of cytoplasm.

Medulla: Demyelination in spinocerebellar tracts. Many cells of inferior olives show disintegration, chromatolysis or vacuolation of the cytoplasm. Marked astro and microglial response in the inferior olives. Nerve cells in hypoglossal nucleus, vagus, n. cuneatus show frequently degenerative changes of various intensity.

Pons: Degenerative changes of rather acute character are found in various pontine nuclei.

Cerebellum: There are striking changes affecting Purkinje cells. There are numerous "torpedoes" (ballooning of the Purkinje cell axons) and bizarre deformities of Purkinje-cell dendrites. Numerous round or asteroid bodies are present, which are brightly anisotropic in polarized light. These probably represent neurofibrillary degeneration of Purkinje cells. Granular layer of the cortex is occasionally rarefied. Very numerous fat-containing reactive microglia in both molecular and granular layers of the cerebellum. In the dentate nucleus, neurons show degenerative changes. Terminal nerve fibers show degenerative bulb formation and scavenger activity of microglial cells.

Basal Ganglia: Substantia nigra shows degeneration of neurons. Numerous reacting microglia cells are distended with melanin, which undoubtedly was picked up from the dying nerve cells. Caudate, putamen, pallidum show advanced degeneration of neurons. Very spectacular neuronophagia of the nerve cells in the caudate. Thalamic nuclei show less intense changes in the neurons; on the other hand, there is very intense gliosis affecting predominantly the medial nucleus. Hypothalamic nuclei including mamillary bodies show marked glial response, neuronal damage is less intense.

Cortex: Including Ammon's Horn, show relatively little pathology with only occasional hyperchromatic or tygrolytic cells. No inflammatory changes can be found throughout. Only few vessels in the inferior olives contain lymphocytes in the Virchow-Robin spaces. This can be easily explained as secondary to the degenerative changes.

Impression: The above-mentioned findings do not give any ground to suspect infectious or inflammatory process. The widespread involvement of various structures does not fit into any of the known hereditary degenerative patterns. Toxic etiology of the condition appears to be most likely. Nutritional factors may be involved in producing toxic metabolites.

15 August 1957

SMADEL to DCG 16 August 1957

Dear Carleton:

I had a very fine time in Europe, a good meeting for a week at the Poliomyelitis Congress in Geneva and then a good touring vacation in Italy. I got back the first of August, but I am now off for a one-week's visit to the Rocky Mountain Laboratory.

I read your correspondence on my return—including the two most recent letters describing your proposed trip around the periphery of the area occupied by the Fore people, and the one which arrived yesterday going into detail about the photography.

One thing begins to worry me a little more than before. What will happen to the records, the material, and the information you carry in your head if the plane comes down in the jungle or if one of the indigenes decides to revert to cannibalism? I think you better finish up the work cautiously (as regards your own well-being) and get the hell back here. It seems to me the simplest procedure for keeping you in one place long enough to analyze all of the data and to write the papers is to put you on as a Visiting Scientist for some months in the Institute of Neurological Diseases and Blindness. This will serve the added purpose of buying time while you decide what your next task should be. I would strongly recommend that you stay home for a year or so and fully recover from your tripping.

Dr. Ness has done a number of copper determinations on the urine specimens. Enclosed are some of the values. While certain of the determinations are higher than the usual normal, the group here is unable to give you any definitive interpretation at this time.

All of the shipments mentioned in your letters have been received. This includes the two lots of urine and the six brains.

Additional histological sections have come through, and the boys are excited about the diffuse degenerative involvement which goes far beyond the cerebellum. A note from Dr. Klatzo is enclosed. Since I am leaving before I have a chance to read the typed version of this, I don't exactly know what Klatzo will have to say. In case he doesn't mention it in his note, he requested the following some time ago:
1. "Send more autopsies of spinal cords."
2. "Send histories of patients from which autopsies were taken—histories that were sent do not match autopsies."

Dr. Shy says he can recall no other lesion with such acute destruction of the nerve cells and associated neuronophagia. At the moment, the inclination is to attribute it to a toxic effect. Shy says that the lesions in the somewhat similar disease found in miners connected with magnesium [manganese] are more limited in their distribution than in kuru: i.e., in the magnesium [manganese] malady they are essentially limited to the basal nuclei.

I haven't the foggiest idea what has happened to the prints we sent you on the 26th of July. It is apparent from even your most recent letter that you have not yet received them. I cabled you August 13 to look for them. Maybe they are lost in the jungle.

Very best.

Sincerely yours,
Joe

ROBERTSON to SCRAGG   16 August 1957

Dear Dr. Scragg,

I enclose a preliminary report on the pathological appearance in Brain I. It had already been cut into slices, which contributed to the difficulty of examination. The second specimen, apparently preserved as suggested, was in excellent condition. Its examination has been commenced. Again, it was macroscopically normal. Embedding in celloidin takes some time, but a report will be sent with as little delay as possible.

While abroad, I took the liberty of showing some sections to Dr. J. G. Greenfield and Dr. Webb Haymaker. The notes comprise all opinions.

Yours sincerely,
E. Graeme Robertson
Department of Neurology
University of Melbourne

Preliminary (Incomplete) Report on Case I.

Macroscopically, brain appeared normal.
Frozen, paraffin, and celloidin sections.
Cerebellum shows most pathological change.
Patchy degeneration of Purkinje cells (probably number lessened: In order cells essential cellular structure has disappeared: silhouettes of nuclei, some cells shrunken and pyknotic).
   Changes appeared more marked in neocerebellum.
Bergmann layer cells look alright.
Molecular layer — increase in glial nuclei: swollen fat-laden microglial cells (gitterzellen) in small numbers, patchily distributed. Appear to be most numerous when Purkinje cells most affected. In one place, marked thinning of the molecular layer with degenerative changes and gitterzellen. Some gitterzellen seen under pia, but no marked meningeal accumulation.
Granular layer — reduction in nuclei — some cells pyknotic. More gitterzellen than in molecular layer, in greatest concentration where grey matter abuts on white. Increase in astrocytic nuclei.
Dentate nuclei — neurones normal. No gitterzellen.
Roof nuclei — some cells vacuolated and shrunken.
White matter — (?) some axons swollen. Few macroglial cells around vessels.
Cerebral cortex — slight abnormality. Some degenerative changes in neurones which might even be like postmortem changes. (?) More horizontal cells in layer V than normal. Axons appear swollen in places.
Basal ganglia — very little change. Few gitterzellen in neighbouring white matter, of perivascular distribution, but concentrations rare.
Medulla — some mucinoid bodies in white matter of medulla. Dr. J. G. Greenfield pointed out the similarity of these to those found in grass sickness in horses. May be artefact. Otherwise no definite change.
   Glial and myelin stains still to come.
Summary: Changes surprisingly slight—most marked in cerebellum—regressive changes with degeneration and disappearance of Purkinje and granule cells, with microglial scavenger activity and astrocytic proliferation. Changes would appear to be of no great age. The process is not acute and does not appear to be inflammatory.

16 August 1957

DCG to SMADEL   16 August 1957

Dear Joe:

Immediately upon the heels of the last verbose letter, another. We have rushed through two hundred feet—two reels—of Kodachrome on the new camera, and are rushing them off to you in the hope that they can be urgently processed and a telegraphic report sent to us as to whether we are recording anything or not. I am a complete novice with this cine machine, but have tried my best. We had a special group of kuru patients on hand, and these I have briefly surveyed. Furthermore, on the tail end of the final film is a congenital tremor, which is another interesting "new" disease (in Melanesians, at least) which must be differentiated here from kuru. Perhaps the few seconds' comparison are enough to show the difference. If we are doing O.K. with the equipment, I can then promise you a complete clinical documentation of the toe-grasping, posturing, reflex patterns, emotionalism, etc. of kuru and full comparisons with the hereditary, familial congenital, non-incapacitating and non-fatal tremor which is found—but rarely—hereabouts. This too we shall report in our next kuru paper.

Acute lymphocytic meningoencephalitis is upon us. We now have a total of five cases with a CSF pleocytosis, fully mononuclear, and unrelated to mumps epidemics or any other cause of benign aseptic meningoencephalitis of which I know. Whether to call the cases encephalitis, or not, is a problem. Four of them would better qualify as mild encephalitides. Two, in today from the same village, have 1200 cells (96% monos) and 550 cells (100% monos) per mm^3 of CSF, respectively. They are two girls who show lassitude, painful neck and head which has just subsided (but no true meningismus, for there is not a positive Kernig or Brudzinski), and marked wasting. I did LPs, more on the basis of their behavior than on the basis of meningismus. We see plenty of the latter, for *H. influenzae* and *N. intracellularis* meningitis are both still with us at Moke-Pintogori. Both fluids were cloudy, the first markedly. Differential counts were checked by direct chamber differentials—using acid and stain, and dried and stained smears of the CSF. If they follow the course of the former three proved cases, they will slowly start to eat again and then regain the lost weight—which is considerable. In addition, two other patients came in with obviously the same syndrome, but so late that I could no longer find cells in CSF; Pandy's were slightly positive, and full CSF chemistry reports are not yet back to me.

However, Joe, these cases bear no relationship to kuru that we can detect. All kuru patients are well nourished, and show none of the cachexia and weight-loss and obvious acute recent illness of these meningoencephalitis patients; and kuru patients and their relatives deny emphatically any disease similar to these newly discovered cases, even when we bring them to see them. We cannot thus far establish any link whatsoever between kuru and this acute CNS infection that we have found in the region (certainly five times, perhaps seven, now). Kuru, on the other hand, is not rare, but is occurring constantly. New cases appear every month; our first estimate that one percent of the affected population was currently dying rapidly of kuru was conservative, and we know now that there are over 150 cases among the 15,000 natives.

[58] We have plenty—really plenty—of some of the best clinical material for cine documentation in the world. With almost 150 active cases on hand to study, all stages of reflex patterns, tremors, emotionalism, incoordination, posture maintenance, ataxia, etc. can be shown in detail—along with studies of our acute encephalitis cases and the congenital tremor cases (important for differential diagnosis), who are really different, as I hope you will see in the film, and and who do not rapidly deteriorate and die.

[59]

Finally, Joe, I must beg your pardon for all this detail. I hope someone at NIH is interested enough in kuru to be willing to take these pleas off your hands, study them and work through a few for us. We know we have unusually good material on hand, Joe, and our study will be reasonably thorough before we terminate it in September or October. Thank you again for all of your help and patience.

Carleton

KURLAND to IMUS   20 August 1957
Memorandum.

I appreciate the opportunity of reviewing the report by Zigas and Gajdusek, and Gajdusek's correspondence on kuru. The difficult conditions under which they are working probably account for the trend the research has taken, for it is clear that additional epidemiologic study would be more productive than their time-consuming therapeutic and repetitious clinical descriptive efforts.

I seriously doubt that this can be a genetically determined disorder. There appears to be a geographic focus and a high degree of familial aggregation of cases, but some condition related to environment common to these families will, I predict, be found to account for such focalization of cases.

I find the genetic hypothesis difficult to accept for the following reasons:

1. If the assumption that the disorder is new—20 years or so in this area—is correct, it is practically impossible to explain the extremely high gene-frequency (at least 25%) in this population of 10,000. For this to occur, there must have been, in a recent generation, an enormous spontaneous mutation rate for a specific gene. This would have to be in terms of several per cent instead of the usual 10^{-5} or 10^{-6} mutation rate.

2. If the assumption that this is a new disease is incorrect, it would still be highly improbable that such a gene-frequency could ever develop, in view of the very serious reproductive disadvantage present in the affected as compared with the unaffected in the population.

3. The age-distribution is from four to five years of age throughout adulthood. The upper limit is not given. Although many hereditary disorders of the nervous system have an appreciable age-span, such as through childhood and adolescence or throughout adulthood, it is unusual to find a single genetic form (recessive or dominant) spread over childhood, adolescence, *and* adulthood.

4. The sex-ratio differs for children and adults. I am not aware of any hereditary disorder due to a single genetic form (recessive or dominant) in which the sex-ratio changes from slightly more than 1:1 in children to 13:1 in adults. If Dr. Gajdusek's observations on sex-incidence in the population are complete, this odd sex-ratio is very much against a genetic mechanism. (A similar preponderance of *males* (12:1) for lathyrism is of interest in this respect.)

5. The occasional case in a non-Fore woman now living in the kuru area cannot be passed off lightly and is very strong evidence for a non-genetic mechanism.

6. Most geneticists do not believe in "anticipation" as described here. One would expect no offspring from affected children in an earlier generation. Obviously, with the wide age-distribution described, some cases occurring in the present generation will be in children whose mothers (affected or not) at least had to reach maturity.

7. A characteristic of most neurological hereditary disorders is that they selectively affect one or only a few of the central nervous systems (motor, cerebellar, or selected portions of the basal ganglia, etc.); the diffuse involvement of almost the entire central nervous system found in the case described by Dr. Klatzo of NINDB is not characteristic of a genetic disorder.

The data on age and sex reveal a selectivity for women and children. I am not acquainted with any such sex-ratio in a hereditary disorder, but am reminded of a similar age- and sex-incidence in the careful community surveys of Goldberger and associates on pellagra. The possibility that adult males might be eating more often away from home would suggest either a deficiency disease or, more likely, that some food substance—perhaps one that is relatively new to this population, or one that becomes toxic in storage and which women and children are more likely to eat—could account for the reported age- and sex-distribution.

I am reminded of the circumstances on Guam, and the manner in which the Chamorros have over several centuries learned to handle and safely consume local toxic food-products such as cycad nuts (federico) and manioc (cassava). These are hardship foods, and during the recent war years were used when other foods were in short supply. Some of the younger and inexperienced people were said to have been poisoned because they neglected to follow the old ritual of repeated washings and soakings which were required to remove the toxic glucosides from these foods. (See attached from Dr. Marjorie Grant Whiting [61].)

It would seem that a more careful delineation of the population affected—with further information on age, social and economic class or caste, occupation, and particularly detailed reports on food and water supplies and food preparation and consumption—would be helpful. To satisfy criteria for a genetic hypothesis, careful pedigree, twin and consanguinity studies with the required parental identification from blood-grouping and other measurements are required.

If Dr. Gajdusek would welcome the assistance which this Branch could offer (subject to approval by the Director), Dr. Marjorie Grant Whiting—a nutritionist from the Harvard School of Public Health, who is now cooperating in our amyotrophic lateral sclerosis (ALS) project among the Chamorros and Carolinians of Guam and Saipan—could be asked to assist him in New Guinea. She is an authority on native foods, and is particularly well acquainted with recognized toxins of the Pacific equatorial islands. (The circumstances related to ALS are quite different from those of kuru; and although I seriously doubt that a good toxin plays a role in the etiology of ALS, I have encouraged Dr. Whiting to carry out her planned investigations.)

We could assign on a short-term basis one of our Epidemic Intelligence Service officers to aid in the study of distribution and population selectivity.

Our geneticist, Dr. Ntinos Myrianthopoulos, would be available for consultation or direct services if desired.

The phenomenon among the Fore people, if properly evaluated and clarified, might be of value in some of the rare but devastating central nervous system disorders which are encountered in the United States. It would seem appropriate to offer to Dr. Gajdusek the benefit of our experience or the assistance of some of our personnel.

Leonard T. Kurland, M.D., Dr.P.H.
Chief, Epidemiology Branch
National Institute of Neurological Diseases and Blindness

DCG to ANDERSON 21 August 1957
Dear Gray,

Enclosed is a copy of a recent letter to Roy Simmons in which I try to straighten out some of the complications of our blood-group studies associated with kuru. The sheets summarizing the schoolboys' serum specimens also apply to virus studies and the kuru-region serum file, and thus I send you copies also. The additional specimens out with this mail are for both "control" serum specimens for protein studies (which Roy Simmons sends to Cyril Curtain) and for Simmons's studies of local blood-groups. In addition are several additional kuru specimens, including one trace-metal blood specimen on Isoisi (F/6) with advanced kuru. [62]

We have used up the trace-metal–free saline sent from Dr. Price in Moresby on the one brain we just took. This is hardening now in the trace-metal–free formol-saline. We need more for subsequent specimens; and if the supply of HCHO you sent Dr. Price has limited him to the two liters he sent to us, then please send him more. However, if he has more on hand he will certainly prepare more for us in reply to my request, which is now going out to him.

Greetings to Sir Mac and everyone at Hall.

Sincerely,
D. C. Gajdusek

DCG to ANDERSON 22 August 1957

Dear Gray,

Enclosed is another list of specimens—this list being of specimens of "control" normal blood for eventual serological survey of this region, but more especially for genetic survey for blood-groups. These include the first Kukukuku bloods we are sending, and a small series of South Fore bloods from just across from them. More of both these series will eventually be coming. Please send all clots to Roy Simmons. One copy of the list, for your records, is enclosed with the specimens coming under separate air shipment. Another is enclosed herewith. Please give it to Roy with the clots.

We are sending, in addition, several urine specimens collected as follows: 24-hour urine specimens collected in bottles and placed in an icebox as collected (where I estimate temperature to be about 10°C!). When the collection was finished, volume was measured, and to entire specimen was added 50 ml of the 20% HCl you had prepared for us and sent to us. Then 100 ml of the specimen was taken and is being sent to you. Obviously, it would have been preferable to collect the urine "on top of" the HCl, for we may well have had some bacterial activity in the 10–20° storage during urine collection. However, this was not done; and it is highly unlikely that we can get a full day to devote to urine collection again during the coming month—with the schedule as tight as ours now is. Thus, I send these specimens along, urging them upon you as the best we can get now on kuru. I know it seems ridiculous, but we are now off patrolling; few if any kuru patients can be kept about the post, and those hereabouts are not capable of yielding valid 24-hour collections. The collections are in acid. The only error is that of not collecting over HCl but adding it to the fully collected 24-hour specimen.

Enough from us now, Gray. Any pathological reports available?

Sincerely,
D. C. Gajdusek

DCG to SMADEL

Purosa—South Fore
On patrol
25 August 1957

Dear Joe,

We are patrolling, finishing up kuru epidemiology—and have just shot films of a group of 8 kuru patients found in this region. Thus, a series of eight rolls of cine film are already off to you (some should already have been received). We are awaiting telegraphic report on whether we have been leaving the lens cap on or some other fool thing. Any comments on technique, improvement in material or photographic technique will be much appreciated. [63] We shall use up the remaining film on kuru in the next few days; and since it behooves us to try to film more of the interesting congenital tremor we have discovered here and of the encephalitis patients as well as the kuru, and since—once we are sure of our technique—we wish to try to get some valuable close-ups of toe movements, stability tremors, reflex patterns, etc., as well as further shots of kuru in the native setting, we could really use further film. [64]

All goes well. A third report on kuru, with much additional material, has been sent to *Klinische Wochenschrift*. We are hoping to get color photos and plenty of black-and-whites published with it. The *Medical Journal of Australia* has accepted without change our second paper, and I am hoping—as I have said before—that the *NEJM* will come out first with the first report of the disease, written two months earlier. I am awaiting arrival of the prints for the Australian journal article, and when we get them I shall write captions and rush them to them. They have agreed on eight pictures!! Perhaps one of the Kuru Hospital and the patients seated about outside of it might be included. The others should be of patients of different ages and sexes in different stages of the illness. For the *NEJM* I leave to you the selection of the best two or three pictures you can find among the negatives I have sent.

The NFIP asked for the first paper to be included in its collected papers, and I have already forwarded the reprint order they sent me to the *NEJM*. I trust the further report in the Australian journal will also interest them. I have no comment on the second manuscript yet from you or Dr. Rivers. The much longer and informative third report I have already sent to the Journal [*Klinische Wochenschrift*], and as soon as I get back to civilization—i.e., our Kuru Center—I shall mail copies out to you and to Dr. Rivers. Our kuru case-series will shortly exceed 200 cases, and in the group we shall shortly have over 50 already dead from kuru; over a dozen brains have been obtained. Two, of children, are waiting to harden before I send them to add to your kuru collection. A further brain collected in special trace-metal–free formol-saline in a special plastic box is hardening also. Would there be trace-metal chemists about who might study this unique specimen thoroughly, along with thorough neuropathology, if I send it to you?? We do not know whether the Melbourne crowd can dig up the proper studies for it and thus would like to know if there is anyone at NIH who would really settle the matter. It has taken a great deal of work to get such a specimen!

I have already had rather extensive trace-metal work on blood done; and although it is not definitive, now manganese and copper are being again suggested—although I am sceptical!!—by our chemists. The immense file of CSF specimens from kuru for trace metals has stood about in a Moresby ice-box unstudied; and I have thus asked for the entire series of CSFs in trace-metal–free containers to be forwarded to you, and have sent a few others directly to you. These obviously are sufficient to settle the matter once and for all as to whether CSF has any trace metal in excess, if there is anyone who can do the study—perhaps both a spectrophotometric screening and a quantitative analysis for anything that shows up.

With the upcoming September patrol of about one month, I hope to wind up this first field-attack on kuru. Our estimate that 1% of the entire population is dying each year of the disease we now know to be low, and probably somewhat over 150 active cases are in the kuru region at any one time. We have not yet had them shoot at us, although they are not always very intrigued by our studies and tell us so in no uncertain terms. A patrol was attacked just within our region a few weeks ago, and we are hoping not to encounter any of the same in our work. Thus far, our Fore and their kuru-affected neighbors have been fine patients, with a greater tolerance for our investigative mien than many a Western group; but they know full well that it's all sorcery, and that it is best to humor our skepticism. Murders in reprisal for kuru sorcery are now rivaling the kuru cases in number in some regions. Thus, kuru plus kuru-reprisal murder account for some 50–75%, at least, of deaths in many regions; and should we add to them the infants who die when their kuru-afflicted mother leaves them behind, we would find, I fear, that the Fore would never die at all were it not for kuru.

Enough for this note. I have pig-fat rubbed all over me from admiring observers (most with great pig-tusks through their noses) who have never before seen a typewriter; and I am hoping that the one who carries the films and this letter to Moke will not open both and inspect the contents with pig-grease–covered hands.

Sincerely,
Carleton

P.S. The *AMA Archives* papers on AICF test are now accepted—unchanged. I am quite pleased. Sir Mac came out in *Nature* with a "follow-up" already.

IMUS to SMADEL
Memorandum.   Subject: Kuru.   26 August 1957

Attached is a review of subject neurologic disorder which was prepared by Dr. Leonard T. Kurland, Chief, Epidemiology Branch.[7] This report discounts possible genetic factors, and suggests a toxic origin.

Dr. Richard Masland has been kept informed of all of the details of Dr. Gajdusek's study. He suggested on July 31, 1957, that kuru was probably not hereditary but due to a toxic or deficiency problem in New Guinea.

Before committing NINDB to any participation in investigations in the field, as suggested by Dr. Kurland, it seems desirable that both Dr. Bailey and Dr. Masland have an opportunity to discuss the problem with Dr. Gajdusek as well as with other members of the NINDB staff.

---

[7]See 20 August 1957 memorandum, Kurland to Imus.

SIMMONS to DCG   28 August 1957

Dear Carleton,

Thank you for yours of 21-8-57. When you send samples directly to me, number them and send a typed key to the numbers. I've never had so much trouble over the months trying to sort out samples with variable spellings, duplicates, and repeats. I think numbers throughout would have saved Lois, me, and Curtain a lot of worry.

I didn't get any extra cells from the 31 convalescent boys, or the 11 of the 13 promised new bleedings from children previously missed in the 1-101 series or No. 82—because you didn't put in a note telling Lois to send them to me. She kept the sera for you, and naturally threw out the clots. This also applies to Curtain.

When you get this lot of results attached, you will have had results for all the samples which have been brought to me—except for duplicates, which I attempted to find to save work on precious sera by repeat testing. Now every sample, both blood and serum, which came to me (including duplicates) went to Curtain. I have no samples, except a few of the last lot.

No. 82-1 and 82-2 were both group O—this was a clerical error.

Congrats on all the material you have in press—it's most impressive!

Thanks for the population notes for the New Britain paper—they will be sufficient. When I get time, I will attempt to finish the paper.

When samples come direct to me I will remove cells, call for Lois to take sera for you, and Curtain may have the balance from her. I have not time to collect and store sera at $-20°C$ for you. Sorry!

Thank you for the kuru list–non-kuru children list, and reference to samples repeated by me. That helps to straighten out the picture. [65]

Remember, when you take duplicate bleeds especially for me: put a note in for Lois telling her the samples are to go to me. This last time she didn't send me any duplicated names.

We have just finished a paper on the Pygmies of Netherlands New Guinea, which was work done years ago. It attempts to compare the various Negrito groups re blood-group data. The paper will go to *AJPA* soon (next week).

I have tested every kuru and non-kuru sample sent to me. If you want blood-groups on any kurus missed in my series, then send the samples to me numbered and branded *kuru*.

Good luck—best wishes.

                                                         Roy Simmons

NB: Samples should be got to me as soon after collection as possible. This is important especially in relation to some blood-group antigens.

SMADEL to INGALLS   29 August 1957

Dr. Theodore H. Ingalls, Associate Editor
The New England Journal of Medicine

Dear Ted:

This is concerned with the manuscript of Drs. Gajdusek and Zigas dealing with kuru, which has been submitted to the *New England Journal of Medicine*. In particular, my letter replies to Dr. Garland's note of the 15th.

I quite agree with you and Dr. Garland that Gajdusek's paper would be much improved by some mention of the histopathological changes found in the brains of patients with kuru. Fortunately, since the manuscript was prepared in New Guinea, Gajdusek has sent to Bethesda the fixed brains from six cases. These are in the process of careful examination by Dr. Igor Klatzo of the National Institute of Neurological Diseases and Blindness. Dr. Klatzo and Dr. Milton Shy, the Clinical Director of NINDB, rightly hesitate at this time to provide a detailed description of the changes which are present. Nevertheless, the lesions which they have found are extensive and are common to all six cases. It is planned that Drs. Klatzo and Gajdusek will prepare a detailed report after the latter returns to this country. In the meantime, Dr. Shy and Dr. Klatzo suggest that the brief paragraph given below be added to Gajdusek's manuscript:

> The preliminary pathological study of six cases reveals a widespread neuronal degeneration. The cerebellum and extrapyramidal system are most severely affected. There are also changes in the anterior horn cells, inferior olives, thalamus, and pontine nuclei.

Drs. Shy and Klatzo are of the opinion that the above material should be used without direct reference to either of them. I would suggest that the paragraph be added as an addendum to the manuscript with a lead-in such as the following:

> Addendum: Since the manuscript was submitted, additional fatal cases have been autopsied. The preliminary pathological study (continue the above description).

Within the next ten days I shall send you six of Gajdusek's photographs dealing with kuru patients. I suspect that you may wish to use several—primarily to present the general idea of the disease, the patients, and the locale. Frankly, the still photographs do not provide the convincing evidence of severe neurological disturbance which is so clearly depicted in several hundred feet of movie film which Gajdusek took of a number of the patients, and which have been viewed by our neurological staff. Nevertheless, one or two of the stills would, I believe, add interest to the article.

Best wishes.

Sincerely yours,
Joseph E. Smadel

cc: Dr. Garland
    Dr. Gajdusek
    Drs. Shy and Klatzo

SMADEL to DCG   30 August 1957
Dear Carl:
Just before I left last week for a visit to the Rocky Mountain Laboratory, Dr. Leonard Kurland (Chief of the Epidemiology Branch, NINDB) came over to inquire about the kuru study. He was particularly interested in the possibility, reemphasized in the last few weeks by the pathological findings, that kuru resulted from some toxin; and he was opposed to the idea of it being dependent upon genetic factors. I asked him to read your manuscript and to let me know what he thought might be done to further the work. Attached are his comments[8] as well as certain

---

[8] See 20 August 1957 memorandum, Kurland to Imus.

information on cycad nuts, prepared by Dr. Whiting. I think you will find Kurland's remarks stimulating and probably helpful, even though you will recognize sentences which are probably unduly dogmatic and undiplomatic. Nevertheless, with the good and bad, I am sending the material to you in toto because I do not wish to lose any of the flavor in abstracting the material.

Attached to Kurland's correspondence is a covering note [26 August] by Dr. Imus, which contains the sound suggestion that plans for future studies of kuru be held in abeyance until you get back here to go over all of the material and to discuss the problems at length with various members of the staff of NINDB.

Yesterday I sent you a copy of my [29 August] letter to the editor of the *New England Journal of Medicine*. I did not add a P.S. to the effect that the editors gave reasonable assurance that the article would be accepted and would probably appear in an issue in November. I am going to have trouble picking out the still pictures for illustrations, particularly in writing legends for the damned things. I had hoped by this time to have had a reply from you indicating which ones you wanted used and giving the appropriate legends. Since your last messages failed to mention the receipt of the stills, I assume they must be lost some place on the jungle trails. Unless I hear from you within the next week, I shall choose the pictures and write the legends; so just don't bellyache about the results.

Last week we received the following from you: five 35mm developed still film; one 8mm developed cine film, two 16mm undeveloped cine films, and one 100-ft. roll of 16mm color film.

I am not sure whether I am going to the Bangkok meeting in November. At the moment, I think the odds are against it. I think you ought to get home, and I would suggest that you skip consideration of Bangkok and also shorten or eliminate your proposed visit to Kuala Lumpur.

Best wishes.

Sincerely yours,
Joseph E. Smadel, M.D.

II.
Bush Correspondence
29 August–1 September 1957

DCG to ZIGAS and J. BAKER Bush camp No. 2—one day away from
Wi'ir, the first "Papuan" settlement
out of the Highlands
29 August 1957

Dear Vin and Jack:

We are in a tight bush camp, in pouring rain—it has poured every day of our patrol and we have only made one walk thus far without being drenched to the skin by torrential rains, and that one rainless walk included several waist-deep fords of streams. Thus, we have not yet been dry. I write in a jungle-built hut which would do credit to [the film] "Walk into Paradise," after two full days in the type of dense jungle which the *Sunday News* readers down south believe comprises all New Guinea forests—to be truthful, the first such jungle I have seen in New Guinea, quite similar to that of the New Britain lowlands.

Until Ivaki village, all was a quiet patrol. We marched from Atigina to Purosa, making the eight miles (?) in under two hours in a terrific storm. On arrival, however, we found another 6 kuru cases; and thus we had 15 new cases at Atigina and Purosa combined, besides seeing our old cases from the region. [66]

The trip from Purosa to Ivaki was again in pouring rain, and we had a good deal of difficulty in getting our cargo hauled up the steep *kunai* slopes to Ivaki. We got chickens and a pig from the Purosa crowd, who were much more friendly and cooperative than I had anticipated. Our cargo-boys—21 in number—were all hired from Atigina and Purosa. At Ivaki we did not get much done, and left early to make Oriei over a trail which is hardly worthy of the name "track." It is a long, steep and hard trail; and we arrived at Oriei rather exhausted, in rain again. Here we tracked down Tawasa (the 14-year-old girl with kuru who ran out on us at Moke) and Mande, who had gone through the syndrome we observed in Nata and Aso and Aoga: i.e., getting dreadfully fat after her kuru began to limit her activities.

Since Mande was on a village diet, was not on drug, and developed her adiposity entirely in Oriei, it seems certain that this is one possible aspect of the clinical picture of kuru. I was surprised that we had not spotted it for our earlier reports. Mande promised to return to Moke for further observation and treatment of *kaskas* [decubitus ulcer] which has developed over her buttocks. Tawasa I left at her village, to call in later. She still thinks she is cured, but I think she has kuru. Amakiora looks more than ever like she has kuru, but she is still trying to live up to the reported "cure." She strives with all in her six-year-old musculature to defeat her tremors and ataxia, but does not succeed. Oma, the small boy of Atigina with kuru, was in about the same condition as when I last saw him.

At Oriei we bought another pig, which broke loose during the night and left us with naught to do but reclaim our hatchet payment. I would not be surprised if the owner did not facilitate the nocturnal exodus. From Oriei we took a trail never before patrolled, that to Kasarai, where there is only one hamlet with somewhat over 100 total population. [67] They are reached by tedious walk on a bush track which, without some cutting and without a good guide, could not possibly be followed. We walked for five difficult hours to gain Kasarai, and found there my old friend Anuma the old *Luluai*, who is certainly their major leader. An impressive second *Luluai* (even less trustworthy, and wonderfully decorated with bird-of-paradise nose-feathers, etc.) joined Anuma in receiving us; they had just built three fine temporary shelters for our party, which weathered the all-night and afternoon storm well. Kasarai, not previously visited, is well out of the way and is the furthest south of the Fore settlements, with Paiti about as far south but across the Yani River. It is the most isolated and "bushy" settlement of the Fore, with our mountaintop Kasokandi the only rival. You must have a guide to reach it! The trail could not be followed without one. We were given a wonderful welcome and loaded with food, a pig, and three chickens. For the small group, this was astounding. Anuma had kept his promise, visited Sorobi's hamlet as he told me he would, and found Sorobi and four others had just died and the Yar people were deserting the site they had occupied because of these deaths (I cannot work out the cause) and have moved even further south toward the Papuan border, but just before the Yani-Lamari junction.

The two days' dense jungle trip from isolated Kasarai to the Yar people has now become three days. The track was not only impossible to find, but impossible to walk without a crew in front to cut, bridge, and bushwhack the way that old Anuma indicated. For the life of us, we could not see how he kept his bearings; but we came through, dropping to well below 3000 feet into real tropical jungle at the rushing river that goes from Urai village down to the Yani. This river we had to ford thrice, and once crossed precariously on an immense tree-bridge. Our cargo-boys were afraid of the current and made but one crossing; we had to have a special trail cut for them through the dense bush. This "road from the highlands to Papua" has never before been traveled by Europeans, we learn. The first night out from Kasarai, we came upon Anuma's "bush camp," which we converted into an immense base, taking down an acre of bush and erecting a camp with our "staff" (now, for this expedition through "no man's land," our component is 26 cargo-boys; 2 *Luluais*, including our guide Anuma; the 2 "Papuans" from Yar whom Aneti brought north for us; our 3 police; 3 Native Medical Assistants; my 6 boys, and one itinerant Kukukuku—Agurio—whom I am adding to the "boys"; and 4 "track-cutters" from Anuma's line).

We had a wonderful afternoon of real swimming with the current (most of the Fore too frightened of the water to try) and today have walked another five hours of dense jungle bush in pouring rain, which has continued from yesterday afternoon without cease. Our second "bush camp" is so completely enclosed that we are walled in on every side. We have met the Yani River—immensely, frighteningly huge at this point south—and left it [to go] over the ranges to meet it again tomorrow when we make Sorobi's group, now led by a man called Pe'i. They have two bridges over the Yani, I am told, hard as it is to believe. We took four hours getting shelters erected here; but all is now cozy and tight, the jungles are alive with game, and the roar of rain rivals that of yesterday's campside river.

Mando, one of our policemen, shot a huge *muruk* (cassowary); and the bird is being *mumu*ed, stuffed with many of the dozens of new native *kumus* [greens] we have learned to eat since yesterday. A huge *kokomo* [hornbill] was shot yesterday. *Kapiok* trees—your Pidgin song *"long as blong kapiok tri"* ["under the *kapiok* tree"] is my only previous acquaintance therewith—provided much of our roofing yesterday, and from their raw and roasted nut-like seeds we made a full meal. Dozens of new wild nuts are about, and John and I may either discover a new edible fruit or die in the attempt in sampling all the acrid but delicious-looking fruits and berries which we pass. We broke into our bush rations for the first time today—although we are still carrying enough *kaukau* from Kasarai to make out today, and tomorrow we should be at the village (I hope!). Thus, with the wild game (we have not yet got a wild pig), everyone is very well fed. We may have three days in the bush beyond Wi'ir, if we attempt to get down to the real population center, So'o; unless I reach So'o, I fear we cannot settle the matter of the southern extent of kuru. I now know that some kuru-area Kimi and some Fore women have married into the Yar village we are approaching! That it has a totally distinct language, utterly unintelligible to all of our Kimi-Keiagana, Kamano-Fore, and Pidgin-speaking crowd is certain. Anuma does a modicum of interpreting for us. None of us speak Motu [the lingua franca of Papua], though many of these Yar people can probably handle it.

An Anson swept down over our first bush camp, banked to view it (we think), and went on. Since this is no place to see Ansons in the sky, we could only assume it was you, Jack and Vin, making our long-postponed aerial survey of this southern region. If so, I hope you saw us. The plane was very low and "searching," judging from the way it banked. However, our camp could not have been easily seen unless from directly overhead—for it was surrounded by tall trees, and the river is bordered by trees of over 100 feet. If it was not you, Jack, who the hell was it? This is, as you know, no place for a small plane! While on the subject, please try to line up with Mick Foley the aerial survey for me just after my return, since a good view of all this country from the air may influence my epidemiological thinking greatly. In fact, it becomes a full essential now, since kuru seems to have no "ecological bounds" that I can find.

We passed below Anuma's village of Abonai (the furthest south the government administration has reached . . .). The Kimi from Misapi village chased them out some 10–20 years ago with warfare. They had kuru at Abonai!!! Abonai is now deserted. No other people are south of Paiti and Kasarai but these Yar people whom we hope to reach tomorrow.

John left Moke with a cold; he has since developed a severe sinus infection, and has been struggling on in the face of rather severe ailment. We underestimated it for a while; it has knocked him out fairly seriously, but we have no choice but to push on now. He seems to be responding to aqueous penicillin injections—which are as painful as can be imagined, he claims—and we started Terramycin today.

Our plan is to study things thoroughly at Wi'ir, and to try desperately to reach the Papuan population center of So'o. All that can stop us is illness and the immense rivers. We hope we can conquer both and get on. We should be "furthest south" within four to five days from now, and start back north, returning on the west bank of the Yani River via deserted Abonai and Paiti villages and the comfortable south Kimi villages. We hope you will send us the supplies I list below via the South Kimi—or, if we are not yet there, to Paiti.

Sincerely,
Carleton

DCG to J. BAKER   Wi'ir village, Yar people
                                                         30 August 1957

Dear Jack:

Continuing where I left off in our bush camp yesterday, I now write from a most astounding settlement of heterogeneous-appearing (and genetically so!!) sophisticated "Papuans," stinking of fermenting sago and with all their conversation shouted against the roar of the Yani below us. [68] Interpreting has been almost impossible; but I now have a word-list of several hundred words, and we are beginning to make out with the aid of the "camp followers" attached to this settlement. These people here call themselves Papuans and are so known to the Fore. They have been settled for at least a decade at the very junction of the Yani and the Lamari, just in the point of land before the junction and overlooking both rivers. This settlement was led by Sorobi, who got his "brass" [i.e., credentials] somewhere "down below"; and they came up to this location well over a decade ago from somewhere down below. [69] They claim the nearest government station is Kikori, and all or at least most of the men have worked at Kerema, Daru, Kikori, and Port Moresby; and in spite of being perhaps as "isolated" a settlement as exists in all New Guinea, they have dragged rather fresh-looking clothing—shirts and shorts (some long trousers, as well) and fine trade-items such as pots, cups, mirrors, glasses, etc. up here into the bush with them.

There could not be many more isolated regions than this, for it is a full seven days of heavy travel, in either direction, to the nearest radios. . . . They left their previous settlement at the Lamari-Yani junction, where their houses are still standing deserted, because of many deaths from "fever" which may be malaria. They have only now built temporary sago-covered shelters here in the bush, north 3–4 hours' walk from the river junction, [in two clearings—one] on the Yani's eastern shore and [one] just across the river. These clearings are already connected by two narrow, immensely long and frightening bridges. They are each suspended by six vines, hang high above the raging torrent, and are only two-to-three poles wide at the base; and on these poles one must secure a precarious footing. One could hardly survive a fall into the stream, I fear. Whether we can get cargo across these bridges remains to be seen. I tried them twice today, barefooted, and do not care to risk them many more times! The people have plenty of sago, this being *as blong sago tru* ["real sago country"] according to everyone; and their new gardens, planted with pineapple etc. are just now planted and fenced. The group is small, and we cannot yet estimate the population. It is probably well under 100 total, although we have seen few women or children as yet. They have built a new "temporary shelter" to help house us, and have given us plenty of sago. They are friendly, and among them I find the "Papuans" who made their way earlier this year to Moke for medical treatment, whom you surely will remember.

The dozen-odd children I have seen all have huge, probably malarial, spleens. They are certainly Papuan types in both physique and behavior, and our [Fore] cargo-boys only struck us as real "savages" after we arrived here.

Mosquitoes and leeches are terrible, and huge insects of every description abound. I will make a collection of those only over six inches long for photography tomorrow.

This is the only population, thus far, south of the Fore. Their leader—replacing Sorobi, who just died—is Aipos (or Pe'i). He is cooperative and intelligent and has worked and visited in Kikori, Moresby, Daru, Kerema, etc. However, in spite of their sophistication, they are living now more primitively and in some respects more dirtily than the Fore; and they needed a great deal of medical attention, which we have started to give them. With them are some Fore boys whom I now trace to Takai village; their fathers were all jailed by Patrol Officer Coleman about two years ago for a mass murder at a place in Purosa which I cannot place but where [Corporal] Homeguei swears the Takai group executed mass murder two years ago. These four boys of 12 to 15 have been adopted into this group; and although they are Fore, they speak this language and live here, now, for good. These Fore are not the only ones [from the kuru area] who are here, for some Fore women have come into this group. Furthermore, one Kimi is here, claiming to be a Yar by birth but Kimi by experience. He speaks both tongues. The entire group look like a bunch of deserters or escaped convicts, and never a more heterogeneous and "atypical" racial group was to be seen. Were it not for the distinct language—absolutely unrelated to Kimi, Keiagana, or Fore—which they speak and which I am now rapidly documenting, I would not believe a word of their tale. Aipos, their leader, looks Papuan, not Highland; yet he claims Kimi for father and Yar for mother and is unable to understand any Kimi. Thus, the entire group is more bewildering and confusing than any I have seen before in the Territory; they are an obvious "back door" to your Highland estate, a traditional "escape" or "refuge" for the crowd of Fore to the north.

We cannot be satisfied by this group, therefore, until we have seen evidence and proof of a Yar population less heterogeneous. This is (strangest of all) another 3–5 days south, we are told, and entails crossing the Yani on an even longer and more precarious bridge than the tightrope we have been surveying here, down at the Yani-Lamari junction where they have just deserted their village. We shall go down tomorrow to see this landmark-of-landmarks and the fine houses and village they are supposed to have left for this "bush camp." Then—*if* we can make enough sago for food; *if* we can get the carriers, ourselves, and the cargo across the Yani; and *if* we can then, after two or three nights of further bush camps, make rafts and go one day further downstream (and I certainly would not risk a minute in navigating the wild torrent rushing by us here!)—we shall try to make So'o, the truly big, permanent, and representative Yar settlement and the next inhabited site below here. We can come back up the river in canoes from So'o, they tell us. If this wild venture proves too much (and some say that current rains will make it so), we shall settle for a study of the Lamari-Yani junction and then start our return. Furthermore, with Abonai abandoned south of Paiti, we are told that there is no trail or route on the western shore of the Yani north; but since we had to cut our way through to get here for the past four days, we have decided to cut our way north to Paiti and then come out along the track I know already to Misapi village, and the south Kimi.

A further revision of all data on these people is needed after further data was obtained today from their real leader, who—strangely—showed up today just as they told me he would. [70] He came from the former village site overland with wife and two daughters, walking slowly because of yaws lesions of his feet. Both children and wife suffer likewise from extensive yaws. His name is Urahau. He brings with him a village census-book, issued from Papuan authorities in Beara in 1954 when he was appointed Village Constable without any census ever being made and without any other of his group being seen by the government authorities. [71] There are no names entered into the book at all—no census figures. No government patrol has ever seen these people.

Urahau confirms much of our previous information. He lists the villages south as So'o, Wa'abowa, Ho'urua, and Beara (the government station in Papua). He calls his people and language Yar. [72] So jotu is the old village site, which he claims they might return to. It is located south —*not* in the junction of the Lamari and Yani, as we were previously informed, but rather across the Lamari, below and south of the Moraei Kukukuku people.

[73]

The census-book listing Urahau as Village Constable states that Aipos was given a Councillor's badge at Beara, which badge he now wears. Urahau holds the book, however. It is also reported that in 1955 Urahau came to Beara to report that the Fore [Bore'e] people killed a woman and man at his village. Papuan authorities noted this and estimated the time of the incident as late 1954 from his tale, closing the matter at that.

[74]

Marriage contacts with the Kimi and Fore have existed for at least one generation, with Aipos himself being of Kimi father and Yar mother. Several Yar people speak Kimi (including Urahau). They have also had a few Fore wives; and one is now present from Kasarai (Iginauri), which is a kuru-affected population. There are four Fore brothers from Takai-Ketabi clan of Purosa living with them for the past two years and speaking Yar well. These boys, aged 10–16 are from the kuru-area center; and the older is already capable of genetic transmission to this Yar group. Fore contact, as well as Kimi, has been extensive, although it is a two- or three-day bush trip from the nearest Kimi or Fore settlement north (Somai and Misapi for Kimi, and Kasarai and Paiti for Fore). Abonai, previously occupied by the Kasarai people led by Anuma, was the closest settlement north but is now deserted.

There is much evidence in the bush of shifting settlement about sago patches—ruins of old houses and occasional garden plants gone wild.

All of the children and adolescents seen were found to have large spleens, extending to below the umbilicus. This included the four Fore brothers from Takai, who have been here only 1–2 years and have not traveled south to visit any other Yar people. Yaws is prevalent and severe. Urahau has at present severe yaws of his feet; and his nine-year-old daughter and one-year-old son are both severely affected, and his wife moderately so. Infected sores (tropical ulcers included) were the only other medical complaint found.

All young adult and older men are familiar with the route south, but all claimed to be unwilling to attempt the trip in this season of high waters; and most are familiar with the routes north to Fore on the eastern shore of the Yani (a bush route requiring guides and cutting) and to the Fore and Kimi people on the western shore of the Yani (through thick bush requiring much cutting and a guide, although always close to the Yani River—unlike the easterly route which cuts inland over the ranges). [75]

Yar, or Te'hei, is a distinct language—totally unrelated to Kimi or Fore, Keiagana or Kamano, and completely unintelligible to all those speaking these languages who were with us. Anuma was the only Fore we could find who could interpret, except for the four boys from Takai who live with the Yar people (they speak Purosa-brand Fore and not Moke- or Okapa-brand!!). Several Yar people can understand Kimi well. We have found no Kimi who speaks Yar as yet. A word-list of several hundred words indicates a total lack of cognates between Yar and languages north, also a different phonetics and phonemics. [76]

Although the men are dressed in shirts and trousers or shorts, and have many trade-items (such as cups, pots, plates, knives, forks, spoons, mirrors, and a few suitcases and knapsacks), the women are all dressed in grass skirts and decorated as are Kimi women—some with *kina* [pearl shells]—but not as extensively. In the women, but not the men, it is obvious that they are a bush people. Women and children ran away on our approach, and stayed out of sight until we sought them out. They accepted medical attention—both sexes—and brought children for injections.

They insisted that there were no stores in So'o, Wa'abowa, or Hoiuru'a'a, thus placing the nearest access to trade-items well over one week away from them—which is remarkable, considering the quantity of such items and their good condition. Their present settlement is insect-ridden, mosquito-infested, and hardly as clean as a Fore village.

[77]

They claim to have had no kuru in themselves or in marrying Kimi or Fore people, or in visitors. Documented kuru thus remains north of here, the furthest southern extent yet known being Abonai (now deserted)—two days' walk north of the Yar people, near the present Kasarai people, who are now marrying into the Yar group.

[78]

DCG to ZIGAS and J. BAKER   Bush camp No. 4 (fifth night in the bush, one
                              day north of Yar people, on west shore of Yani
                              River; one day—we hope—south of
                              abandoned Abonai)
                              Sunday night, 1 September 1957

Dear Vin and Jack:

The enclosed envelope contains two letters previously written during our trek south through the bush. We obviously got to where we were heading, but not as far as we wished to go (i.e., So'o); this I hope to do later from the Papuan side. In these times of high waters, it is perhaps impossible—at least, most difficult.

The enclosed two previous letters need no prompt perusal. There is, in addition, a short discursive summary of data we collected on the Yar people. This I hope you will carefully save and read and show to Mick Foley, who may also be interested. Any differences in facts from sheet to sheet represent later accumulation of information. The final summary is thus-far definitive. I have four Yar guides and bush-cutters with us now, and I may even further modify this information as I have more opportunity to talk with them.

The present letter demands immediate attention, and thus I urge you to attend to it—putting off the sealed sheets herewith enclosed, until you have time to browse.

We hope to be at deserted Abonai tomorrow, make another bush camp there, and arrive at Paiti the next day. On Wednesday we should be at Misapi, and Thursday at Uvai. Since we did not get on to So'o, and since we have so little time left to complete kuru work, John and I have decided to head for Lufa from Uvai; the route via Mani is direct and covers the West Kimi (Lufa) epidemiological-patrol area. This patrol I must squeeze in, in order to complete the epidemiology on this side of the kuru area. We therefore write, begging you to send out supplies to meet us at Uvai on Thursday, or at Mani if we have already left Uvai. [79]

Bush camp No. 5
2 September 1957

We managed to find Abonai, and had more difficulty cutting our way through its deserted and overgrown gardens than through much of the virgin bush. Rather than stopping there, we have come on and are camping tonight—in rain, as usual—in another *saksak* [sago]-covered shelter. We made use of what is probably the last clump of sago we shall find, on our trip north, to make a camp. Tomorrow, with luck, and still cutting a trail, we should reach Paiti and the beginning of a maintained system of trails. Previously, Paiti was the furthest point away from Moke in my thinking of your sub-district; but now, for the past week, to reach Paiti has meant "being back home in civilization."

We are hoping to reach Paiti early enough to get messengers off today toward Moke with this letter for you. They will also be carrying some mail to post for us, and blood specimens from "down south." These are already too old, and should be immediately refrigerated (not frozen) and shipped to Melbourne as soon as possible. These bloods must certainly be shipped off soon. They may have some importance both in genetic and in antibody studies. I may also include with them some bloods on kuru cases, if I can get them at Paiti in time.

[80]

Our crew of over 30 (probably 40 with the boys and hangers-on) have made out on the meat, rice, and sago we brought, and on what fresh food we hauled with us from Kasarai, for the past eight days. They are hungry for *kaukau*, and so are we. Every conceivable bush food has been used to supplement the diet, but they "missed" tonight. They picked a wild *limbum*—a tree from which the center core is eaten—and although it tasted good, it was the "wrong one" and everyone is just now getting over rather violent attacks of vomiting. Homeguei shot two wonderful birds-of-paradise this afternoon, and we just missed a wild pig which Kosinto swears he wounded with his arrow. With the cassowary and several *kokomos* and several *kokis*—and now these *kumurs*—the diet is being supplemented well by fowl.

Please rush us the supplies, if you will, and tell us of any plans you and Vin have. Except for the possible survey-flight first, I am ready for the Kukukuku trip the day I get to Moke, and that can be the day you get back from the Kainantu race meeting, or the next.

Sincerely,
Carleton

III.
Excerpts from Field-Journals Kuru Epidemiological Patrols 3–14 September 1957

3 September 1957   Paiti village, South Fore

[1:3–3:35]   For the past two years I have kept a personal journal in a haphazard, irregular, and half-hearted fashion. My Gidean phase of journal enthusiasm subsided as I finished reading all of Gide. The most exciting times to have kept a journal well—while traveling throughout Australia, hiking in the Tasmanian mountains, working with the Flying Doctors among the Cape York aboriginals, traveling in New Guinea and New Britain in 1956, and, most important, during these present kuru studies in the Eastern Highlands—have all been missed. It is with a constant desire to keep a journal, but with such a tight schedule that I never do, that I have lived these past 18 months; and I feel this omission has been something of a loss to me—loss of richness in the experiences, for journal-keeping tends to add condiment to them. Again, I make a start.

We[9] have at last arrived at Paiti after a six-hour hike from our Bush camp. This eighth day of our trek through previously "unvisited" country was track-cutting all the way. Since we left the Oriei Rest House, we have been over roads and country which no European has previously traveled....

Agurio, a Kukukuku lad, sits beside me on the patrol boxes as I type under a *kunai*-grass–covered shelter just erected beside our hut and overlooking the cloud-filled forested valleys of Paiti and Somai to the west. Behind me rushes the Yani River, its noise dimmed by our elevation above it and by a shielding hillock....

Agurio came to Pintogori in response to my call for Kukukukus, a call made to establish further contacts in addition to those provided by Waiajeke, the only boy working with me who is a Kukukuku. Agurio arrived with a dozen other Kuks from Moraei and a delegation of Agakamatasa Fore, their escort. We had our new "Kuru House" filled with Kuks and their lice for the two days before my departure on this patrol. I tried to win their favor with friendship and payment—payment being held to be part thereof hereabouts, I find—and won at least Agurio's firm following, for he did not return via Agakamatasa to his home with the others when they left us at Purosa, but joined us when we turned off for Ivaki.... Although he looks frail and

---

[9]European Medical Assistant John (Jack) Berkin accompanied DCG and his entourage of assistants and cargo-carriers on this patrol. Patrol Officer Jack Baker remained at Okapa Patrol Post during this period.

needed help in the crossing of the Urai River (which we crossed after leaving Kasarai Village)—presenting a picture of sensitive, tender frailty—he makes a fine performance on rough trails, swings a bush-knife expertly in cutting the track, and climbs to staggering heights like a monkey in the giant-leafed *kapiok* trees to cut leaves for roofing and bedding as well as an occasional young leaf for a spinach-like food.

4 September 1957   *Misapi village, South Kimi*

We spent a morning in the clouds at Paiti, where we managed to examine over 100 people and find a few cases of yaws, residual after my more thorough medical inspection and yaws treatment here in July. In addition, hidden in the village huts, the NMAs found other sick. One was a man with severe right-lower-lobe lobar pneumonia. With much difficulty we persuaded the people to carry him along with our patrol to Misapi for continued treatment, for the sulfadimidine and penicillin therapy which we have started will certainly be insufficient unless continued for several days.

Clouds moved in and away all morning, yet the boys managed to wash all of our clothing; and during one parting of the clouds, I managed to get 30 feet of cine film of two kuru cases—Aurika and Esita—with Paiti hamlets and gardens for background. *Luluai* Wanto, while giving the impression of cooperating and understanding much that medicine has to offer, joins with his people in hiding sick cases from us and argues with them against taking the sick with us for a few days of treatment. Enough argument and discussion wins the point, and with no more coercion than that which I exert to force medicine down the throats of skeptics in the U.S.; to use all means short of any actual force is fair. We managed to treat all the sick.

Anua, my 13-year-old Ketabi-Purosa boy, has been ill and ailing for days in the bush. Two days ago, at Bush Camp No. 5, he and Wanevi and several of the cargo-boys, searching the bush to add to our monotonous diet of sago and rice and bullied beef rations, decided to use, in addition to the species of *limbum* (Areca palm) which they usually use for food (eating the soft friable central core, which tastes very good raw but better cooked) another species prevalent in the bush about our sago-camp. With sago-palm present we did not make use of the *limbum* leaf for roofing our camp shelters, as we would have otherwise. They ate the *limbum* with relish; and about one to two hours after consuming it, they became violently ill with retching, intense nausea, and vomiting. This vomiting persisted long after their stomachs were empty, and the retching and vomiting—coming in sudden spells—made a half-dozen of the boys seriously and severely ill for many hours. They were ill all night, many continuing to vomit periodically until near morning; and in the morning, some 12 hours after having consumed the *limbum*, they were dehydrated, exhausted, and had an acetone odor to their breath. They have slowly recovered. Wanevi is now well, but Anua and two of the *limbum*-poisoned cargo-boys are now ill with an acute febrile illness which has affected 10 of our party. They have headache, fever, intense malaise and exhaustion, and some vomit; but there is nothing specific enough to warrant making a diagnosis. That they may have anything from Murray Valley encephalitis and dengue, through the host of arthropod-borne tropical viruses not yet identified in New Guinea, to more common bacterial infections is certain; and in the bush we can make no diagnosis. Since two are coughing somewhat but have clear lungs, and a few have sore throats—without exudate or pharyngeal injection—I am giving all penicillin and sulfadimidine (or sulfamethazine) therapy at present.

[3:46–4:17]   Anua was still well enough this morning to start our three-hour hike to Misapi; but in the middle of the journey, at the point where we cross a large stream in which we swam and washed, he could no longer walk and a litter was made to carry him to Misapi. Here he still has a high fever, but no chills and nothing specific in his clinical picture.

In the episode of *limbum* poisoning, it should be noted that all who ate large quantities of the *limbum* were violently ill, while a few others who tried a small amount were not ill. The *limbum* was consumed raw, as usual.

We are all suffering severely from the cold of these high lands, having been too long—in the brief week—in the lowland jungles and their heat. We did not notice the heat but did sleep without a blanket or with one blanket in wall-less shelters, whereas here in the Highlands we find five blankets are needed in a closed house to keep warm. It will be hard to get accustomed once again to the rain, wind, and cold of Moke-Pintogori. Perhaps we shall have worse as we proceed now with the circuit of Mt. Michael, which we are attempting. Tomorrow we shall find how many of our cargo-boys will stay with us, and replace any who wish to leave with the friendly Misapi Kimi.

6 September 1957   *Amusa Rest House, South Kimi*

[4:25–5:24]   Anua developed a picture of typical meningitis later in the day at Misapi, and he was already on high doses of aqueous and procaine penicillin in oil and sulfadimidine when the diagnosis became apparent. He had headache, neck pain, positive Kernig and Brudzinski, and all signs of extreme meningismus by mid-afternoon on the 4th, after arriving at Misapi. His cerebrospinal fluid could not be examined for lack of proper needle, and I hesitate to do a lumbar puncture with the poor regular needles we have with us. It is conceivable that he has a meningoencephalitis of virus etiology, but he must be treated as though he had acute bacterial meningitis. He kept the first several doses of sulfadimidine down, but has vomited all since; and when he arrives at Uvai, to where he is now being carried, I shall start injectable chloramphenicol since sulfa therapy is obviously failing in the presence of his emesis. . . .

He is drowsy and somnolent, but shows no other encephalitic symptoms; and he remains conscious and coherent—without convulsions, pareses, paralyses, or any other signs of central involvement other than the difficulty in rousing to response. Thus, three of our boys have now developed meningitis, and I cannot fathom the factors responsible. . . . I have taken an "acute blood specimen" and plan to take later a convalescent specimen for virus studies, which might give a clue; but the etiology is probably bacterial, for that is what we have usually been finding. However, since he developed it just after eight days of exposure to bush south of where he is at home, and in the malarial forests he never knew before, during which period we were constantly cutting trail and clearing camps from virgin bush, it is very possible that he has an arthropod-borne virus meningoencephalitis.

The other boys pay little attention to Anua's illness, but are solicitous of his needs. Though they do not take much notice of him lying by their cook-house fire critically ill, yet they will help him out to urinate, feed him, and inform me of any change in his condition. However, their games, their ribald and loud humor, their *singsings* do not cease a moment in deference to his illness. We carried him the two-hour trip from Misapi to Amusa by stretcher, and will carry him for the one-hour walk to Uvai this afternoon via a good trail along the Yani River.

6 September 1957   Uvai Rest House, South Kimi

[5:40–6:7]   Anua is now on intramuscular suspension of chloramphenicol; and although still vomiting, I know he is getting broad-spectrum drug and feel safer with him. He still has much meningismus, is acutely and critically ill, but is less febrile and appears somewhat better than yesterday. I only hope we are winning. I believe it safe to carry him to Mani tomorrow. The trip back to Moke would have been an arduous 2–3 days by stretcher without medical supervision, and no one is there—with Jack, Vin, and Lucy all at Kainantu for the race meeting. Thus, even distant Moke, our one outpost of civilization to where we have addressed pleas for further food and medical supplies, is deserted; and we head for Lufa tomorrow, unreplenished in trade-items, medical supplies, or the usual patrol luxuries such as sugar. We live well on the Kimi foods, and at each place they have had a huge *mumu* prepared. When hungry, nothing could be more welcome than freshly *mumu*ed *kaukau* and *pitpit*. Sweet bananas are plentiful, and we have obtained chickens.

[7:13–8:3]   The typewriter is without one of the shift keys. I suspect one of the curious natives, who have never before seen such a machine, has plucked it out when I left it unattended outside the house today. However, I may well have dropped it, packing the machine away. Two lads ran all the way back to Amusa to hunt for the missing key, and the Kimi dignitaries showed grave concern when I reported its loss; but it is probably lost for good. It had been loose for ages. To have brought the typewriter has required the struggle of one cargo-boy over these trails—John and I have hardly been able to drag ourselves over them, especially while we were trailblazing in the south—and the trip has put a good deal of heavy strain on the machine. The midday sun seldom takes out the dampness which the typewriter accumulates. However, without it I could not have our records up to date, nor have completed the case histories on kuru, the epidemiological and expedition reports, nor—and this pleases me most—started a journal again after this lapse of months, almost years.

Obviously, I am having trouble getting into journal-writing, and find myself writing either a cheap travelogue or a self-conscious diatribe; but it does serve to organize my thoughts, direct observations, and bring an added pattern to the day. As the "crew" *singsing* across the rain-soaked field and the rain drips off our *kunai* thatch into huge puddles, I work with this machine, aided by the luxury of a kerosene pressure-lamp instead of the old wick-lamp I brought on past patrols. In this entire 30,000–60,000 population center of Kimi, Keiagana, Fore, Yar, Iagaria,

Auiana, and Kukukuku linguistic groups, John and I are today the only ones who have seen anything of the world other than the New Guinea island. Few have seen the coast during their lives, and most of those few are with us now (none are literate or "civilized" yet). During this entire two weeks, I have not missed "civilized" company any more than I did on the last 18-day patrol or during the month of patrolling with Melanesian natives in New Britain. These people are sufficient unto themselves; they have a complex and organized-enough society to provide a cultural entirety devoid of contact with the outside world. Language and culture may close much of it to me, but I am at home with them.

[8:15–33]   I came into the Fore on March 14th for a brief visit. Kuru has now kept me in so-called "uncontrolled" regions for almost half a year, much of which time I have been out on patrol. I have abandoned my French and Russian authors; my correspondence, which has fallen to naught but the voluminous scientific exchange about kuru and our three kuru papers, has afforded me little time to follow world news, literature, or home events for these six months. I am a bit shocked when I consider how little all this bothers me and how little is my anxiety to leave. I know full well that I shall be walking the streets of Paris, Rome, London, New York, and Washington again, and that from these places the New Guinea jungles and the "savages" are but remote museum pieces or subjects for arty films and literature—hardly the humans with whom I now live and sleep. To me they are, as were my friends of New Britain, among the warmest and closest friends I have had. I respect, admire, and love them—and know that once I part from them, I may never see them or hear from them again. I am in no hurry. But kuru work, at any rate for the first field-stage, is nearly done; and I must be on my way—long and devious though it may be.

7 September 1957   Mani Rest House, South Kimi

[9:13–11:12]   Anua is very ill and the long stretcher trip over the ranges is not too good for him. He has had a total of 3.0 gm of injectable chloramphenicol, which we gave in 1.0 gm injections at 12-hour intervals; in addition, we have salvaged some 500 mg of drug from the injectable suspension residues. This is all we have; and since he vomited much of the sulfa, we again turn to it. He has lost much weight, but ate a bit this afternoon and is now apparently afebrile. I think he is recovering, and certainly hope so—with all broad-spectrum drug, aqueous penicillin, and most sulfa now exhausted. Furthermore, we shall now wait here at Mani for the supplies [ordered from Lufa Patrol Post] to arrive; we are drinking tea and coffee without milk or sugar, which is a real hardship for us both. Besides, the rest will do Anua good.

The population about Mani is immense, with almost 2000 Kimi censused from this point. We fully examined 1350 of a previously censused 1450 people today—a very good turnout for medical inspection, exceeding the census in some lines—and found well over 10% with something to treat, but no critically ill patients. In fact, we have brought with us more severe illness in Anua than we have found. Supplies of oily penicillin are still holding out; thus we have enough to treat the yaws, abscesses, tropical ulcers, infected scabies, and other wounds which form the majority of the problems. OMPA in infants and young children should also respond well to 1,000,000 units of penicillin in oil.

One case of kuru has occurred in the Mani area, and I have surveyed only one-quarter of the hamlets. It was imported by a Kimi woman from the Amusa kuru-affected region, but developed here. She is one of the rather frequent Kimi "recoveries" from kuru. The apparent importation of most of these cases suggests support for the genetic predisposition hypothesis of pathogenesis; but the frequent "recoveries" here, in the Kimi, suggests emphatically an extraneous environmental factor operating additionally. Furthermore, the rather rare newly discovered phenomenon that kuru has appeared in a few non-kuru–area Kimi women who lived with or visited the Fore, speaks strongly against genetic pathogenesis. Thus, we remain confronted with an unsolved problem of great magnitude.

Anxiety for Anua, the dwindling store of medical supplies, and lack of supplies to replenish all the comforts and luxuries we have enjoyed for the first ten days out of Moke-Pintogori of these 16 days of patrol—all serve to create anxiety, since we have at least 10 days of journey before us. So does the knowledge that Moke is deserted, with Vin and Jack away for the past two weeks and Lucy certainly out at Kainantu by now. In addition, the report that [Patrol Officer] Ian Burnet has burned down his house at Lufa, and is out in Goroka, also leaves the second of our two emergency outposts of "civilization" quite "uncivilized." However, the peaceful beauty of Mani and the dramatic panorama of peaks and valleys, of forests and steep *kunai*-covered hills dotted with large Kimi villages, and of precipitously hanging gardens—all these add reassurance and tend to dispel our concerns.

8 September 1957   Mani Rest House, South Kimi

Anua remains critically ill, and our total of 3.0 gm of injectable Chloromycetin and 0.75 gm of outdated Teracyn are all the broad-spectrum drugs we have. Sulfadimidine and penicillin did not do the trick. He remains with a splitting headache, extreme malaise, somnolence, and irritability and shows at least 15 pounds of weight loss. He tends to lie motionless, and sits up only with great difficulty. Meningismus is gone, or nearly so, and fever has subsided. Today he ate a bit and did not vomit. . . .

Supplies arrived from Ian Burnet today, via one of his native policemen, since Jack asked him by radio to get them to us. They were scanty, but enough to pull us through to Lufa. Were we to arrive at Lufa on schedule (i.e., today), they would never have reached us. Our delay accounted for the arrival of the supplies. They were sent via the northern route around Mt. Michael; since we are circling the mountain on the south, had we been on schedule they could not have overtaken us. Ian and Jack admitted that neither of them knew just where Mani was, and that their maps were of no help. Jack's supplies arrived a few hours later, coming via Ke'efu and Henegaru with one of his police; and they contained all we had asked for, including sugar, powdered milk, rum, wheat meal, etc.—and only with their arrival did we realize how nonessential they were, how much of a luxury. What did *not* arrive was urgently needed drugs. Jack and Vin had spent a week trying to get in by Landrover to Moke, but rains and washed-out bridges had made the road

3–14 SEPTEMBER 1957   *139*

impassable. They arrived on Friday to find Lucy alone and struggling to assemble our cargo. However, they further found Moke devoid of all drugs, with no aqueous or oily penicillin, sulfa, or broad-spectrum drug at the hospital. Vin promptly left the next morning to try to remedy the situation, but we now know how disastrous it would have been to send Anua back to Moke, with no medical personnel and no drugs there; we, at least, on patrol, had some.

[11:18–12:30]   We lined and examined the remaining 3 of all 11 censused groups (or clans) of Mani today and found the usual OMPA, yaws, tropical ulcers, occasional leprosy, many infected scabies, and sores of all sorts. . . .

We are poised ready for an early takeoff tomorrow morning, hoping to get away before 5–6, if possible; but the problem of raising the cargo-boys and our crew that early is ever great, especially after the pig feast this evening. The Mani natives were understandably reluctant to haul all the needed *kaikai* [food] up to our camp; and our cargo-boys, police, etc. were mighty worried about the possible inadequacy of food. I finally sent Manto down to rouse some natives, threatening to take the food from the gardens if they did not bring some up. A good supply of sugar cane and *kaukau* and a bit of corn, pumpkin, onion, etc. finally arrived. John and I have feasted on these vegetables tonight with the remaining pig-leg. The requests for a pig were at first of no avail, until Manto arrived with a "conscripted" beast in tow. He was followed by an aged woman covered from crown to toe in mud, mourning for the death of her pig—with tears in her eyes. Her aged husband refused to state a price for the pig, and was taunted by some of his clansmen to refuse any payment. Obviously Manto had used considerable duress; and I could not let our crew pressure me, as they tried, into taking the pig. I simply turned it back to the man and his wife, and shamed the Mani crowd with tales of how well small Kasarai with its mere 100 people had fed our patrol while 2000 Mani Kimi could not. Shortly thereafter, another pig arrived; and this group disclaimed any duress on the part of our police and were willing to contemplate a *kina* shell or a large axe, selecting the large axe in payment—which, by local trade-scale, is a very good price for the pig. I doubt that we got the pig without duress, but shaming them a bit made the Mani group quite determined to supply one; and they most graciously pointed out that since we came "to look out for their skins," they were going to feed us properly. Our crew is thus feasting royally today, and I suspect we have gained more goodwill than ill will by the entire incident.

Actually, we are "invaders" into these people's quiet (at least, now that warfare has ceased) valleys—uninvited and probably unwanted. However, I have tried not to permit the use of force, and have used less coercion than most government patrols would, I am sure. Jack Baker's experience in Tarabo (where they frankly told him that since he was an "easy-going" *kiap* who did not *kalabus* [imprison] people right and left, they saw no reason for feeding him and his patrol—which they did not do, until he had his police dig their own food from the gardens and conscript a few pigs) is a situation we may easily get into, but one we deserve; I hope it can be handled by argument, discussion, and persuasion short of force—as has been possible thus far. The Kimi are hospitable, friendly, and cooperative at every turn. That they do not always climb 2000–3000-foot mountains with loads of food from gardens miles away at our every call is certainly only sensible and much to their credit.

9 September 1957   Hegeteru, Kimi

[13:1–5]   Anua had a rough stretcher trip, and exhausted four cargo-boys transporting him here. He arrived "washed out," but soon rallied and now appears to be on the way to recovery. Vomiting has not resumed and he is sitting up a bit, eating a bit and conversing a little.

[14:17–40]   Today I learn that kuru definitely does stop at Mani, on the Okapa side of the Divide. The people in this region (over 1000) have never experienced kuru. The over-2000 people about Mani have no kuru either, except for one case (that being one of the disturbing and questionable "recoveries") which has occurred at Raro clan in an "imported" woman from Amusa, a [Kimi] region of sporadic kuru with Fore intermarriage. However, in discussing kuru cases occurring elsewhere, I am informed of the important history of three Kimi women from villages with Fore intermarriage and sporadic kuru who were taken as wives by three Lufa men who live and work at the Lufa station, far beyond the limits of the kuru region. Two still live in Lufa, in good health; but the third has died at Lufa of kuru!!! This case was a woman named Urai, who married and came here to Lufa and then returned with her husband to visit her family—and, while back home, developed kuru. She returned to Lufa, walking, with mild early kuru; and at Lufa this mild early kuru progressed slowly and typically to severe ataxia, requiring a stick to help her about, and then to severe total incapacitation and death in a period of several months. Thus, although this case may well have been "contracted" in the kuru region, it illustrates the point that kuru, once started, may progress typically even when far removed from the kuru-region environment. Although this by no means proves genetic predisposition, it is a case which would add evidence in favor thereof.

[15:3–17:1]   Kuru is thus "boxed in" on the south and west. The north and east present problems which we must now investigate, but I think I already know the situation there fairly well.

In restarting this journal, I had every intention of steering clear of writing a research workbook or a travel diary. As a result, I have formed an unhappy mixture of both. The personal is never submerged; it does get pushed aside a bit when my enthusiasm for the work and the bare dramatic facts of the travel loom as interesting as they do now. I fear that the too subjective approach and the too self-analytic tend to distort the fact—that work and enthusiastic dedication, coupled with a dramatically changing environment, tend to overwhelm the strange underlying motivations that led to them. In the face of such factors, the motivations themselves begin to lose significance and interest. They have brought us to the time and place. It is now the *event* that has the stage.

10 September 1957   Gono Rest House and Mission, Lufa, Kimi

We arose at 5:30 a.m. and by 6 were on the trail from Hegeteru to Mengino. The trip was over a fine trail, but the mountains were steep, and we descended and ascended to cross three rivers—each bordered by high ridges—before we made

Mengino. I walked very fast and made the trip in three hours. John took four, but walked well. At Mengino, I found no admission of kuru; and we stayed on this high ridge only for about three hours, during which time the boys killed one of the chickens we were carrying and prepared a boiled chicken and soup which we carried on with us in the afternoon.... The steep ascent and an undulating descent was again arduous, but I made it in a raced two hours—outwalking all the boys, Kosinto, and the others. Many of our "crew" are complaining of foot and knee pains, and claim we have done too much walking. This comes as a surprise to me, for it is from highland natives that the complaints come! However, I suspect that when weeks of this sort of walking are involved, the lack of shoes begins to have its effects. At any rate, our "crew" are weakening.

... I spent an hour or more upon arrival talking to American missionaries Mr. and Mrs. Laughlin in their house on this windy ridge almost 7000 feet high, but finally began to wonder where Jack was. The *haus kiap* is well above the Mission, on a narrow ridge with a drop of several thousand feet to either side. A messenger arrived to tell me that Jack had hurt his foot and was lying on the trail some way back, unable to walk. Mr. Laughlin and I started back and, some 20 minutes from the Gono mission, met Jack being carried by some of our cargo-boys on a native-material stretcher. He was acutely ill, febrile, had intense lymphangitis and edema of his entire left-lower extremity and lymphadenitis in the left inguinal region, with moderate visible swelling and venous engorgement of all superficial veins of the entire left-lower extremity. The small-toe ulcer which he mentioned to me and dressed at Mengino was obviously the source of the trouble; but I had never previously seen so acute a picture of toxemia, probable bacteremia, septicemia, and lymphangitis. We rushed him to the mission house, where I gave him 4.0 gm sulfadimidine orally and 1,000,000 units of penicillin (500,000 aqueous and 500,000 procaine penicillin) i-m and placed him on hot soaks. He had extreme superficial venous engorgement, exquisitely tender inguinal glands, and marked swelling of leg and thigh. An acute thrombophlebitis is undoubtedly accompanying his picture.

I came up to our camp at the rest house, treated my own small foot ulcers and leg ulcers, and went back down to join Jack and the mission family at a fine American-style dinner with apple pie and cheese, ham and chicken, raisin muffins, etc. Jack was much better, but still acutely ill and obviously in no shape to walk tomorrow. Thus, I left him with the mission family and came up to camp to finish work and sleep on our windy, cold ridge....

With Sinoko, Kosinto, and many of the cargo-boys all with complaints, I remain one of the few well ones; but my leg ulcers are nearly critical. Thus, I keep my jubilations cautiously in abeyance, for I cannot tell whether or not they will act up, as John's have done. I cannot understand why John's small-toe ulcer should have so suddenly become such a malignant affair. There is always the possibility that the oil-penicillin he poured onto it to dress it just before leaving Mengino was a pure culture of some penicillin-resistant or -insensitive gram-negative organism, and that the $1\frac{1}{2}$ hour's exertion with such an inoculum on the ulcer is the cause. Then again, the ulceration itself may have had virulent-enough organisms in it for such a reaction to set in upon strenuous exercise. I only think of the oil-penicillin contamination because I have been decrying the NMAs' practice of opening a 10ml bottle, contaminating it by their technique—which is atrocious—and letting it incubate in the warm patrol box for a few days thereafter. Such an opened bottle John used to dress his wound.

[17:12–15]   Anua walked much of the trail today, slowly and falteringly, but he preferred walking to being carried. He is still on sulfadimidine, which he is not vomiting, and on penicillin. It looks as though his meningitis may be conquered.

11 September 1957   Lufa Patrol Post

[18:18–32]   Ian [Burnet, Patrol Officer at Lufa] has burned his house down completely, losing most of his personal belongings—as well as one of the best houses in the Highlands. He is living in the office where I am now camped. A stack of recent *Time* magazines has me enthralled, since I am out of touch with the rest of the world and a half-year behind in the news. The account of Edith Hamilton being made an Athenian citizen was enough to awaken me back to a lost environment. Ian has rescued a few books from the flames, among them Joseph Conrad, American verse, Macaulay's speeches, and some Scott Fitzgerald—the first "literature" I have seen in this book-poor country of "colonials" and adventurers who appear never to read, or to read stuff that I cannot waste my time reading. Thus, even if I sit out a day or two here, I shall find the time well spent, catching up. I have already moused through all the maps and patrol reports of the Lufa patrols, and found all I can from these documents that pertain to kuru geography and epidemiology.

12 September 1957   Lufa Patrol Post

[19:31–21:25]   While we were south of Abonai in the Yar sago country, the poisonous heart of *limbum* palm was the major environmental hazard we encountered. All of our carriers and other natives suffered heavily from leeches, however, which were extremely numerous and aggressive. All of our party had bleeding feet each day, during and after our treks. John Berkin had less leg protection than I had, often having to remove the leeches from his feet, while I had no trouble because of jungle boots and jungle trousers which I donned after discarding my shorts. Another hazard was afforded by the native bees. In cutting tracks, we often penetrated to trees, stumps, and old logs in which there was a hive of native bees. Twice I was severely stung. I was astounded that, although we were cutting what appeared to be a new track, in country wherein I had my doubts that he knew his way at all, old Anuma, our guide, would often suddenly stop, point ahead to a tree or log, and say that in that one he recalled a hive of bees which we should avoid. Mosquitoes were never a problem, but swarms of other native Diptera and some Hymenoptera were always disturbing at our bush camps—each species seeming to have its particular time of day to swarm.

The *kusa*, a stinging-leafed plant, became a great problem as we moved north from the Yar people. This low plant has a leaf which, when touched on the underside, stings severely, producing an erythematous, edematous reaction which is intensely painful for hours, and which no amount of bathing or rubbing relieves appreciably. The upper leaf surface and the stems can be handled with impunity, but the slightest contact with the undersurface does the damage. The natives beat themselves with *kusa* as a counterirritant for many illnesses accompanied by pain or severe malaise, or for any severe local pain. When they inadvertently touch it on the trail (they are most expert in recognizing it from afar and avoiding it, although they occasionally walk through patches of it), they rub the injured portion with the stem and leaves of other plants. This they claim relieves the sting; and John, who was often severely stung, agrees.

In addition to *kusa*, there is an even greater hazard in the leaf of a rather tall, large-leafed tree which produces a really dangerous sting and can be totally disabling. When felling clearings, the natives generally recognize this tree and never cut it. However, Kosinto came into contact with this leaf over an area of some 100 square-centimeters of his lower thigh, and developed sudden massive inguinal lymphadenopathy, lymphangial edema marked by local erythema and edema, and such intense local burning sensation and inguinal pain that he could not walk for a few hours and was in misery for two days because of the sting. John tells me that the same tree, or one closely related, exists in the Queensland bush in Australia.

Thus, bees, flies, leeches, stinging plants and trees, and poison *limbum* heart were the major problems of our crew. As for myself, the *kunai*-grass before one descends from the highlands is always a problem, for though I try not to, I grab at it at times, cutting my hands; and when I try to avoid it, I lightly brush against a razor-sharp edge and get deep cuts. Those on the eyelids I fear the most; for along trails in high *kunai*, facial cuts are frequent and those about the eyes could be serious. In the lower jungles, the problem became a vine with reverse-hooked thorns, associated with long similarly hooked fern-fronds. Along the frond rib, on the underside, are a row of strong, curved, bent-back, needle-like hooks which fasten strongly on clothing and grab at skin even lightly brushed against them. These often stopped me short on the trail; and in clearing the trail, it was this "wait-a-while" vine and the similar fern-fronds which we had to clear most meticulously. I was pretty well scratched up by these two species. The final hazard was vines of every sort which, when not assisting us as a hand grip for support, were tripping us. However, in spite of all these discomforting features of the trails, the New Guinea bush is fairly harmless. We saw not a single snake in the weeks of patrolling, nor is there a really dangerous animal in the country except for the sharks off the coasts and the crocodiles in the streams nearer the coast. Barefoot walking, even for our tender feet, is possible everywhere; and with caution one can walk barefooted for miles through rough bush without injury to his feet. Much of the bush is almost stoneless, and wherever stones appear they are rarely sharp. There are sections of the country where this is not true, especially so-called "limestone country," but not so in the areas I have thus far seen.

Ian [Burnet] should be back from Goroka with his jeep in a few hours, and I expect Jack down the trail from Gono shortly. With luck, we shall get out to Goroka in the car in which Ian arrives, get our work down and soon head for Moke and the mass of correspondence, medical problems, and writing I expect to find awaiting me there. I hope also to get the Kuru Hospital into swing again, for I suspect that our usual kuru-patient census of 15–30 is down to 0–10 since my absence. Finally, we must get off on the Moraei patrol soon, and thus bring this first field-attack on kuru to a close. The difficult western, eastern, and southern boundaries will then be established, and the easy-to-patrol but thus-far-neglected exact northern limit can be quickly determined. Roads, good trails, and rest houses, good transport, and a proximity to "civilization" should make the northern portion a cinch.

[21:35–22:16]   The language spoken in the Lufa patrol-post area we traversed after crossing the Divide from the Okapa side is still Kimi, for Kosinto and L/L Haneo speak to and understand the natives easily. However, they seem to use the term Kimi only for the Okapa-side people, so that what to call the linguistic group as a whole is a problem. I shall ask Ian. . . . Our Kimi are certainly strongly influenced by the Fore; and although dress is strikingly different (long anterior *bilum* hanging from a thin belt for the men; and fern leaves or other leaves, stuck in fresh each day, for an anal covering—which all Kimi wear), the language has many similarities to Fore. . . .

Besides similarity in spoken language, place-name similarity is intriguing. Thus, with the Fore, Kimi, and Keiagana, the same place-names recur constantly. Just as we are not confused by reusing given names over and over again, geographic specificity makes the duplicated names unambiguous to the natives. They always know what general geographic location they are talking about.

1:30 p.m.

[22:25–39]   Ian's house was not—as I was first told by exaggerated rumor—burned by cigarette-started fire, but by a wood-burning heater which was used to heat the house. He and a Goroka European Medical Assistant had returned from a month of patrolling, lighted the stove and sat about it on an evening; and the roaring fire, aided by the strong wind, drew fire out the vent to the outside woodwork of the house, which ignited.

I found the Lufa native who had married Urai, the Kimi woman who died here of kuru. He confirms the story—adding that she developed kuru only a week or so after a return visit they paid to Amusa region, and remained there only a month or two before they both walked back to Lufa together, his wife not then requiring any support. Her kuru progressed to dysarthria, total loss of speech, and full incapacitation and death *here at Lufa!* An NMA now in Goroka is supposed to have seen her. I shall try to get his further cross-checking confirmation of the story when in Goroka.

14 September 1957   Numpuru Rest House, Frigano area

[23:2–24:5]   John and I drove into Goroka, a four-hour drive of about 45 miles, in the same jeep that brought Ian Burnet back to his station the day before yesterday. We arrived in Goroka at about 8:00 p.m., still in time to have a fabulous steak dinner at the one cafe in this highland metropolis of about 500 white population. . . .

We booked into the hotel by phone, got their one remaining room, and found dining in the cafe an assortment of Gorokans and visitors—none of whom we knew directly but almost all of whom we knew indirectly, it later developed. Thus, two Americans were first to be introduced. One was inebriated, incoherent, and spoke with such dysarthria that for a while I had difficulty believing that I was simply unable to understand him. He was apparently traveling about the world idly, as he had done for years, and seemed to have but one interest: cine photography (which, from his behavior and remarks, I judge cannot bear markedly sensitive editing nor very exceptional craftsmanship).

The second American was Father Bondar, a Slovak from Pittsburgh who is the Goroka priest and has been ten years in the Eastern Highlands of New Guinea. He is the priest who Vin had told me wanted to see me; and we were soon conversing garrulously. He is a working-man's priest, of and from the people, and a first-generation American born from Slovak parents. In spite of ten uninterrupted years in New Guinea, he gave the impression of having just left the United States—from the prominent way he revealed his Americanism with every remark and gesture.

Seated with these two Americans was a young chap of 23 or 24, serious and intense, who discussed cine photography so intently that I had but to conclude that he was part of the [Australian] government film unit for which we had prepared and waited in vain in Moke two and three months ago. It turned out that he was one-half of the team: the director. A middle-aged ethanolemic gent seated at another table proved to be his cameraman. The two informed us that they were ready to proceed by Landrover to Moke-Okapa the following day to film kuru. I had to tell them that it would do no good, for no one was there to work with them and that no—or very few—kuru patients would be on hand now. However, I promised to meet them at their boarding quarters after we moved into the hotel; and that, John and I did. Father Bondar, the "film pair," and a sundry crowd of unidentified (more likely identified but unheeded by me) others crowded into a small room where, over whisky and wine, I held forth on kuru for about two hours. The "director" and Father Bondar and the others kept me on the subject, a feat requiring little stimulation on their part; but the two inebriates (the cameraman and the American adventurer) kept drinking toasts of praise to me for my enthusiasm, work, integrity, dedication, etc. These were directed at trying to silence me and get the gathering down to serious drinking and idle chatter more conducive to drinking; but I and the others remained intent on the subject of kuru, and the two euphoric cronies left.

[24:32–47]   Sinoko (to get a chest film on him) and two of the Yar people who had accompanied us on patrol were the only people I took into Goroka with us. All the boys had so often pestered us about the trip to Goroka that I became somewhat peeved and left them behind, rationalizing that there were no accommodations in Goroka for them. Actually, I could have boarded all of them that I could carry in at the hospital, so it was a poor excuse. I fretted most of the way into Goroka, angry with myself for not having brought three or four of those who had never seen Goroka into the town. I was most foolish to have let myself be so disturbed, and not take this opportunity to widen their travel horizons.

The trip back, mostly in darkness, took over four hours; and the seemingly endless low-gear climbing up the twisting, steep, and treacherous road to Lufa dragged on and on. Upon arrival, I was delighted to receive a jubilant welcome from the boys, with no sign of ill will for my having left them behind.

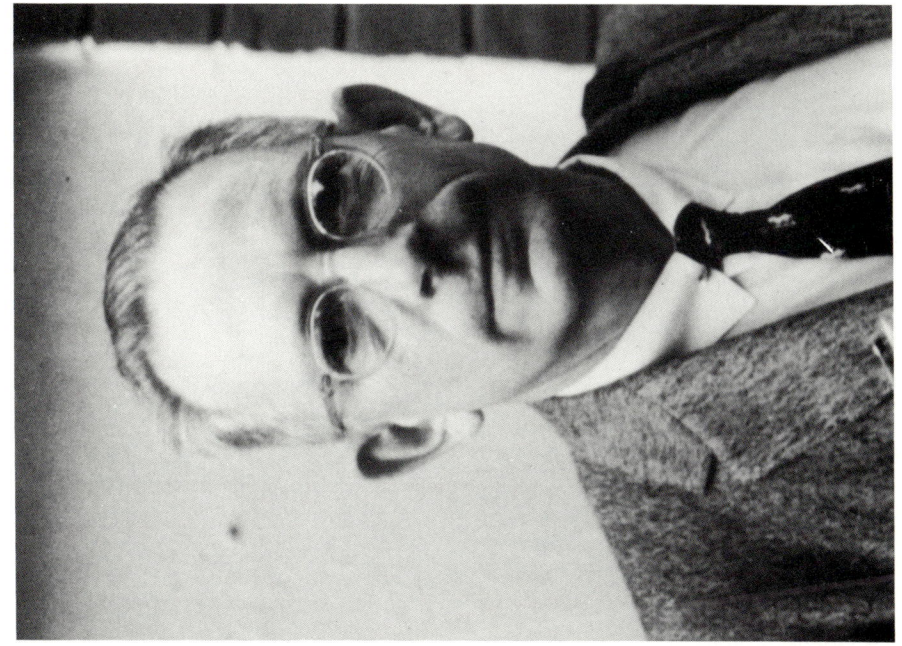

Figure 2. Joseph E. Smadel, M.D., c. 1959. (Photo courtesy of Ted Woodward.)

Figure 1. Sir Frank Macfarlane Burnet, c. 1957. (Photo courtesy of the Walter and Eliza Hall Institute of Medical Research.)

Figure 3. Vin Zigas, Jack Baker, and DCG in the only residence in Okapa in early 1957. Light was supplied by a Tilly kerosene pressure lamp; the dining table was used for desk work and microscopy, as an examination and treatment table, for lumbar punctures, and for autopsies. Human brain and viscera are in the enamel plate and wash basin on the table, and on the windowsill are specimen bottles for fixed tissue. Village census-books are on the folding field table in the foreground. [DCG-57-NG-IV-7]

Figure 4. DCG and Jack Baker in front of the *kiap*'s residence, the first permanent residence constructed at the Okapa Patrol Post. Andi stands in the shade of the porch in the background. [LH-57-PNG-36]

Figure 5. Six of the boys from the kuru region who served as translators and assistants in the kuru hospital and epidemiological work, photographed at the Okapa Patrol Post. *Left to right*: Morieto, of Arora village, Auiana group; Taka, of Aga-Yagusa, North Fore group; Masasa, of Aga-Yagusa, North Fore group; Tiu, of Agakamatasa, South Fore group; Waiajeke (Haus Kapa), of Moraei, Simbari Kukukuku group; and Anua of Ketabi-Purosa, South Fore group. [57-487]

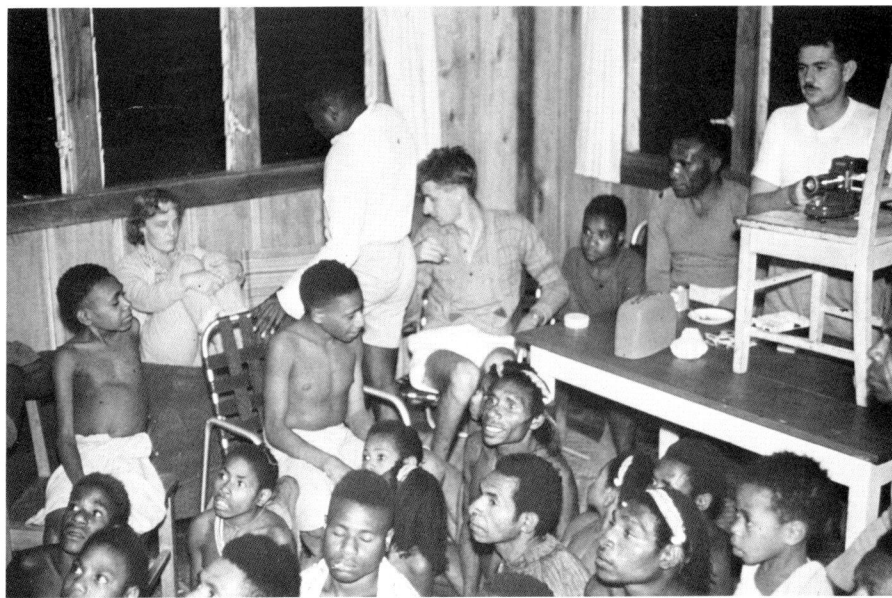

Figure 6. Kuru research workers assembled with visiting natives to watch a battery-operated slide show in the *kiap*'s house at the Okapa Patrol Post. Jack Baker at the projector; Miss Lucy Hamilton, seated on bed; John Berkin, on chair; Boy assistants of kuru research team seated on chairs, from left to right: Wanevi, Tiu, and Masasa. [57-524]

Figure 7. Kuru research team at *kiap* Jack Baker's residence with visiting Gimi warriors (drinking tea in background). Around the table *clockwise, left to right*: District Officer Bill Tomasetti (back to camera), John Berkin, DCG, Jack Baker, and Lois Larkin (back to camera). [LH-57-PNG-142]

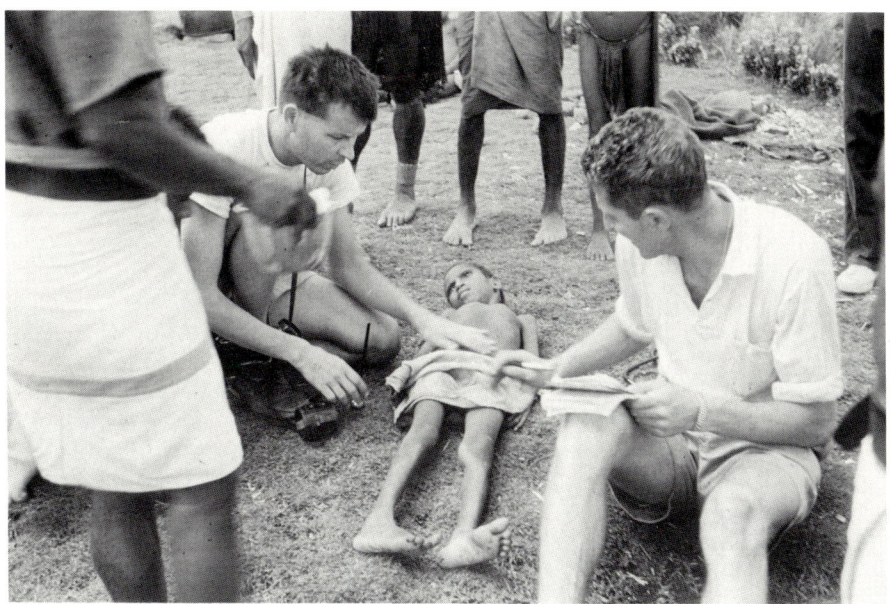

Figure 8. Dr. Vincent Zigas (*right*) and Dr. Carleton Gajdusek examining a South Fore child victim of kuru at Pintogori Kuru Hospital, Okapa Patrol Post. [57-369B]

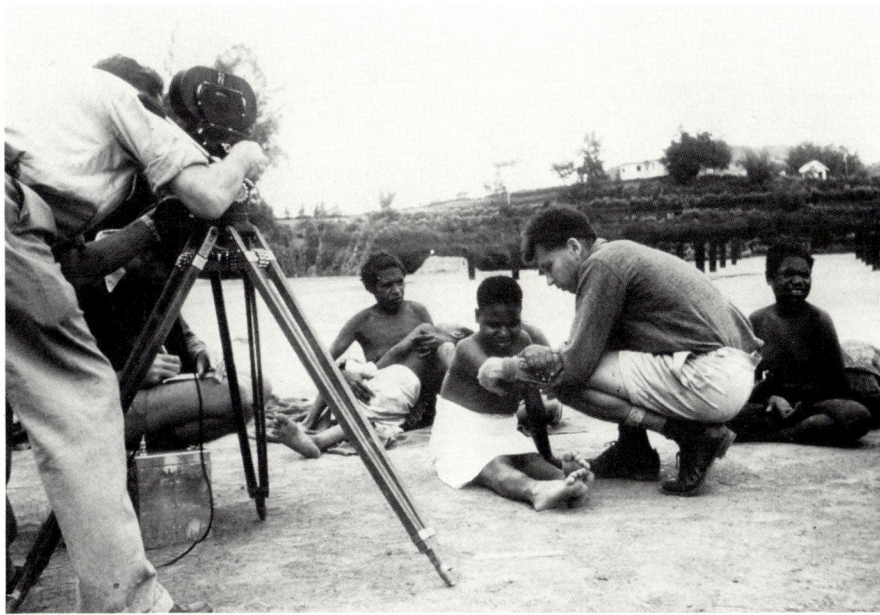

Figure 9. Dr. Gajdusek examining child kuru victim Nata for the Australian film team that visited the Kuru Research Hospital, Okapa Patrol Post, in September 1957 to make a medical film documentary on kuru. [LH-57-PNG-224]

Figure 10. Tosetnam (*left*) and Anua (*right*) at DCG's residence, *haus kuru*. [DCG-57-NG-XLIII-4]

Figure 11. Wanevi, before and after his braided and greased hair was cut. [57-354 and DCG-57-NG-XXXVI-24]

Figure 12. Pintogori Kuru Hospital, seen from *haus kuru*. Wasanamuti hamlet of Moke village lies in the bamboo grove on the ridge behind and above the hospital site. The two small buildings on the left were cooking houses, and the new kuru ward is in the center; on the right are the residences and cooking houses of the native medical orderlies (*dokta bois*). Taka and Masasa pose in the foreground. [DCG-57-NG-V-27]

Figure 13. The Subdistrict hospital at Kainantu, Eastern Highlands District, constructed by Dr. Vincent Zigas in the mid-1950s. Administrative office is in the foreground, surgery is in the center, kitchen and utilities in the rear, and ward on the right. [57-852]

Figure 14. Kosinto (*standing, far right*), and (*right to left*) John Berkin, Lucy Hamilton, and Lois Larkin, with faces painted for a *singsing* at Henegaru village, Keiagana. [57-805]

Figure 15. Lois Larkin and Kosinto at Pintogori, Okapa Patrol Post. Miss Larkin was a technician in virology from the Walter and Eliza Hall Institute of Medical Research, Melbourne. She came to Okapa to help in the collection and processing of tissue specimens for laboratory studies, principally for microbiological purposes. Kosinto was a translator of Fore, Keiagana, and Kimi languages into Pidgin, and was employed at the Okapa Patrol Post. His Keiagana village of Ke'efu lies on the Fore-Keiagana border area, where most people are bilingual. Because of warfare, his people had fled when he was a small boy to join Kimi-speaking groups where they had found sanctuary. He remained there throughout his childhood, learning to speak Kimi fluently. [57-463]

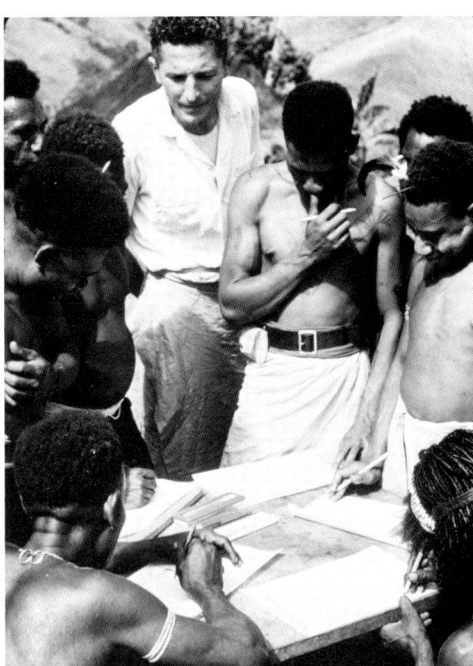

Figure 16. Dr. Vin Zigas looking on as Fore men and youths draw pictures at Abomatasa Village, North Fore. To his left, holding white pencils, stand Native Medical Orderlies Tarangau and Tiu. [57-336]

Figure 17. Jack Baker carries his dog Kuru across a river en route south to Awa villages from Okapa on an early kuru epidemiological patrol. [57-861]

Figure 18. Jack Baker resting on the stones in the Puruya River, Simbari Kukukuku, during the patrol from the kuru region through the Kukukuku area and south to Papua. Kuru assistants Tiu and Waiajeke are with him. [DCG-57-NG-XXVII-6]

Figure 19. Kuru assistants during the patrol through the Kukukuku region. *Front to back*: Tiu, Taka, Masasa; Wanevi stands to the right. [DCG-57-NG-XXV-2]

Figure 20. Five boys in Henegaru village, Keiagana, aged approximately 5 to 10 years, carrying camera equipment on a kuru research patrol. Boys of this age readily volunteered to help carry supplies for patrols between friendly villages. [57-727]

Figure 21. The visiting Adelaide medical team at the Ke'efu, Keiagana Rest House. Dr. Harry Lander examines a youth, with Dr. Donald Simpson looking on. Members of the Papua and New Guinea Constabulary stand by on the left, next to *Luluai* Apekono of Aga village, Keiagana. Three of the kuru assistants (*seated, left to right*): Taka, Waiajeke (wearing a hat), and Tiu. [DCG-57-NG-XLIV-35]

Figure 22. The South Fore village of Agakamatasa (village of highest incidence in the early years of kuru study), showing men assembled before a men's house (*wa'e*) overlooking a row of eight women's houses (*umu*), below and to the left. A bamboo fighting stockade is beyond them. It was erected to prevent surprise attacks from the ridge below by the Kukukuku people from across the Lamari River valley. The forest around the village has been cut for gardening. The village stands on a knob atop a steep-sided ridge, affording strategic advantage in warfare. The Agakamatasa people had fled from Mugaiamuti and Wanitabi villages to the slopes of the Lamari River to get away from kuru, which had more than decimated them there—but the disease came with them. [57-169]

Figure 23. *Luluai* Igaga of the South Fore village of Agakamatasa and his villagers. The *luluai* was a government-appointed leader of the "census units" which the Australian administration tried to establish to define social units of people who lived, worked, and shared garden sites and land with each other. Such social units usually had 100–200 individuals, and rarely exceeded 300 persons. The traditional fight leader was often chosen as *luluai*, for he had some authority over many of the young men, but there was no traditional leader who held authoritative sway over other aspects of life and who could direct or command the services of all the men, women, and children. [57-170]

Figure 24. Six adolescent Awa warriors of Agamusei village in their boys' house. The earth floor is covered with debris from chewed sugarcane. Bows and arrows stand against the wall beside the boy on the right. The braided wristbands and armbands they wear indicate their initiation grades. [57-890]

Figure 25. A Kimi family beside their mother, who has advanced kuru, outside the woman's house. Five of the patient's children cluster about her as she is supported by her husband and her sister. [DCG-57-NG-I-1]

Figure 26. A kuru victim from Moke village, North Fore, at work in her sweet potato garden. She must keep herself from toppling over by using her digging stick for support. [LH-57-PNG-239]

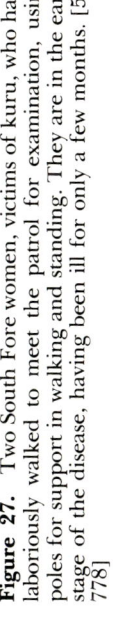

Figure 27. Two South Fore women, victims of kuru, who have laboriously walked to meet the patrol for examination, using poles for support in walking and standing. They are in the early stage of the disease, having been ill for only a few months. [57-7778]

Figure 28. Nine kuru patients from a total population of about six hundred in four nearby villages assembled on one day at the rest house in the Purosa valley. They include seven women, one late adolescent male (*left*), and one prepubertal boy (*foreground*). All these patients died of their disease in less than a year. [57-573]

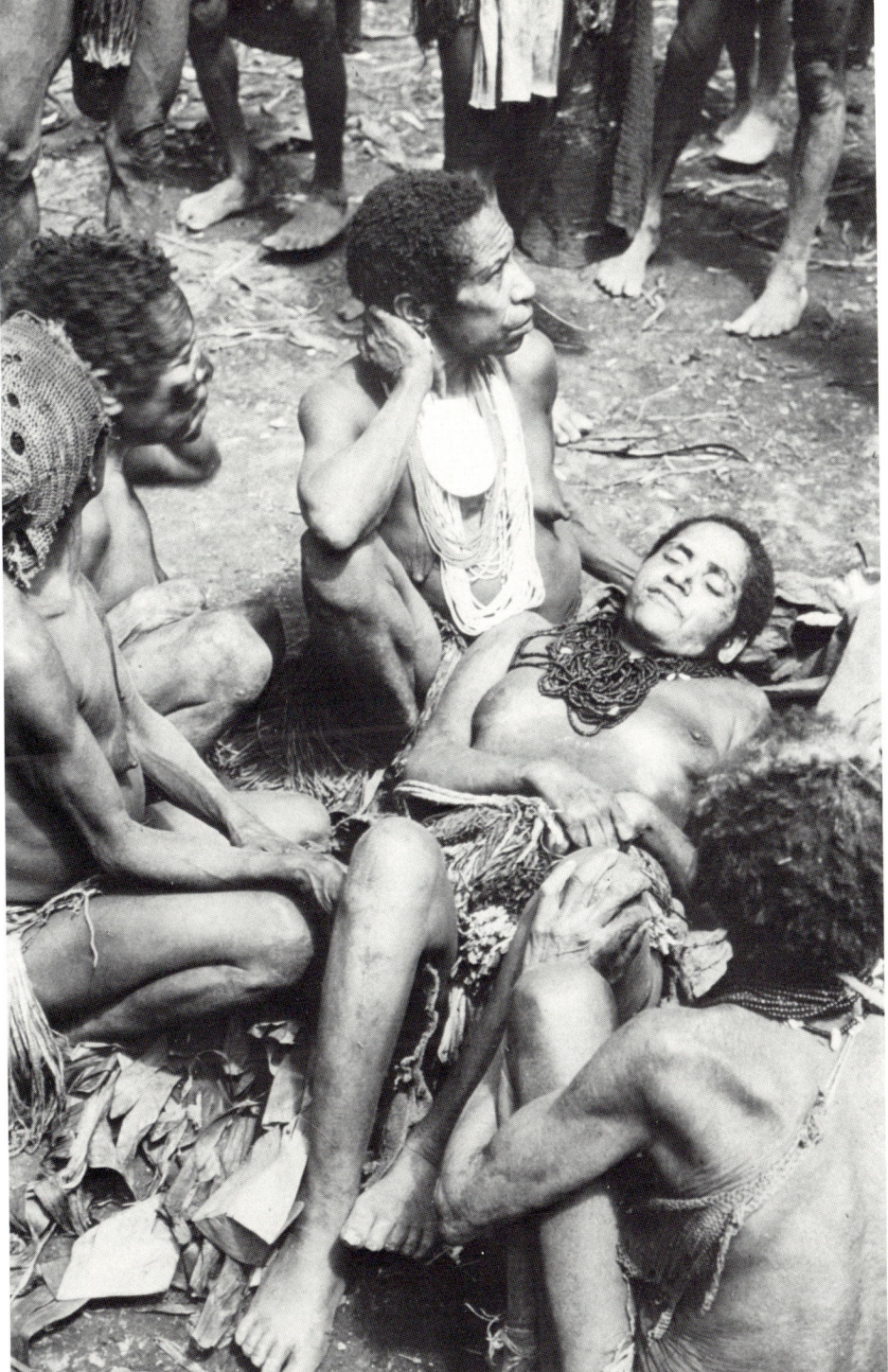

Figure 29. Kinswomen mourning an adult woman who has just died from kuru. Four women cluster closely about the body while children, whose feet are visible in the background, look on from nearby. Male kinsmen had assembled for ceremonial distribution of sugar cane and other foods, marking the death. [57-783]

Figure 30. Eight preadolescent children with kuru, four boys and four girls, on the porch of *haus kuru*. They were all assembled for twenty-four hour urine collections while on treatment with calcium versenate and dimercaprol for heavy metal excretion studies. All these children died within six months of this photograph. [57-507]

Figure 31. Kuru victim Yakurimba, a boy of seven years from Agakamatasa village, South Fore, arriving at the kuru hospital at Pintogori after two days of being carried by his agemates on a stretcher across the ranges to Okapa Patrol Post. [57-197]

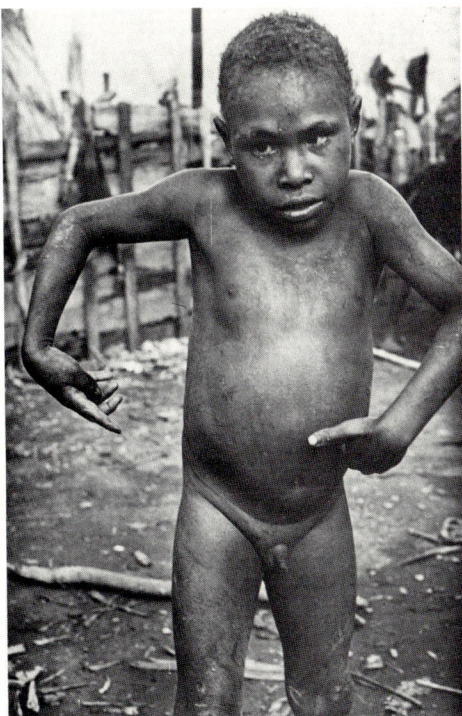

Figure 32. Yakurimba, caught by the camera during an athetoid jerk precipitated by his attempts to remain standing. [57-117]

Figure 33. Yakurimba held in his father's arms. [57-165]

Figure 34. A kuru victim of Ibunarai hamlet, Urai village, South Fore, in the terminal stages of rapidly advancing disease. She is no longer able to sit without support and is unable to articulate responses to questions, although she hears and understands them. She is obese from over a month of sedentary life accompanied by overeating, probably from the bulimia associated with the hypothalamic lesions seen in kuru patients on postmortem examination of the brain. [57-460]

Figure 35. An adult male Fore victim of *tukabu*. A few hours earlier this man was beset by a group of murderers who tried to avenge the kuru sorcery it was thought he had performed. They had beaten his renal areas with stones and bitten his neck over the left jugular area. His attackers fled when his screams aroused his fellow-villagers, and he survived the assault, unlike most such *tukabu* victims. [DCG-57-NG-I-30]

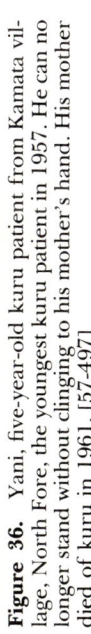

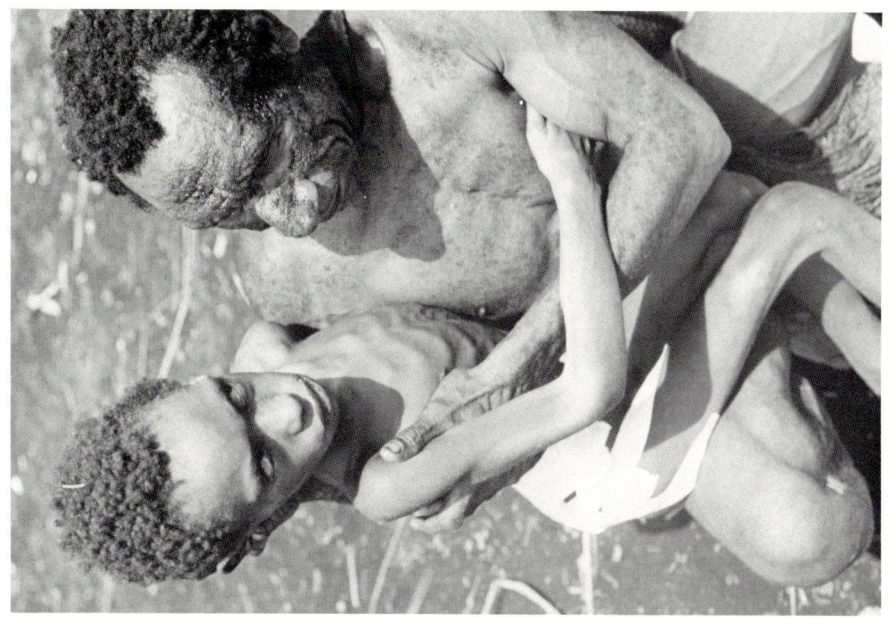

Figure 36. Yani, five-year-old kuru patient from Kamata village, North Fore, the youngest kuru patient in 1957. He can no longer stand without clinging to his mother's hand. His mother died of kuru in 1961. [57-497]

Figure 37. Kuru patient Amakiora in the terminal stage of the disease. He is held by his father. [57-1135]

Figure 38. Carriers crossing the Aziana River between Agamusei village of the Awa people and Ureba village of the Auroga Kukukukus on a rattan vine suspension bridge built by the patrol. This kind of bridge was introduced into the area during the course of kuru epidemiological patrolling. [DCG-57-NG-XX-49]

Figure 40. The Kukukuku patrol pauses at a small tributary of the Wugamuwa River to erect a bridge that will enable the carriers to cross with their cargo. Jack Baker is on the far shore. [DCG-57-NG-XXI-7]

←

Figure 39. Carriers crossing a bridge that spans the Wugamuwa River between Tchaiorogoro and Anji villages, Wantekia Kukukukus, en route to Anji. [DCG-57-NG-XXV-19]

Figure 41. The Kukukuku patrol camp at Tchaiorogoro, Wantekia. Temporary shelters are being roofed with *kunai* grass. A flagpole bears the Australian flag. [57-966]

Figure 42. Women and children stand at some distance to watch the Tchaiorogoro camp being built. [DCG-57-NG-XXIII-25]

Figure 43. Two sentries stand guard along the trail near Ureba village, Auroga, where they have been watching the Kukukuku patrol approach their village. [DCG-57-NG-XXII-6]

Figure 44. Jack Baker leads a patrol up *kunai* slopes toward Kasokana village in the North Fore. The Lamari River Valley is in the background. [DCG-57-NG-XIII-6]

Figure 45. *Haus kuru*, DCG's residence, under construction at the Okapa Patrol Post; the frame is nearing completion. [DCG-57-V-1]

Figure 46. Fore and Keiagana people from the Okapa area carrying *kunai* grass and poles for construction of the *haus kuru*. [57-244]

Figure 47. Men, women, and children of Agamusei village, Awa linguistic group, assembled at the camp of the Kukukuku patrol to barter sugar cane, bananas, sweet potato, taro, and *pitpit* in exchange for trade items brought by the patrol. [DCG-57-NG-XX-18]

Figure 48. Sergeant Malekor supervises the distribution of food to carriers for the Kukukuku patrol. A share of sweet potato has already been given to each man. The tubers are disappointingly small to those who are accustomed to *kaukau* in the Okapa area, and the ration is sparse. [DCG-57-NG-XXVII-22]

Figure 49. Carriers cross an elaborate bridge across the upper Lamari River between the villages of Irakeia and Mobutasa of the Awa people. The bridge was constructed by the patrol from a fallen tree. [57-879]

Figure 50. Carriers negotiate the slippery stones at the headwaters of the Puruya River, en route to Iwane village, Simbari Kukukukus, from Tchaiorogoro village of the Wantekia Kukukukus. [DCG-57-NG-XXVII-4]

Figure 51. Kuanimugu hamlets on a ridge as seen from Igopiji village, Auroga Kukukuku. The *kwal anga* (men's house) crowns the summit between the two groups of *ambel angas* (women's houses). [DCG-57-NG-XXIV-35]

Figure 52. Members of the Kukukuku patrol visit the almost-deserted village of Ureba (Auroga), above the first camp among the Kukukukus. [DCG-57-NG-XXII-4]

Figure 53. Wantekia men and boys in their *kwal anga* (men's house) in Anji village. The second man from the left is the village headman, who wears a blue plastic ring given to him by DCG and Jack Baker. Reed sporans are suspended from the roof above them and fighting shields, bows, and arrows rest against the wall behind them. [57-989]

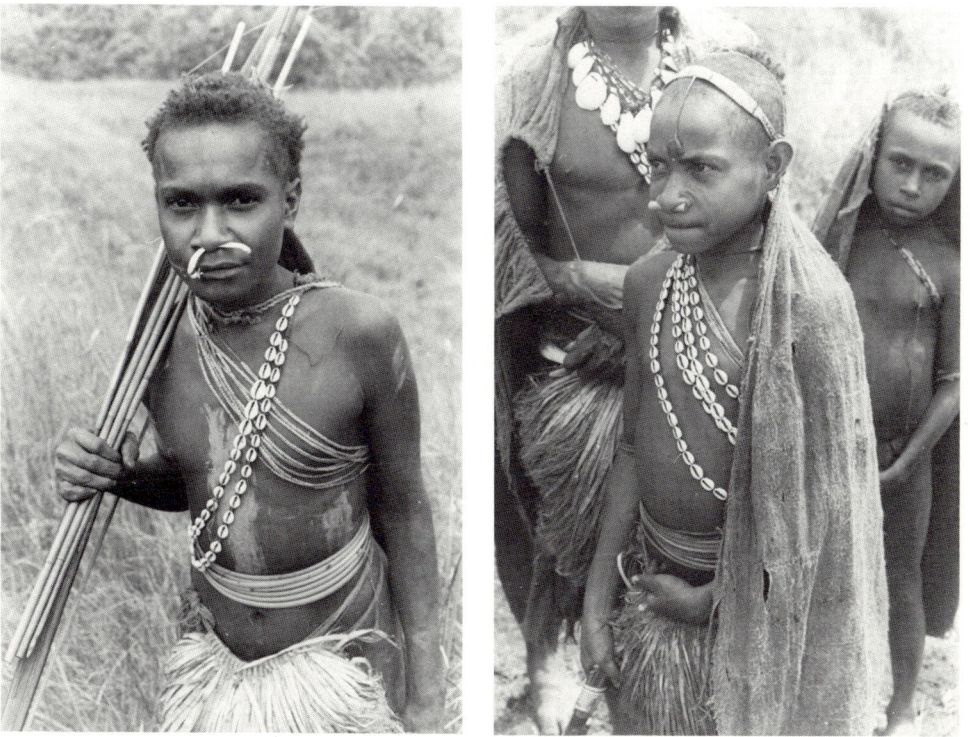

Figure 54. Two second stage initiated youths (*kawatnye*) at Tchaiorogoro village make cautious first contact with the patrol. [DCG-57-NG-XXI-27, DCG-57-NG-XXI-40]

Figure 55. First and second-stage initiated Tchaiorogoro youths. [DCG-57-NG-XXVII-26, DCG-57-NG-XXVI-6]

Figure 56. Four adult Auroga warriors showing a range of responses on their initial visit to the Kukukuku patrol camp. [DCG-57-NG-XXI-44, XXIII-7, XXI-4, XXIII-23]

Figure 57. Yar Pawaiian women of Weme village making sago. First they prepare pulp from the pith of the sago palm tree trunk by beating it with adze-shaped beaters (*top left*). Then they carry the pulverized sago in net bags to a washing trough, where water is poured over it (*top right*). The trough is made of a frond from the sago palm. At its lower end on the right, a sieve made of sago leaf-fiber netting catches the chaff and permits a sago starch suspension to be squeezed from the wet pulp (*bottom left*) and to flow into a collection trough made from a folded sheet of heated bark of the sago palm (*bottom right*), where the starch sediments and the water overflows. [DCG-57-NG-XXXI-34, 57-1054, DCG-57-NG-XXXIII-6, DCG-57-NG-XXXIV-23]

IV.
Bush Correspondence
6–11 September 1957

I. BURNET to DCG   6 September 1957
Dear Dr. Gajdusek:

I have been in contact with Jack Baker on the radio and have been told to expect you on Sunday. Neither Jack nor I is very sure of just where Mani is, but I expect my Constable and Interpreter will find you eventually.

My own supplies are a little bit short, but there should be enough going forward to see you to Lufa without any trouble.

I look forward to meeting you again.

Yours faithfully,
Ian Burnet

HAMILTON to DCG   Saturday
                                                      [7 September 1957]
Dear Carleton:

Received your letter last night. Your first letter arrived Thursday morning and was put away until I arrived up here from Moke at about 5 p.m.; I began getting the things you wanted together as soon as I found the note. Then Jack and Vin arrived when it was pouring with rain, while I was rummaging in your drawer for socks. They were both out of sorts; they had had a very difficult trip, and from what I gathered had to push the car as far as it carried them. With all the rain, I really wasn't expecting them, as the bridge has been washed out for some time and the river's too high to ford. Jack said he had tried every day since Sunday. I had given up hope of going to the race meeting, and was getting prepared to go to a *waswas* [baptism] at Keefu with the rest of the Moke crowd.

The outcome of Jack's and Vin's arrival was to stop everything. The next morning, Jack radioed Lufa for things to be sent in from there.

I had hoped you would be back this week, as I wanted to talk over my work with you before you go to the Kukukuku. I have finished about as much as I can here; I have got pretty well all the foods the people eat here and have checked them with what all the kuru patients at the hospital eat, and with Kosinto, Afakanu, and a lad from Tarabo of Keiagana, and with some Kimis and the Kukukukus who were in, and with some Kamano people who are living at Moke. It now remains for all this to be put into typing, so that others can read it. However, I am not sure whether you would like me to do further investigations as to the food eaten in areas outside the kuru area. I think it might be interesting to do some investigations at Agayagusa.

I have made a map of all the Moke ground-names and investigations of family histories of everybody in Moke, and entered the kuru deaths on the map. They seem fairly evenly scattered throughout; and it is hard to say where they might have picked it up if it comes from the soil, because they have their gardens all over the place—that is, on different grounds from where they live. In addition, they seem to move their houses from place to place quite frequently, judging by the numbers who have died—or, I should say, have contracted kuru—on land where there are no houses now.

Jack [Baker] and I leave today about midday to go in to the races. Vin left yesterday morning when he found that there were no new kuru cases here and hardly any drugs. I have moved all my things from Moke. I am undecided what to do; and if you are going straight out on the Kukukuku patrol immediately after getting in, you won't have much time to go over my stuff—but I do not want to leave without your having done so.

Haven't heard anything from the botanist in Lae since I sent the last lot of specimens in. Jack Reid is going in to the races too, so I might just go back to Lae with him for a few days and check up with the botanist.

Read your letters with interest, and wish that I were with you to see all the strange things you have been eating—though I don't relish the thought of all the hiking in the rain and the precarious bridges. The wicked-looking insects arrived, and I gave them a whiff of ether. Are they eaten?

Sorry you are not coming to the races with us. Cheers to yourself and John.

Lucy

J. BAKER to DCG and BERKIN   7 September 1957
Dear Carl and John,

Quite a foul-up! Your note was not given to Lucy till after five Thursday in pouring rain. Vin and I got in 6:30 p.m., just after dark, to find Lucy wandering around—soaking wet—trying to organize gear.

Read your note and, working on your announced itinerary, decided that it would be quicker to radio Lufa to send gear in from his [Ian Burnet's] side, since we couldn't get it to you till Saturday night. This was done.

Last night—Friday, 10 p.m.—your second note arrived indicating you may wait for gear, so am sending it despite my opinion it may be too late to be of any use.

Re your boots, etc., I suggest that you go by car from Lufa to Goroka, do your necessary shopping, and come home via Kainantu.

Have a happy birthday in Goroka. The *kai* [food] at the pub is A-1, much better than sago.

The drugs you require are not available. This Aid Post Orderly is treating everything from tertiary syphilis to minor abrasions with zinc cream. *No* antimalarials, *no* aqueous penicillin, *no* oily penicillin, *no* sticking plaster, and so on down your list. Trade-goods etc. are there, plus a few home comforts.

Yours,
Jack

P.S. Anything you require to buy in Goroka can be placed on my account either at Burns Philp or Buntings stores.

DCG to J. BAKER, ZIGAS, and HAMILTON

Mani Rest House
Central Kimi
8 September 1957

Dear Jack, Vin, and/or Lucy:

We are behind schedule. Anua developed a violent meningitis, did not respond to sulfa and penicillin and vomited most of the sulfa, and only now seems to be rallying—our having plotted a therapeutic attack based on the total of 3.5 gms of broad-spectrum antibiotics of various sorts we had in our dwindling supply. I hesitated carrying him along, but knew it would be better than chancing the long trip back to Moke; and now that we learn we had more drugs than Moke, we are most grateful that we decided as we did. He is by no means cured. He has lost well over 15 pounds in a brief illness; but I think we may now get by, his vomiting having ceased. We waited here at Mani from where over 1500—nay, 2000—people are censused, and we have managed to see over 90% of the censused group in the past two days of medical inspection of the dozen-odd lines. Our penicillin has just held out; and thus, Kiap, we can now report that the two kuru-epidemiology patrols to this region have medically inspected your Kimis fully. In certain lines we managed to round up more than are censused. We cross over to Hegeteru (Kimi) on the Lufa side tomorrow morning.

I have been keeping maps, field-notes, and compass-sights, which will help in getting a first-draft map ready when I get back and go over the data with you. Four Yar people are still with us; you will know about them if you have had time to read my verbose accounts in the previous letter. These sago-eaters are "putting on the dog" for the savage Kimi, dressing in long trousers and sport shirts; one, disdaining the high bridge over the formidable Yani at Uvai, simply jumped in and swam across in spite of the current. They are, in spite of their appearance belying the fact, in many respects "more primitive" than the Kimi or Fore; and their women dress primitively, show none of the sophistication of the men, and run off into hiding in the bush like first-contacted people.

Please ask Mick Foley to dig up, as soon as possible, all maps and government patrol data he can get from the Beara Patrol Post in Papua (Gulf District) and all accounts of patrols north to So'o. I think none have gone beyond, to the home of the Yar people we have with us.

Your supplies arrived today from two directions. Thus, a policeman, taking the "wrong" route from Lufa, reached us (he came via Frigano, Kagu, and Henegaru and, had we been on schedule, would have missed us completely). Ian [Burnet] sent us enough to get by in a pinch; but it was sugar we most wanted, having gone a week on black tea and coffee without sugar—which to both our palates is "revolting" (in deference to Lucy's tongue). We are sending Aneti back with a large blood collection documented by the enclosed sheet. This is a most important collection; and since sterility could not have been perfect, and since 24 hours' transport (unrefrigerated), at least, is required, incubation of what bugs are present will take place. Therefore, good and continuous refrigeration after receipt is very important. Even keeping them iced en route to Kainantu is advisable from now on. It is a very good collection, and I would be most happy if most of it will be saved. The Yar bloods are, I trust, already all off to Hall Institute. If not, please rush them off, for on the entire patrol we have managed to get only these three dozen Yar specimens. We can always bleed the four Yars who are with us now a second time, however. Here are Mani bloods from Amenetu clan; no protest or difficulty bleeding anyone was encountered, and the natives evinced some disappointment when we ran out of bleeding-containers. [81]

We are well supplied now, the supplies having arrived; and since Sinoko did not come, we shall keep the three NMAs we have with us for the rest of the trip. We shall be in Hegeteru tomorrow, in Mengino the following day, and, with luck, in Lufa on Wednesday—certainly on Thursday. Please let Ian know by radio of our delay but planned arrival and route. It is the "southern" and not the "northern" route.

We shall want to make Goroka for shopping, and then either come back via Kainantu or return to Lufa and walk back with our boys along the "northern" route. We shall let you know, via Lufa radio, of final plans. Please, Lucy, hang on until our return, for I want to go over your stuff with you in detail and will take time for that alone when I get back.

We are really all set now and will patrol in luxury. It was really essential to stay here for two days, for we could never have seen the over-1500 people in one day. Anua has, furthermore, picked up for a further stretcher trip, and thus we should be able to leave at daybreak tomorrow to cross the Divide.
[82]

We cannot take any further blood specimens, since all bleeding-equipment is now exhausted. That is good, for the Kimi collection—should this lot "survive"—is now done. We shall have plenty of "survey" bleedings to do back in the Fore and Keiagana before I leave, but that can wait. If you have contact with Kainantu, or if Vin can find one anywhere, do all you can to get your hands on a scale suitable for weighing schoolboys and other children. Even if it is not fully accurate, we can calibrate it against postal weights in Kainantu or some such thing.
[83]

Carleton

DCG to BERKIN (at Gono)   Lufa Patrol Post
                                                                  11 September 1957
                                                                          8 a.m.

Dear John:

I passed the mission station at 4:59 a.m. by your watch and made the walk to Lufa in exactly three hours. I was walking strong; I made four miles an hour most of the way, I suspect, and I made no stops. Thus, it is a good three-hour walk; and four hours should, preferably, be allowed for it. There are three short cuts, cut off the main road through heavy bush to regain the road; but these are short and need not be taken. The "gorge" is crossed by a good trail which requires, as reported, about 15 minutes to descend and reclimb. From the other side of the gorge from you, there remains 30–45 minutes further walk on to Lufa along which the trail deteriorates considerably. This, then, is the route. Nothing hard.

I arrived here in a rush to get a radio sked [scheduled radio consultation] with Goroka and with Moke, to find that Ian [Burnet] had left yesterday for Goroka and plans to return either today or tomorrow. That was fine, I thought, and rushed to the radio transmitter to make the sked myself—only to find no battery; and a rapid search revealed that Ian has taken it with him to Goroka. Thus, no Ian, no sked!

A note from Ian states that he will return today—at the latest, tomorrow. Thus, it is hard to decide what to do. The native constabulary executive at this post, a corporal, says he can feed our cargo-boys and entire crew well; and, since it is a good day's cargo-trip for them here, I suggest you plan to send everything here as soon as convenient. I shall await the car and Ian—and if it comes today, try to hold it until tomorrow to go out with you to Goroka in it. You must not walk today! The thrombophlebitis and lymphangitis and lymphadenitis, and the septicemia that you have had, may flare up if you do. If you do not mind being carried, perhaps—if the weather is fine—it would be advisable to be carried in. The road is good; hip motion or foot motion would hurt your condition, but a stretcher trip would not. A three-hour trip, made into a slow six-hour walk in good weather would be OK. Otherwise, if you are up to it, come in tomorrow. Since I think that even tomorrow is too early to walk on your thrombophlebitis and even if you come tomorrow you should be carried, I suggest resigning yourself to being a litter case and coming in as such. Today, if weather permits; otherwise tomorrow, leaving early. Should the jeep arrive today, I think we can hold it. If it does not come until tomorrow, you will be here in time for it.

Moke has it all over Lufa, except that they are alike cold and foggy. I have not been able to see more than a few hundred feet through fog since leaving Gono at 5 a.m.

As for our cargo-boys, police, NMAs and boys, and the Yarevana "Papuans" we have with us, I would send them all in to Lufa when you are ready to leave. If Ian arrives early, I may walk back today and either find you there at Gono or meet you on the road. There is really no hurry, John; and if you are not feeling well, ride it out a few days at Gono and I shall show up again either today or tomorrow. If, however, today turns out to be fine (and now nearing 9 a.m. it is still a thick soup through which I can make no forecasts, and I am huddled in Ian's dressing gown—feet freezing, my clothing being all wet) and you are fine, then come in today, sending in all our cargo and crew. Should you do so, come via litter and do not try walking. I gave Sinoko further penicillin for you. Take two injections of 500,000 units today: one already given, I presume, and another at noontime or thereabouts. That should be sufficient. Keep up with sulfa, taking two pills (1.0 gm) at a time for five doses today.

The books up at the Gono *haus kiap* where I slept (Homeguei will bring them to you) can all be returned to their respective L/Ls. Please "sing out" for the others, putting into them a note identical to the brief note in each I left behind. If you see the L/Ls yourself, ask each, with the aid of an interpreter, whether they cannot—if they try hard—recall a single past recovery or death from kuru (*uruna*) in the region. They all know what it is. If they report any, record name, relatives, location, etc. All I inquired of said "No; emphatically no!"; I am sure they will say the same to you. However, press the point; insist that they search their memories, try them out on whether they recall the Kimi woman from Revesavipi (from our side) who died of kuru at Lufa four years ago who was married to a *turnim tok* [translator] here. This should take no more than a half hour or so; and, since I do not know just which L/Ls I have already interviewed, any you can see should be questioned. I feel rather sure you will have very little to record, and it will not take long. Finally, ask a few if they know of or have seen any congenital tremors in this region. That is all.

Sorry to load you with work. The chances are even that I may be back today, for everyone says Ian will arrive shortly. If he does, I shall probably come in to Gono to come back out with you. Boys can run the trip in three hours, and there is plenty of traffic on the road. Thus, send notes often and freely. Let me know when you get this. It goes off at 9:15 a.m.

My gratitude to our wonderful American hosts. I have already lost their last name, so please record their full names for me; I would like to record them in my address file. In the event I do not see them again, thank them heartily for me. Good luck.

Carleton

V.
General Correspondence
4 September – 4 October 1957

ANDERSON to DCG   4 September 1957

Dear Carleton,

I have just received two samples of urine from Tovepi and Atogori, and they will go across for amino acid analysis.

We have also received two bloods on some of the (?)mumps cases and the list of 101 people. Four paired sera are being examined at the Royal Children's Hospital for mumps antibody, and I will let you know when the results come through. I also have your letter of the 22nd August; Lois is looking after all the samples for Simmons of C.S.L.

I am sending herewith one more pathology report, and that completes the reports available to me. CSIRO has still not yet told us what results they have obtained with heavy metals in blood on a quantitative basis, and as I know they are very busy I do not like to hurry them. [84]

Could you please send me the ages of the patients supplying the bloods which you sent to us for heavy-metal analysis of the 12th August?

Kavuiompa and Anona were put through suckling and adult mice and eggs, but yielded no virus.

Yours sincerely,
S. G. Anderson

BURNET to DCG   10 September 1957

Dear Carleton,

I hope things go well with you. The stimulus to write was contained in a letter and reprints that I had from Telford Work at Poona. You may have heard this already but it was new to me: that they have isolated a new type of arthropod-borne virus in India related to Russian Spring-Summer and producing a severe generalized disease with haemorrhages. The reference is "Kyasanur Forest Disease: A Preliminary Report on the Nature of the Infection and Clinical Manifestations in Human Beings," *Indian Journal of Medical Sciences*, II:619, 1957.

The clinical symptoms include severe muscular pain and prostration; enlarged lymph glands; papular eruption on the palate; and bleeding from nose, gums, stomach, or intestines. Apparently the virus is fairly readily isolated from blood and CSF.

Lederberg has settled into the Institute for a month or two and is being extremely useful.

With kind regards,

Yours sincerely,
F. M. Burnet

SMADEL to DCG 12 September 1957

Dear Carleton:

Dr. Bailey, Dr. Klatzo, and I have just looked at your Kodachrome movie film which was taken at Moke and Kemiu on August 7 and 8. The photography is excellent. The color is very fine, except for one short sequence which was overexposed. Whatever the difficulty, it was promptly resolved by you.

One generalization which must be remembered by all who transfer from stills to movies is worth repeating: don't pan! It is almost impossible to swing a small movie-camera for a panoramic view if it is held by hand. Furthermore, even with a tripod, the beginner almost invariably swings the camera too rapidly. This produces a jiggling effect when the film is projected, and may produce seasickness in the audience. If you must have a panoramic view, use the tripod and rotate the head one-quarter to one-half as rapidly as you think it should be rotated. If you hold the camera in your hand, operate it for ten seconds, stop, move across one-half a field, operate for another ten seconds, stop, etc.

Take all future movies at 24 frames per second!! It is almost impossible to transform 16-frame movies into 24-frame movies; and practically all good sound-movies are taken at 24-frames at a cost of about 16 cents per foot. However, such a product is not suitable for general use or for teaching; it is essentially limited to home use.

Yesterday I sent you a carbon-copy of my letter to the editor of the *New England Journal of Medicine*. I am sure that you could have written more intelligent legends for the three photographs I sent the editor. However, only one of the four patients illustrated in the photographs has a clinical record here. Let's have the rest of the clinical records, beginning with those of the fatal cases.

Dr. Klatzo showed colored slides illustrating the CNS lesions in the patients. The degenerative changes are really interesting. He will write you an informal note giving you a tentative description.

Everyone here continues to be greatly interested in the disease you uncovered. The leaf we received from you has been identified by Dr. Bernice Schubert, a botanist with the Department of Agriculture at Beltsville, as belonging to the family Araceae, genus *Homalomena*, species as yet undetermined. She was fortunate in being able to consult Dr. Jacques Barrau, a botanist with the South Pacific Commission who is an expert on plants of the New Guinea area. He states that the roots of the plant in this family are often used in native medicines in Malaya and Indonesia. It is interesting that alkaloids are not known to occur in this family.

Dr. Schubert asked if it would be possible to receive from you one or two complete plants, in order that she may have a flowering spike and a fruit.

Dr. Evan Horning is interested in undertaking toxic studies on the plant which is reputed to cause tremors similar to those of kuru. He suggests that you obtain one and one-half to two pounds of dried leaves. In order to do this, approximately five to ten pounds of wet leaves should be collected. If the weather is dry enough, the leaves can be spread out and dried by sunlight, after which they may be placed in the container for mailing. The customary way of drying the leaves is to place them on a screen six inches above 100-watt light bulbs. If the weather is too damp for drying the leaves in the sun, perhaps you can arrange some way to heat them; they should not be heated to a temperature which would scorch the leaves.

It may be helpful if a pound or two of the fruit of the plant could also be sent along with the whole plants referred to above. Send them back by air express converted to a government bill of lading. Special tags which will expedite the passage of the material through Plant Quarantine are enclosed for you to tie on the packages.

Best wishes.

Sincerely yours,
Joseph E. Smadel, M.D.

P.S. Dr. Horning has also just requested one or two pounds of the rhizomes, if it is possible to obtain them.

KLATZO to DCG   13 September 1957
Dear Dr. Gajdusek:

I had the privilege to receive, through Dr. Smadel, six brains from the kuru cases which you sent to the National Institutes of Health. My associate Miss B. Garry, M.S. and I have been studying, during the past few months, the pathology of these cases. By now you must have received our preliminary findings on the first complete case (Yabaiotu) we have had the opportunity to examine. The incomplete survey of the remaining cases reveals that all of them show pathological changes within the CNS; however, the intensity of the process varies from case to case. The complete and thorough workup of all the cases will undoubtedly take some time, since a great variety of techniques must be used, and topographically all the anatomical units within the CNS should be screened as completely as possible.

The brains generally arrived in a good state of preservation—except Natameia and Aranaka, but even these two brains are still usable for histological studies.

If I may express a suggestion, it might be better to fix the brains somewhat longer (3–4 weeks) at your center (brain immersed in 15% formalin, suspended by the basilar artery, corpus callosum cut), with possible exchanges of the formalin several times. The other thing I wanted to ask you is whether it would be possible to get more complete spinal cords and dorsal root ganglia (also a few more peripheral nerves), to have the pathological material completely covered.

We have received the clinical histories on four cases (Kinao—brain not available, Yabaiotu, Natameia, and Yoeia) and we are anxious to have the histories on the remaining three cases (Inome, Mulinapa, and Aranaka). This clinical information is very important to us for pathological investigation and for future publication of these cases.

I am afraid I am unable to give you any useful leads as far as the etiology of this disease is concerned. It seems to be definitely a new condition without anything similar described in the literature. The closest condition I can think of is that described by Jakob and Creutzfeldt. However, in cases reported, the neuronal degeneration was most intense in the cerebral cortex; also, no cases involving children or adolescents are mentioned. The etiology of Jakob-Creutzfeldt condition is entirely unknown (only about 20 cases have been reported). Since there are no known hereditary disorders resembling even remotely the kuru condition, I am inclined to suspect some toxic metabolism responsible for the process.

We sincerely admire the splendid work you are performing despite the adverse conditions, and we are looking forward to meeting you in Bethesda.

With best regards.

Sincerely yours,
Igor Klatzo, M.D.

BARRAU to DCG   Washington
                                                14 September 1957
Dear Dr. Gajdusek,
Being in Washington for a few weeks, I received thru the New Crops Research Branch, U.S. Department of Agriculture, the botanical specimen you sent to NIH. The plant belongs to the family "Araceae," most probably to the philodendroideae, philodendreae. Without the inflorescence, it is rather difficult to tell you what is the species; but I am sure it belongs to the genus *Homalomena*.

Several plants of this genus are used in native medicine in India and Indonesia. The natives there said that it is a stimulant and an aphrodisiac. Moreover, the plant has been reported to have an irritant action and is said to be used as dart-poison.

Its use in New Guinea was not known. I should be very grateful if you could send me more information on the locality, where you found the plant to be used, the method of preparing the leaves, if any, and more details on the effects resulting of the consumption of the leaves. A complete specimen with flowers will be also extremely useful but I hope to be able to collect the plant myself as I plan to visit New Guinea once again during the next few years.

I am sorry to ask you so many questions but I am really interested in the ethnobotany of Melanesia and this *Homalomena* is something new in this field.

With many thanks in advance.

Yours sincerely,
Jacques Barrau
%Plant Introduction and Collection Section,
South Pacific Commission,
Noumea, New Caledonia

P.S. I hope you will forgive my poor, basic English. I am French.

DCG to SMADEL   18 September 1957
Dear Joe,
I returned yesterday from a month of extensive walking and patrolling—some through country not previously visited, to hamlets and settlements never before visited by Europeans—and on this patrol have completed some 1000 to 1500 miles of mountain-climbing and walking (often with bush-cutters to clear the trail) trying to track down kuru epidemiology since March 14th, when our study started. To return to the extensive mail from NIH was a wonderful thing, and I cannot thank

you sufficiently for all the fine assistance that Dr. Carl Baker gave me while you were at the polio conference in Europe; for the supplies, the film, etc. from the NINDB; and, especially, for your own help in selecting the photos for the *NEJM* article, and adding the pathological report. Finally, the wonderful work of Dr. Henry Imus and Dr. Leonard Kurland is the first really sharp critical evaluation of our work we have received in such detail, and I am most grateful for it.

Dr. Kurland, of course, does not know that most of our time is and has been spent on epidemiology. Entomological surveys; extensive botanical and geological surveys; trace-metal analyses of food, smoke, fire ashes, soil, etc.—all have been arranged for and are still being carried out. For three months Lucy Hamilton, an expert on native diet and nutrition for the entire territory, has been in full-time collaboration with us; she has prepared with us extensive lists of all the food, paints, native products, etc. which the Fore infants, children, men, and women—the kuru victims and non-kuru victims—come in contact with. This work has been done for the hamlets and houses with current active kuru, and contrasted with similar situations and populations without kuru. The Fore, all of whom are in the "kuru region," have been compared in this respect with surrounding non-kuru populations; and those linguistic groups with kuru in one part (i.e., adjacent to the Fore) are being compared to those portions of the same groups without kuru. Botanical, zoological, entomological taxonomy on all species eaten, involved in occasional pica, chewed for salt, used in fires, used in sleeping houses (the smoke), and used in cooking are all being done. Thus far, we cannot find a single clue.

Three months of intensive work by Lucy, seven months by myself, and the added linguistic-anthropologic help by Dr. Smythe, who is an expert on Melanesian languages, has netted us not a single clue as to what might be "poisoning" the Fore—not a single sign of a dietary deficiency. Experimental approaches (such as removing patients from all contact with native products, as far as possible, for months—with continued progression of their illness) have been tried. Epidemiological case-plotting, past and present, has shown cases who develop into severe and fatal kuru after a complete exodus from the kuru region, and other cases apparently starting in kuru-free regions after marriage into such regions from kuru populations. However, there is an increased number of "recoveries" at these borders, all still fitting the hysterical hypothesis; rare and semi-documented cases from low-kuru-incidence regions get kuru in the kuru-region, but none from absolutely kuru-free populations have yet been found with a case of kuru during residence in the kuru region. Studies of all toxic, possible toxic, and medicinal herbs, minerals, and plants is under way. But, without a trace of increased or special exposure in either the kuru patients as opposed to normal Fore, and, in addition, without a trace thus far detected of the real differences in food habits or environmental exposure between kuru populations and kuru-free populations, we are thus far totally stalled on the toxic approach.

Case-plotting geographically and chronologically within a small set of hamlets (tribe or clan) as to ground, gardens used, living and sleeping houses, eating patterns, etc.—all fail to show any accumulation of cases or simultaneous cases. Familial incidence is conspicuous and high; but, as a rule, siblings and mother tend to be sick at widely different times—often with illness years apart and often with place of onset miles apart. Thus, Joe, all of our work is designed to prove current or antecedent infection, or to demonstrate a current or a past toxic exposure via skin or any body orifice. Nothing yields a clue to us thus far. Two cases in a single village occur in distant, remote hamlets of the same village, in two individuals coming into contact rarely. When the disease strikes the same "family" or household again, it is living at a different site; often different food-preparers, etc. are involved, and it is often years after the death of the first in the family. A few exceptions—with parent and sibling having the disease within a few years of one another, and a rare simultaneous case (one instance)—have yielded no added information on careful study thus far.

Thus, we are well aware of the unlikelihood of such a phenomenon with such astounding, unprecedented incidence being genetic—but what can we do? I am constantly playing with the hypothesis of antecedent infection, perhaps a mild encephalitis, giving rise to a post-infectious process. Even our AICF test has been used on the cases without help, but we cannot yet find a suitable clinical entity via case-study, anamnesis, nor by study of all illness in the region for over half a year now. I am naturally most anxious to have further neuropathological reports. Your preliminary reports are so impressive, so contrary to preliminary reports on one brain from Melbourne (a further two have not yet been reported), that we are astounded—yet, of course, much encouraged. We at least have something in the neuropathology to anchor on, and our ward full of dying cases at the moment demonstrates to us every moment that there must be CNS pathology visible; there is far too much evidence of extensive neuropathological lesion clinically. In recent months we have proved to ourselves that the lesion has to be diffuse, and that cerebellum and basal ganglia in addition must be involved. This now is clearly evident from examination of our patients; and, at the moment, with a film team from the Australian government finally here, I am trying to show this clearly in cine film. This you will shortly see; and together with the longitudinal studies of single cases which our early cine efforts have made possible, there should be no problem demonstrating most of the clinical features, and some of the epidemiological features, of kuru to research audiences of potential collaborators. The film is designed not for instruction but for consultation with research associates, to get added advice and opinions such as Dr. Kurland has supplied.

We have now tried stilbestrol, testosterone, and cortisone further without help. BAL proved of little or no value in repeated courses. Tranquilizers are en route and will be tried further. Strep studies on the nasopharynx of patients, and extensive blood-group genetic survey of kuru patients and all kuru-involved and surrounding kuru-free populations are well under way.

I am sending today two more fixed brains from children with classical kuru. They have stood for over one month in fixative. Also, fixed tissues on these patients, including cervical cord, pituitary, adrenals, kidney, liver, lungs, heart, muscle, fascia, etc., etc. We have now plastic and glass containers on hand; we also have specially prepared trace-metal–free formol-saline made at Hall Institute for us, using analytical reagent NaCl and trace-metal–free HCHO and water. One such brain I shall send to Melbourne. Another such—along with other tissues similarly collected—I hope to send to you. When the trace-metal–free brain arrives packed in plastic bags, please make certain that it does not get shoved into routine neurohistological channels but rather gets appropriately handled for trace-metal studies. All previous specimens have been fixed in usual autopsy-room HCHO (the provenance of which I do not know) and kept in metal tins. Only this specially handled specimen will be really adequate for trace-metal studies, and we hope you will be able to get the lot on it.

In a week we shall be off on the final Kukukuku patrol, and then I shall plan to close down our field program and start homeward. A conference first in Port Moresby, a line-up of follow-up work: there will not be anyone in the field to collect further specimens, etc. once we close down, nor to assemble cases, track them down, etc. However, I am arranging all so that in 12–18 months a follow-up epidemiology could be done quickly and with profit, based upon our extensive current epidemiological files.

Will you let me know whether you selected any photos for the *MJA* and sent any in to them, or whether you have sent them only to the *NEJM*. I telegraphed to you today, asking for frame edge-numbers from the negatives of whatever pictures you selected, along with negative numbers. Since I have a record also, I can easily write captions and rush them to the journal. If you have done so yourself, OK. [85]

We are working on and have been continually seeking a toxic factor. No one will appreciate having to study toxicologically the 500-odd foodstuffs eaten by our Fore kuru patients (as well as all other natives hereabouts); and thus far we can find nothing peculiar to either kuru patients, kuru-affected populations, households with currently active kuru, or the age- and sex-group predominantly suffering from kuru. Without such a "lead," we are faced with the impossible task of toxicologically surveying some 500 species and varieties of animal and vegetable life of which the Fore partake. All will soon be taxonomically identified, and a search through our lists and commentaries by competent Pacific Island toxicologists may be of help—and by botanical pharmacologists. However, who has seen anything like kuru following any known toxin thus far? Manganese poisoning, perhaps; but we cannot anchor on heavy metals as yet, either from laboratory study of tissues or from local epidemiological study.

[86]

The new shipment of tissues which is now off to you will have with it, insulated in plastic in case of fluid leakage, the stack of case-reports mentioned above and the third kuru review (which has the more recent data and which, we hope, will provide a chance for publishing a definitive detailed regional epidemiological map and plenty of pictures). This is all we shall write while here, although I have material on some 50 children (for a review of kuru in childhood) and data on everything from Fore genetic blood-group studies to clinical lab studies and treatment trials on kuru to work up later, back in the U.S., for further reports. Obviously all is of minor importance as long as etiology remains a mystery. As with EHF, we are thus far licked. Worse than with EHF, I can find no toehold from which to start infectious-disease or toxicological study. Further neuropathology may be the key matter, and obviously we are looking forward to further reports. How possible do your NINDB men consider a mild—usually clinically inapparent or minor—undiagnosable antecedent acute infectious disease from the neuropathological picture?

As far as deficiency: Fore diet is excellent, patients are well nourished, showing no sign of any dietary deficiency of any sort and occasionally even fat and strikingly well developed and nourished at the onset and well into their disease. Dietary supplement does not influence their course, nor do vitamins in moderate dosage. Outsiders can spot only a superfluity of foods of every sort here, in contrast to much of the rest of the Highlands. Water sources give us no clue.

[87]

My greetings and intense gratitude to Dr. Klatzo and Dr. Milton Shy for their help and collaboration.

Please let me know whether or not you will be in Bangkok, for I shall be passing through at just about the right time for the meetings, as things are turning out.

Greetings to Betsy and all others in Washington who may remember me.

Sincerely,
Carleton

[88]

DCG to WINTON 19 September 1957
Editor
The Medical Journal of Australia
Dear Sir:

Enclosed you will find the signed assignment-of-copyright form which you requested we sign and return to your office. Enclosed we are also forwarding to you a set of photographs for illustration of the article. The map which is Figure 1 of the paper was already supplied to you with the manuscript and your office indicated that it was satisfactory. You suggested that we keep the number of photographs down to about 8. We are, however, submitting a total of 12 in the hope that you will publish at least the first 7 (lettered A through G inclusively) and one of the remaining 5 (which include photographs of children with the disease and are designated H through L inclusively). However, we believe that all contribute significantly to presenting the picture of this "new disease" and urge that you include as many of the last 5 pictures as you can.

[89]

We know that photographic illustration is expensive, but believe that this disease, now being described for the first time, warrants full and inclusive documentation. It is a new disease, occurring exclusively in Australian Trust Territory; and we believe that a full photographic coverage of it will add considerably to the quality of this report of kuru in your Journal.

Please inform us of the receipt of this letter and, again, at a later date, of the receipt of the finished glossy prints for publication from America; for we are somewhat isolated, and such notification assures us that there has been no loss of mail.

Sincerely yours,
D. Carleton Gajdusek, M.D.

[90]

DCG to SMADEL 21 September 1957
Dear Joe,

Just after a stack of letters each of which had to be packed into a real package rather than an envelope, you no doubt would like a rest rather than more mail.

However, now that a great shipment of tissues (including three brains) is off to you, I will discuss herein a few remaining things of kuru-interest before packing up for another long patrol—the long-postponed Kukukuku patrol. [91]

We have just finished today shooting 3000 feet of 35mm cine film on kuru with the Commonwealth government's professional cine team. We had waited for them for months and months, and went ahead with our own filming, as you know. However, they finally arrived; I assembled two dozen kuru patients, and planned a rather extensive clinical survey of kuru which we shot in four days. It lacks any chance of showing progression of the disease in single patients and also shows nothing of kuru in its native village setting, since all was filmed here at our Research Center. However, it will be a professional rendering of the clinical features, and we tried hard to show many of them. I shall eventually get a 16mm copy of the entire film off to you. It is made for advice from research collaborators and not as a teaching demonstrating film. We hope that you will see things we did not know were in the film. However, this 3000 feet of film (actually 3200) will make 30 minutes showing. It is shot at 24 frames per sec., and copies are taken from a 35mm master film. It is black-and-white. This material can be coupled with the 16mm 24-frame/sec. material we have shot with the camera you supplied, and which we shall continue to shoot. We shall concentrate on kuru in the native setting, and on serial horizontal studies of individual patients at various phases of their disease. Thus, the 16mm material we shoot, amateur though it may be, can, we hope, be coupled with this professional material to make one rather exhaustive and comprehensive film covering everything from ecology, native setting, social setting, the research project, and especially, the clinical features of kuru. Other congenital disorders, differential diagnosis with the contrasting congenital tremor, and such things as effects of cortisone, stilbestrol, and testosterone therapy can all be demonstrated as well. Thus, this 3200 feet of 35mm film will be a great addition to kuru documentation. [92]

After the Kukukuku patrol, we shall have a Kuru Conference in Port Moresby. Actually, field research has been restricted to Vin and myself and to our nutritionist Miss Lucy Hamilton. Dr. Smythe was with us for a month working on grammar, and his preliminary grammatical analysis of Fore has just arrived and is of much help in our work. However, fully half the time, Kimi, Keiagana, Usurufa, Kanite, and other languages are the problem. Our last patrol revealed that kuru involves all of the small and narrow belt inhabited by the Kanite peoples rather than part of it, and the same applies to the Usurufa tribal groups. These people are thus entirely involved in kuru—not only partly, as we had thought.

However, since they number under 1000 total (each group) and live only in a narrow belt all of which "borders on the Fore," and since they all marry into Fore, this does not alter our thinking any. Iate (a cultural group closely related to Kanite and Keiagana) is also involved in kuru, we find. Thus the kuru-region is a slight bit larger than we had thought, but we have established fully the extremes of its borders. Only one case has yet been found outside of these "limits of the kuru-region": That is a case in a woman who developed her disease while visiting her home in the kuru-region but who walked to Lufa with her husband—an arduous four days of mountain-climbing—in very early illness and died at Lufa after three months of slow progression of her illness in this densely populated but kuru-free region. She thus demonstrates that very early kuru can run a full course to incapacitation and death in a totally kuru-free region in a cultural group devoid of kuru. This woman was brought in by marriage as only three kuru-region women have been to this group. The other two are living and well. Moreover, she lived for over a year in the kuru-free region of Lufa (west of Mount Michael) with her husband after leaving her home, and noted kuru first after a return visit to her kuru-region home, after only a week or two back in kuru region.

Such cases make it most difficult to think of infection or toxin; yet, Joe, that is what we are thinking of and concentrating on—but, unfortunately, with no leads. I am asking Lucy Hamilton to prepare a collection of those plants and native products which for some reason or other we have wondered about, suspected, or been disturbed about. These will go off to you for taxonomic and toxicological analysis of whatever sort the Smithsonian and others at NIH can get for us. The full taxonomy of Fore diet is being done, but we have no expert on *toxic* plants, etc.

Sincerely,
Carleton

DCG to SIMMONS   22 September 1957

(Roy, please excuse the carbon-copy, but the original is so illegible—because of poor ribbons—that I sent the carbon instead)

Dear Roy,

I have been completing the kuru epidemiology and finding the racial, ethnic, and linguistic—as well as ecological—limits of kuru by extensive patrolling, much of it through previously unvisited and "uncontrolled" territory. We find kuru involves somewhat more people than our original estimate: some 17,000 to 20,000 people living in the kuru-region—with some one percent of all these currently affected with kuru, and dying. All but a very few of our first series of patients (since March) are now dead. Finally, neuropathological reports are beginning to arrive. From the USA (National Institute of Neurological Diseases and Blindness, of the National Institutes of Health) we have reports of extensive neuronal degeneration in the cerebellum and anterior horn cells, with pontine nuclei, inferior olives, thalamus and extrapyramidal system also involved in the first six brains studied; and further studies are in progress. Toxic factors are still urged upon us, but we cannot find any signs of them, try as we may. We are probably just missing them in our study of the pica, diet nutrition, soil, smoke, body paints; and yet, genetic predisposition (at least) looms ever more likely from the field epidemiology and clinical studies.

A large shipment of blood specimens on the Kimi, from the very central Kimi tribe just beyond the border of kuru in a kuru-free region where there is a history of only one case in the past (and that "imported"), went off to you. Thus, you have a fine, racially pure non-kuru Kimi group—if the specimens arrive intact. In addition, I sent a series of specimens on the Papuan Yar people, which we managed to contact by exploring south of the Fore in the sago country. These are a sparse population; and you have, among the twenty-odd specimens we sent to you, specimens from over half the individuals seen in a region patrolled with a week of hard walking.

We had to send specimens back to Moke by the carriers, who walked twenty-four hours a day to get them placed in an icebox there and thereafter taken out by jeep for air shipment to Australia. One box of venules remained, by being overlooked in the Moke icebox until our return. Thus, those twelve or so venules will be one-to-three weeks old by the time they reach you. Being valuable specimens, we rush them on. They have remained refrigerated all the time, and it is still possible that some work with the cells, washed from the clots, will be possible; if not, discard them. I have sent these on to Lois to prepare for you. Enclosed you will find further copies of the lists documenting the Kimi and the Yar ethnic blood collections.

Also enclosed are the lists documenting the F Series of bloods, numbered 1 to 106, which I have just sent out to you. These are our definitive adult Fore blood collection, for comparison with kuru patients. They are from adult Fore—all racially pure, without intermarriage, and all from a group of small adjacent tribes in the very center of the Fore linguistic and cultural areas, in the very center of the kuru region. Since every Fore is a blood relative of a kuru case, and since every Fore within a clan or tribe has some blood relationship with all others in the clan and tribe—because of intensive in-group marriage—we cannot avoid "relatives." However, in general, there is only one specimen per family group (father, mother, children). These specimens are for blood-group study, normal Fore serum studies (Curtain et al.), and for infectious-disease serological survey at a later date. I have identified subjects by sex and estimated age, place of clan affiliation, and the affiliation of mother and father of patients. All are pure-blooded Fore from central Fore territory, and from central kuru region; *none are kuru patients.*

We leave in a few days on the long-promised Kukukuku patrol, and we may make an attempt to walk through from the highlands to the Papuan coast. Obviously, we shall be with peoples which, if we can get specimens, will never have been previously studied. Furthermore, since we are patrolling on this patrol the "fringe" of the kuru region, and just beyond into kuru-free populations, there is a great deal that may eventually be learned by comparison of all these peoples. Kuru has made it essential for us to have sound ethnological, linguistic, and marriage-pattern documentation of all the peoples we study; and this data—vast and voluminous as it is—fills my notebooks and can be supplied, once we call a halt to our kuru fieldwork. I shall call a halt here, in order to get this out by the waiting jeep.

Sincerely and gratefully yours,
Carleton

P.S. I shall try in Moresby to dig up reimbursing funds to cover these expenses. Lois reminds me that we are indebted to you for all the supplies.

DCG to BURNET 23 September 1957
Dear Sir Mac,

I have your letter of September 10th and a letter from Gray of September 4. These two need answering, although I have just sent Gray a rather lengthy report.

I am taking the liberty of sending with this letter a very poor carbon-copy—all I have—of a long four-page letter to Joe Smadel in which much of the current kuru quandary is discussed. Since it is still highly possible that an unusual deficiency; an unusual toxin in food, smoke, or pica; or some unforeseen antecedent infection is involved in kuru pathogenesis—in spite of the increasing evidence for heredofamilial pattern—I think that these musings are still in order. I am sorry that they are so close to illegible. They were written at a time when our stationery and typewriter were at one of their periodic "lows." If you have not the time to attempt deciphering, Gray may be able to find the time and requisite ciliary muscle energy.

I am poised on the eve of our final kuru epidemiology patrol—one which will really be extensive, we hope. At the moment, I am trying to complete therapeutic evaluation trials of testosterone, Sertesin (a reserpine tranquilizer), stilbestrol, astronomical doses of vitamin B1, and continued use of cortisone and of BAL—the only two drugs which have given even a slight suggestion of perhaps (?) some palliative effect. We have also accumulated a new group of patients—our original series whom we collected in March, April, and May have very few survivors—willing to stay in our Kuru Research Center and, perhaps, willing to die here. Thus, we are continuing to try to line up pathological materials. I telegraphed to Gray the astonishing pathological reports from the NIH in the U.S. on six brains from kuru they had examined. Extensive neuronal degeneration in cerebellum and extrapyramidal system with similar changes also in basal ganglia, pontine nuclei, anterior horn cells, and inferior olives is a rather dramatic report; but it remains preliminary and totally a first description rather than a definitive report. Gray has telegraphed back that on August 16 he sent Dr. Robertson's preliminary pathological report—the first kuru brains we obtained went to you at Hall—but I have not received it. This may be due to Dr. Scragg's absence from the Territory. Since this delay in transmission through Moresby is inevitable, and since reports of anything which might influence our work can only do so if they get to me and Vin Zigas (no one else has yet come to work on kuru other than Lucy Hamilton, who is helping us with epidemiology considerably), could you ask Gray to send copies directly to us? Even now, for I have no idea when the Moresby report will be opened and transmitted to us.

Obviously, such extensive neuropathology surprises us, after preliminary reports of no gross lesions of any sort. However, even these are preliminary studies, and I fully appreciate how much work laboratory study of a brain entails. Neuropathology is no simple matter! The findings make further collection of pathological material essential. It is not easy—of that I can assure you. However, we may be able to get further brains; and any pathological report may influence greatly what tissues and organs we take, how much time and effort we put into getting more and more autopsy material, and how we collect the material (trace-metal–free, fixatives, perfusion or not, etc.).

If Gray requires further trace-metal–free blood specimens, please ask him to send up a few more trace-metal–free bottles, as we now have none on hand.

On returning from the Papuan border-region after an extensive patrol south, in which we established the southern extent of kuru-affected populations, I visited Lufa and saw Ian [Burnet]. We had only a brief time together, but an enjoyable one. He told me of his exciting Karamui patrols. We managed to locate one kuru death—in the east—occurring right at Lufa. This in a Kimi woman from the Fore-Kimi border region, where kuru is sporadic. She came out of the kuru region to Lufa in marriage, stayed a year or so, and then returned to her home with her husband, a Lufa man, for a visit. While back home in the kuru region, but only after a week or two at her old home, she noted the early ataxia of kuru. She remained at her familial home for only 1–2 months and then, in very early kuru, walked the arduous hike of three days back to Lufa without support of a stick or help—thus she was really an early case. Back at Lufa, her disease ran the usual typical course of kuru in this previously kuru-free region. She progressed over 3–4 months to dysarthria; marked ataxia; inability to walk, then to sit; and finally to decubitus and neurological degeneration and death. This is the only case of kuru we have yet found on the Lufa side.

A "new disease" in Melanesians (and also a kuru-differential diagnostic problem), a really marked congenital tremor of familial nature, is present in the Kanite and Fore people; and we have now found three cases. It is a non-incapacitating yet dramatic disease, and tends to emphasize how unlike this sort of congenital tremor the peculiar inconstant tremors of kuru are. We have made cine records of this differential diagnosis problem, contrasting it with kuru.

On this last patrol, I found kuru for the first time outside of the Okapa Patrol Post–administered region—but only this case in Lufa (by history) and a large series of cases in the Henganofi Patrol Post region. Now that these new cases are mapped, it is evident that they fall only in a narrow belt immediately adjacent to our previously documented kuru region and associated with the kuru region by direct marriage ties. Thus, all we have done is enlarge the kuru-affected population as known to us by another 1000–2000, and enlarge the "kuru region" a bit. I still estimate that, at the moment, 1% of all members of the affected populations are dying of kuru; and the total "affected population" seems now to be closer to 20,000 than 10,000.

A half-year is enough on this frustrating problem for the time being; and I shall call a halt to our field-study after this coming patrol, and try to gain distance and perspective before deciding just what else I may be able to do about kuru. It is a problem I do not want to leave; it is good, and obviously offers a chance to get some leads on neurological degenerative disease if intensively enough studied—with some "breaks." We have not had any "breakthrough," however, and it is time to move off and look from a distance.

I worked this week trying to assemble [clinical] material, and demonstrate it in 3400 feet of 35mm cine film which the Commonwealth Government cine team shot for us. This film, once I edit it in Moresby, I shall send to Melbourne promptly, in the hope that from it you and others at the Royal Melbourne may see in kuru things we have failed to notice or detect. In addition, we are shooting 16mm cine film regularly, which shows more of the local setting and kuru in the native village, and also individual patients as their disease progresses. This is being processed at the NIH. As soon as I return to the U.S., I shall edit this 16mm film and send a full-length copy to you at Hall for the same purpose.

This, then, Sir Mac, should bring you up to date a bit on kuru. [93]

Intensive epidemiology covering place and ground of onset, of progression, and of death of all documentable kuru—past and present—in the Moke tribal region in the center of the kuru region is nearing completion. Lucy Hamilton has been a great help in getting this together. From it, it is already evident that throughout all hamlets occupied by the Moke tribe, kuru has been sporadic, preferential in none, selecting none in any given year or two in preference to others. Gardening patterns and food-consumption patterns do not seem to differ in households with active kuru and those without. This intensive local epidemiology is revealing in that it clearly shows how difficult it is going to be to find the environmental variables associated with kuru pathogenesis.

Thank you for the information about the new virus from Telford Work. It certainly sounds like one of the so-called hemorrhagic-fever group; and Omsk Hemorrhagic Fever is known in Russia to be a variant of RSSE, and it is quite possible the other hemorrhagic fevers also are. In my EHF review (in German from Hall Institute), I mentioned this relation of some hemorrhagic fever to RSSE. It is most encouraging to find the syndromes outside of Russia—from the point of view of being able to study them, of course, for the diseases are certainly no boon to humanity.

I wish I could be around Hall for a while while Lederberg is there. I certainly hope you and he have time to thrash over all the problems of mutual interest.

Proofs of the *American Journal of Internal Medicine* articles on the AICF test can still reach me here for about one month. Thereafter I shall have to send you a new address. It can be c/o Dr. Robert Traub, Institute of Medical Research, Kuala Lumpur, Malaya. In other words, I shall be underway again, I hope.

Thank you, Dr. Burnet. My greetings to Dr. Gottschalk, if he is back, and to everyone else at Hall.

Sincerely,
D. C. Gajdusek

DCG to SCRAGG 24 September 1957
Dear Dr. Scragg,

In the past six weeks—nay, eight weeks—kuru has been such a rush of work for me that I have been unable to keep up with kuru correspondence. However, there are many things which I should bring to your attention, and I trust that you will pardon my so doing by use of carbon-copies of two recent letters. One is to Dr. Smadel, [Associate] Director of the National Institutes of Health in the U.S., where we have been getting extensive "logistic support" and neuropathological collaboration. Another is to Sir Macfarlane Burnet at Hall Institute. I am sorry that I am forced to send such poor carbon-copies, but already the extent of writing and correspondence which our kuru work and reports to all our collaborators entail is eating into the time I should devote to direct study of the disease.

As you will learn from the letters, Miss Lucy Hamilton is still working intensively with us; and her assistance in preparing an intensive study of a small representative kuru-afflicted community is of utmost importance. If kuru is purely a heredofamilial, genetically determined disorder, its unprecedented incidence for such a lethal combination of genes is hard to admit. The now-apparent extensive neuropathology which I have learned of from the NIH reports (preliminary reports from U.S. neuropathologists) tends to emphasize the need for an even more intensive search for possible toxic, deficiency, or antecedent-infectious pathogenesis. Such has been our approach, for from the beginning we are and remain reluctant to accept the genetic predisposition possibility which seems to force itself upon us. If genetic, the Fore are doomed. Thus it behooves us to deny the genetic evidence and keep trying to locate toxic, deficiency, or post-infectious factors—or anything else in the environment which might lead to such a neurological degeneration.

We have found acute encephalitis in the Okapa region. We have studied it and collected specimens for study of it both at the Hall Institute (by Anderson) and in the U.S. Nothing has yet identified the agent responsible for acute meningoencephalitis here. However, it is completely impossible to associate this meningoencephalitic illness with kuru in any way. Kuru patients have not had the disease, and patients with the illness recover without any sign of kuru. They are being followed. We are continuing to collect specimens for every type of study available to us. Streptococcal studies underway at the Department of Bacteriology in the University of Melbourne and Fairfield Hospital show nothing significant thus far. Intensive genetic blood-group study of kuru patients, of non-kuru Fore, and of all surrounding kuru-affected and kuru-free populations are now in progress.

Vin Zigas and I wish to know whether the Department of Public Health is interested in ordering reprints of the two kuru papers—both preliminary reports—which are now in press and will shortly appear. Furthermore, we shall soon send you a copy of a more definitive paper which we hope to publish soon. If you will let us know whether or not the P.H.D. desires to order reprints, we shall be able to place the order or send order forms to you and thus avoid the possibility of missing an order.

I am hoping to be in Port Moresby in late or mid-October for our Kuru Meeting. Vin Zigas and Lucy Hamilton are making the same plans, and it would be good if Dr. Bill Smythe could be there as well as Dr. Price. In addition I should like to suggest that Mr. Jack Baker, Patrol Officer of the Okapa Region (whose administration is so intensely involved in kuru and on whom much of the responsibility for support of our studies lies) be likewise included in the meeting. Our patrolling schedule does not make the setting of an exact date possible. I shall get in touch with you by radio before the proposed meeting, suggesting possible dates and leave the final decision to you. Thank you again for your help. All drugs have now arrived. The filing cabinet for filing-cards never did, but cards are here—although not in the quantity which we may require. Therapeutic trials of tranquilizers and further trials of stilbestrol, testosterone, cortisone, and BAL are all underway; and we are adequately stocked at the moment.

[*remainder illegible*]

BURNET to SCRAGG 25 September 1957
Dear Dr. Scragg,

We have just finished preparing more trace-metal–free formalin to be sent to Dr. Gajdusek. This is time-consuming and expensive work, and—as the Institute is now taking no part in the organization or responsibility for the work on kuru, which seems to be being channelled more and more by Gajdusek to American centres—I think the time has come for us to withdraw from what has been from the beginning a most unsatisfactory situation.

I know that there is still some work being done by the Chemical Industry section of CSIRO, and Dr. Graeme Robertson has some interest in the neurological side. Any withdrawal by the Institute does not, of course, involve these workers. We have provided facilities of various sorts for over six months, and feel that under the circumstances this is all that can be done.

We are not therefore prepared to fulfil requests from Dr. Gajdusek in the future. I should like it to be quite definite that this in no way diminishes our willingness to cooperate with the Administration in any matter within our field of interest and competence.

Yours sincerely,
F. M. Burnet

DCG to INGALLS 25 September 1957

Associate Editor
New England Journal of Medicine

Dear Dr. Ingalls:

I have a copy of Dr. Smadel's letter to you of August 28, referring to the paper by myself and Dr. V. Zigas on kuru. I assume that you have already received, in addition, a selection of photographs of the kuru region and of kuru patients from which to select illustrations for the article.

I am writing to request a few minor additions and changes in the paper, since it is possible that I may be unable to do any galley-proof correction prior to publication in November. This letter is being forwarded to Melbourne (Walter and Eliza Hall Institute of Medical Research), Australia, where definitive copies of a map showing the location of the kuru region are being prepared. It will be mailed to you from Australia along with glossy prints of the map which I hope arrives in time to be included as Figure 1 in the article. Geographic localization is of much importance in kuru epidemiology. Furthermore, the map will greatly aid the reader. The map is itself self-explanatory. If a further legend is desired, use:

> Figure 1. Map showing location of the Kuru region in the southern portion of the Eastern Highlands of New Guinea.

In addition, I would like to request that the pathological findings which Dr. Smadel supplied to you, in his letter of August 29th, be placed within the body of the paper rather than as an Addendum, as he suggested. In the revisions I previously mailed to you (July 26), I already brought case figures sufficiently up to date to embrace all cases thus far examined neurohistologically. Thus I suggest including the same description Dr. Smadel furnished to you—with the introductory word "However"—as an added three short sentences which can be inserted into the text on page 4, just after the sentence which ends on the next-to-last line of that page: i.e., "Histological and neurohistological findings will be reported (Insert) *in detail* in a further communication. (Insert) *However,* the preliminary pathological study of six cases reveals a widespread neuronal degeneration. The cerebellum and extrapyramidal system are most severely affected. There are also changes in the anterior horn cells, inferior olives, thalamus, and pontine nuclei."

Dr. Smadel will shortly inform me of the film-negative edge-numbers of the photographs he has selected from our kuru film-file for the paper. From these I can identify the photographs and promptly write captions which may be used should they differ significantly from those Dr. Smadel has submitted. He has attempted the task under the disadvantage of not knowing the case depicted, and it is possible that pictures may have been inappropriately described.

Finally, I should like to insert the following brief text at the very end of the paper. The pathological findings just recently made available to me in no way change any of the facts we report, but they do influence somewhat the stress and interpretation thereof. It thus behooves us to recapitulate our strong suspicions of toxic or post-infectious factors, in spite of our inability to demonstrate them. Thus, please insert, before the bibliography and after the last word in the current manuscript on page 8, the following brief lines:

> The extensive neurological degeneration found on histological study strongly suggests some toxic factor. Thus far search for such factors in diet, pica, skin painting, smoke and

other sources has failed to reveal any organic agent which might be responsible for the disease and trace-metal studies on blood, urine, cerebrospinal fluid, and other pathological specimens from kuru patients and on foods, soil, water and fire ashes have failed to indicate any heavy-metal intoxication. If the degeneration of kuru is a post-infectious phenomenon, the antecedent illness must be so mild or subtle as to escape detection by the natives and ourselves.

I am sorry to bother you with these further changes, but in the light of this added data, which greatly improves the paper, I believe, they become mandatory.

Sincerely yours,
D. Carleton Gajdusek

P.S. Since completing your letter, mail has come in with several letters from your office. We are glad to learn that the paper will soon be published, and trust that the map will arrive within two to three weeks after receipt of the photographs from Dr. Smadel, along with this letter. By that time I hope to have captions for the photographs to you. The change in title suggested is satisfactory to us. Perhaps, however, the word *Natives* should be inserted after the word *New Guinea* in the title, as suggested by you. Thus, it should read: "Degenerative Disease of the Central Nervous System in New Guinea Natives: The Endemic Occurrence of 'Kuru'."

Enclosed you will find the two cards you sent to be filled out by Dr. Zigas and myself.

DCG to SMADEL   27 September 1957

Dear Joe,

We leave for the Kukukuku patrol and an attempt—one of the few made—to "cross the island." We hope to track down populations south of the Fore, back down to the Yar people with whom I previously made contact, and thence on down the Purari River to the Papuan coast. It is fully "unexplored," and we cannot guess yet how we shall succeed.

[94]

Lucy Hamilton will be collecting the medicinal plants and drugs of the Fore, and also all the plants, varieties, and species of food-plants which we have reason to suspect. These she will send off to you while I am on patrol. Things eaten, especially by women and children (usually kuru patients), and items which our kuru patients seem to have eaten more regularly than other people—as well as items we have reason to suspect in their being peculiar to the kuru region or, at least, changing in variety throughout the highlands from region to region—will all be sent for:

1) botanical (zoological, in the case of food insects and animals) identification;
2) comments, by whoever is qualified, on the likelihood of toxic alkaloid or other toxic products being a species variety of the genus involved;
3) comments on any known pharmacological or toxicological properties of the specimen or related specimens; and,
4) suggestions as to which might be advisable to pursue more intensively by physiological and toxicological and local ethnological food-habits study.

In addition, Lucy will get off to you, via the Bureau of Plant Introduction, a shipment of the kuru-like-tremor–producing plant, as requested. It is all hearsay, for we have not seen it used. However, all informants agree on the information. We shall try to get a complete Fore ethnobotanical file off to you. A crew of entomologists are soon to be with us for a week or so trying to identify all food-insects, which women and children particularly eat in large numbers. However, I suspect that Smithsonian sources will be of more help eventually than our visitors. They will be asked to make a survey of ectoparasites in dwellings of kuru patients, and to comb rodents living in the dwelling thatch for fleas, etc. Furthermore, I hope to ask for an intensive search for all biting insects about the houses, gardens, and haunts of the kuru patients—to get some idea if we have missed rare mosquitoes (they are known to us at this altitude) and to get information as to what species of fleas, ticks, sand flies, etc. bite in this region.

I have given Lucy the addresses to which to send the specimens: formaldehyde-pickled things directly to you at NIH; unpickled botanical specimens via the Department of Agriculture, Office of Plant Introduction, from the address cards which you sent me.

I have received word of the pathology from Dr. Klatzo, and am most excited and anxious to learn if further work bears out these results. Three well-fixed brains are off to you. I have managed to keep four terminal patients around this month, and we may (?) be able to get the brains on some of them, at least. It is most difficult; and most terminal patients must, on the insistence of villagers and kinsmen, be returned to their home ground for their death and burial. Thus, we miss most possible posts; and those we get are a difficult politic.

Last week, one of our friends from Aga came in to report that his clansmen—against his advice—had eaten his grandfather. Such recent, nay current, episodes of cannibalism are not unusual here; but it is highly unlikely that all of our kuru patients have eaten human brain or ectoderm and thereby gained human neuroectodermal sensitivity, a remarkable concept which Anderson discussed with me many months ago. It is so unique a concept, and such a romantic one, that I almost wish cannibalism was more prevalent than it is. Ritual payback murders, usually for kuru sorcery, are a regular occurrence; and the tribesmen do not like to be interfered with in their vendettas. A gang are now in the calaboose for "putting bows in" [shooting at] the police from our post who went to investigate the last murder. A day's walk from our Research Center, the upper Lamari people held down a patrol with arrow-fire for some time last month. We now head for restless areas and are proceeding with due caution. But thus far, all this sounds far more "dangerous" than it has been; for our Fore, Kimi, Keiagana, Usurufa, and Kanite friends and patients have been friendly, helpful, and cooperative, and—except for "trustworthy"—seem to live up to the Boy Scout's oath rather well. I shall write to you from patrol.

Sincerely,
Carleton

P.S. Lucy has found a few additional plants reputed to produce "tremors," and these we shall also collect for identification. However, the plant you have already received is exclusively a Fore drug, making it more interesting. We cannot pin the use of it, however, on our kuru patients, and we cannot yet verify the rumor as anything more than legend; we have not actually seen it consumed.

[Enclosed in the above letter to Smadel was the following extract from a letter (2 September 1957) from the Australian Museum.]

In regard to the spiders, Mr. Musgrave reports as follows. The spiders forwarded have been identified as follows:

    128. Obana. "Spider, small, black with white markings. Found in bush. Eaten only by small children. L. Hamilton."

The glass vial contained two large spiders: *Cyrtophors mollucensis* Doleschall, 1857. A species ranging from Java, Amboina (type loc) to New Guinea and Australia. A smaller spider, *Leucauge* sp., too mutilated to be identified with certainty. These two species are members of the family Argiopidae (orb-weavers).

    150. Kogogora. "Spider, small. Found in bush and garden and *kunai*. Eaten by women and children only. L. Hamilton."

The glass vial contained three specimens representing three species, all members of the family Argiopidae. The largest of the three is *Argyope aemula,* Walckenaer, 1841; the other two species are members of the genus Araneus (Epeira).

It is interesting to hear of spiders being used as an article of food by the natives of New Guinea. La Billardiere in 1799 relates that the natives of New Caledonia eat certain members of the genus Nephila (a large orb-weaving spider) in large quantities. He gave the name of *Aranea edulis* to one of these New Caledonia Nephila, although today it is known as *Nephila edulis.*

BURNET to DCG   27 September 1957
Dear Carleton,

Thank you for your letter, which I found of much interest—though so far I haven't managed to get very far in the carbon of Smadel's letter!

I feel you have done everything you could on the problem at the moment; and as people in America are well interested by now, I decided that we would withdraw now from any further kuru responsibilities like supplying trace-metal reagents, etc. It is a lot of work for Holden, and could better be done by those receiving the specimens for study.

We shall continue to be most interested to hear how results work up—the picture seems to me to have steadily firmed in favour of a heredofamilial disorder throughout the period of investigation.

I had a letter from Ian [Burnet], telling me of having seen you and taken your photograph for the Institute's archives! He was a bit unhappy, naturally, about having his house burnt down.

There is no further word of the AICF papers, beyond the note that they had definitely been accepted. We shall let you know anything that eventuates.

I understand that Dr. Graeme Robertson has sent you a copy of his report in the last day or so.

With best wishes,

                                                                         Yours sincerely,
                                                                         F. M. Burnet

CURTAIN to DCG   27 September 1957
Dear Carleton,

Herewith further results and the promised pictures. The sheets include the corrected names of the first series of sera.

I was rather upset by the finding that Ivuti (one which you said was not kuru) showed a high β-globulin, and was beginning to wonder if this was a characteristic of the region (that is, some "normals" had this type of globulin disturbance). However, thirty of your normals arrived in due course and have been electrophoresed. These, although differing in some respects from the normal European pattern, do not possess the striking abnormalities observed in your kuru series. I feel, however, that we must do as many normals as possible. Information on electrophoresis of the sera of New Guinea people is most meagre. I wrote to Walsh of Sydney Blood Bank, who told me that his limited observations showed that many coastal natives and Solomon Islanders had high γ-globulin. I believe that there is a Nuffield Foundation–sponsored Sydney Anthropological group somewhere in the Western Highlands. From press reports, they have apparently been doing some serology. Do you know anything of them?

Looking over the 60-odd kuru sera, I can find four groups: the first with a marked elevation of the β-globulin, the second with an elevated γ, the third with an elevated α_2 and β, and the fourth with a pattern fairly close to the European normal. I might point out here that, in actual concentration, there is no diminution of the albumin in any of these sera.

I have run approximately 60 haemoglobins, and found them all normal (that is in pH 8.6, I = 0.03 veronal). John Owen at the University has repeated 26 of these on starch, with the same result. The people are the same as those on the serum lists.

I have a good supply of samples, about another 50 to be run or calculated, and will push on as fast as I can.

Yours sincerely,
C. Curtain

SCRAGG to BURNET   4 October 1957
Dear Sir MacFarlane,

Your letter of 25 September was received along with one from Dr. Gajdusek and two copies of letters which he had sent, one to you and one to Dr. Smadel.

Needless to say, we are sorry to hear of your decision not to fulfil any more requests from Dr. Gajdusek. I do, however, agree that it has been a most unsatisfactory situation from its inception.

I feel the six months in question should be adequate and that Dr. Gajdusek should now be reasonably informed, and that further repeated widespread investigations as you have done are no longer necessary.

As you will gather from Dr. Gajdusek's letters, he is planning to leave the area within a month. Prior to his departure he will be coming to Port Moresby, and a general discussion will be held on kuru. I will inform you of the findings of this meeting.

My impression is that even though a great deal of work has been done, we still have no idea of the aetiology or a possible cure for the condition.

The condition does not appear to be within the scope of your interest and competence.

Dr. Zigas will no doubt continue with field-studies of the condition; and we may, at a later date, discover something.

Thank you for your reassurances of your willingness to cooperate with us.

I have recently visited Maprik and Kundiawa, and I can foresee the development of research centres at both of these sites. We are, however, still awaiting the appointment of a Director of Medical Research, and feel that these projects must await his appointment.

I will forward you a copy of the report of the group discussions to be held on kuru, and inform you of any future progress in the field of research.

Yours sincerely,
R. F. R. Scragg

VI.
Excerpts from Field-Journals Kuru Epidemiological Patrols
26 September–9 November 1957

26 September 1957   Okasa Rest House, North Fore

[54:2–55:19]   The Kukukuku patrol, ambitiously contemplated as a "cross-island" Kukukuku-Papuan coast patrol, is at long last under way. We walked down to Okapa from Moke this afternoon, having dispatched our cargo by the treacherous track from the Okapa Patrol Post in three loads of the Land Rover and sent the 50 cargo-boys (whom we selected this morning) down the trail unloaded. We arrived to find that Sergeant Malekor had established a fully equipped, comfortable, and well-arranged patrol camp here at Okapa Rest House.

I have browsed over maps, records, and kuru charts—and have tried to bring our work up to date by making epidemiological data cards from dozens of scribbled notes, and by entering on the patients' records stray blood counts, urine analyses, and CSF examinations which have remained scribbled on bits of paper in our disordered rush. It is the first time in weeks that I feel myself achieving full control and getting "up to date" in our expanding, already overburdened kuru program.

Jack [Baker] and I are alone. Vin, who was to be with us, is limping about with a ruptured knee-cartilage which he must soon have removed. He first had trouble with the knee years ago, in the steep Goilala region where he started his work in the Territory. On the Kasokana-Abomatasa patrol which he made with me, he had much trouble with it—twice being immobilized on the trail with a locked knee. Thus we are to miss Vin, and I regret it. However, I must agree with Jack that were he with us friction would inevitably be high. Vin and I have continued to work well together, but Vin is made ever more impatient by the inevitably slow progress of research and the apparently endless investigative inquiries. Where he looks for kuru study to be finished, I can see it only starting. Thus, with the program a bit out of his control and hands, expanding beyond his "organization," he has withdrawn from it considerably. However, during our last epidemiology patrol down to the Papuan side and around Mt. Michael to Lufa and back, he came in to Okapa to supervise the kuru patients and did a vigorous and excellent job of conscientious study—which was the only follow-up we had on our major, closely studied cases during my long absence.

Jack, however, has noted our frequent disagreements, spats, and sharp exchanges. Actually, they are more in the nature of "in-family" squabbles, for I consider myself very close to Vin—much akin to him in interest, temperament, and background—and I am very sympathetic to his need for independence, dominance, and direction in anything in which he engages. He has been an excellent colleague; I could not want a better one, nor would I have exchanged him and his vigor and enthusiasm for anyone else. . . .

Tiu, Masasa, Taka, Wanevi, and Waiajeke are the five *mankis* I have taken on the trip. Anua, who is convalescing after his meningitis which occurred during our last patrol, and which nearly brought tragedy to it, is home in Purosa on two weeks' "leave." . . .

Because of the recent attacks on the patrols in the North Lamari region, the totally uncontrolled nature of the region we are to traverse, and the presence of much warfare, we have been required to take 10 of the Royal Papuan and New Guinea Constabulary from Jack's contingent—and some borrowed from Kainantu—with us. Malekor, in all his efficiency, is in charge.

[55:28–36]   I lined the cargo-boys at Moke today, gave them each 0.5 gm of chloroquine-phosphate malarial prophylaxis, and took sterile clotted blood specimens from each for blood-group genetic studies and as "pre-bleed" virus and rickettsial serological specimens to be compared with "post-bleed" specimens collected after the trip. This may show us what, if any, inapparent infections have occurred—at least give us a serological lead on this long expedition into regions foreign to our carriers. I certainly hope we do not have clinical infections to boot.

[56:6–28]   Just before leaving Moke, I discharged Yosetme, currently our youngest kuru patient (M–5). He has deteriorated much since we brought him in from Tunuku several months ago, in spite of intensive cortisone therapy. Yesterday Kuroeva was carried back to Ke'efu in terminal condition, with little time left to live. Intramuscular testosterone has not helped him; however, his treatment was discontinued over a week ago. Although our neuropathological reports suggest a toxic process, we remain stalemated and we cannot shake off the evidence for a heredofamilial pattern pointing to genetic predisposition. However, an environmentally, ecologically determined toxin, deficiency, or infection could easily create a false impression of genetic predisposition. Our whole effort is to find the environmental variables involved—but thus far without success. . . .

27 September 1957   Yakeia village, Awa

We managed to get our carrier line of 68 carriers, 10 police, 6 *mankis*, 1 *haus kuk boi* (Andi), Kosinto (who accompanies all of my patrols), and Tarangau (our NMA) on the trail by 6:30 a.m.—having arisen at 5:45, lined and enlisted 22 more carriers from Ilafo and Okapa to add to the 46 we brought down with us from Moke, and issued Nivaquin (0.5 gms) to all *mankis*, police, new carriers, and others who had not yet received it. After seeing that they took it, we started out on the long hike to Yakeia.

[57:15–36]   I arrived at Yakeia at about 2:30, and Jack came in about 3:00. We were both in good shape and without complaints from the long stiff walk; and thus it is quite evident that we are in considerably better condition than when Jack, Vin, and I made our first Fore patrol to Agakamatasa and Awarosa from Moke.

Wanevi stays close beside me on the trail, and he and I lead the patrol most of the time. It is all I can do to keep up to the speed of the natives while I hike without a load, while most of them are carrying heavy cargo. Wanevi carries his *bilum* filled with his blanket and extra *laplap*, some *kaukau*, and some soap, in addition to my Leica and a telephoto lens, and manages to be more stable on the trail than I and even gives me a hand at times when I can not maintain my position on the steep slippery slopes. . . .

The Yakeia people are a different group from the Fore or any other we have been with, speaking a totally different language unintelligible to our Fore, Kimi and Keiaganas, and Kamanos on the trip. We have one Okapa Fore lad who can translate for us, however.

I shall devote tomorrow to trying to learn the marriage patterns, the linguistic relationships, and the tribal affiliations of these people—and also trying to determine whether they have ever had any kuru.

28 September 1957   Yakeia village, Awa

[58:43–59:17]   Jack and I argue—as Tomasetti and I did—about the advisability of taking blood specimens for genetic studies on the Kukukukus and others. Tomasetti and Jack both believe that any subsequent illness might be attributed to the procedures and thus cause trouble. Their objection is theoretically sound, but I do not know of an instance in which such has been the case. Without worrying about this, I have already bled 30-ml specimens from two dozen Kukukukus who arrived at Moke-Pintogori; and I am not worried about any reactions to the bleeding when we turn up at Simbari and Moraei, from where my subjects came. The fact is that I have done so much bleeding of primitive people that I am in all probability a bit overconfident, and Tomasetti and Jack are reasonably suspicious. However, I think I am a better judge than most of the clime of acceptance, and I am certainly not going to bleed for the research study unless the clime looks good. We have left the matter hanging, and I am left with only the anger which both Tomasetti's and Jack's objections have evoked in me. It is a peculiar response—stemming, I suspect, more from wounded pride than anything else.

[59:39–44]   All in all, the disagreement is minor, although it may result in failure to gain a fundamental biochemical insight into the race we are with; and I shall do all in my power to see that selling the idea of a blood collection to the Kukukukus (and to Jack as well) is successful. I suspect the problem will be greater with Jack than with the Kuks.

[60:14–33]   Today's valiant attempt to figure out the tribal relationships has failed. Our translation is wretched and totally unsatisfactory, and I cannot get consistent accounts. Word-lists show that the language is distinct from those with which we are familiar. However, we get various accounts that it is identical with that of Mobutasa across the Lamari River, and others that it is distinct therefrom. . . . They react to the name *Awa* as the name of the language, but at other times call themselves *Fore*—adding that they are "different" Fore from the Fore we know at Moke, Okapa, etc.

30 September 1957   Mobutasa village, Awa

[61:1–62:46]   This morning I have worked on compiling a word-list of the language. The people here are either more intelligent or, what is more likely, more anxious to communicate; and thus, in spite of the language barrier, I am getting incomparably better informants than I got at Yakeia. Other places with the same language as Mobutasa are listed here as: Amoraba, Agamusa, Tainoraba, Yakeia. I am still unable to get a name for the language, although Coleman has called it Awa. The word-list reveals a few similarities with Fore, but most of the words are distinct. These people have more members who speak Fore than we found at Yakeia—or at least more who admit that they do. They remain more cooperative, friendly, receptive, and happy than the Yakeia people were. On my first walk through their large village—in which 149 have been censused—they greeted me with smiles and food (sugar cane, bananas, etc.); and children and men accompanied me to the Rest House on another ridge, outside of the somewhat-sheltered bamboo grove in which the settlement is located.

Today the information I collected included an absolute denial of any past or present kuru, and a denial of any marriages outside of the Mobutasa group except for a few of their women who have gone to the Fore in marriage. . . .

Mobutasa people have brought us ample food, and the variety corresponds to that which we usually get from the Fore. One pig has been given to us; and Jack let the *luluai* shoot it with his rifle, placing the barrel near the pig's head. The *luluai* was much pleased with the results of the shot, and has been parading about the village with his villagers ever since—strutting, with the gun on his shoulder.

[62:17–63:2]   The cultural similarity to the Fore is so marked that I am surprised to note the marked differences between the languages. Obviously this language, Fore, and most of the other surrounding languages with which we have dealt would lend themselves well to a linguochronological analysis; and the relationships between these languages would be of interest to us in our work.

We are now five days out from Moke, and thus far have been most successful. The walks which threatened to be extremely difficult have passed as pleasant exertions. Weather has been good and trails fine—at least, under the dry conditions we have encountered. Our food has been excellent, for Jack's cook Andi is along preparing stews, bread, roasts, etc. Mobutasa was at war earlier this year with Agamusei, which we shall visit tomorrow. At Awarosa in April, the visiting *luluai* of Mobutasa complained to us that men of his village had received arrow wounds (some severe) when they were attacked by the Agamusei people. Now, only six months later, the Agamusei leader is here visiting us at Mobutasa and all appears to be peaceful once again.

6 p.m.

Mobutasa's *kros* [feud] with Agamusei is not as quiet as I thought it was a few hours ago. The Mobutasa *luluai* explains that no one from his group has dared cross the range to Agamusei since the arrow-shooting earlier this year. They still fear a fight. The entire matter stems from disagreement over an Agamusei woman, bought in marriage by Amoraba people, later taken by a Mobutasa man without a mutually satisfactory new exchange of bride-prices having been arranged. In a discussion of the matter, the Mobutasa *tultul* was wounded in the hand by an arrow from the Agamusei; and this has added cause to the quarrel. Mobutasa men and Tori (my Mobutasa linguistic informant, who is about 15 years old) will accompany us to Agamusei tomorrow, under the protection of our native police; and Jack will try to straighten out the matter then and there.

[63:22–65:3]   Yesterday we traveled from Yakeia down a precipitously descending *kunai*-covered ridge and near-vertical slopes to the Lamari shores, where two inclining trees formed a bridge over the Lamari—a span of 50–70 yards, I estimate; a small gap between the two trees (some 20 feet) was spanned by bamboo and vine. One had to climb the trunk of the first tree to start over its overhanging branches toward an immense tree leaning some 100 to 200 feet over the stream from the other side. Our cargo-boys could not cross this bridge as it was, and our native police worked a full two hours before it was strengthened and improved enough to trust cargo loads over it. Many of the cargo-boys had to be helped across by our police, who were less afraid than they of the raging torrent below. A nest of tree-ants (Jack calls them green ants) hung in the branches of the first tree forming the bridge. When I first came upon the bridge, I did not notice the ants; and while surveying it from the trunk of the first tree, they commenced to devour me. It was a hectic ten minutes before I could rid myself of all the biting Hymenoptera. To record the dramatic crossing we took motion pictures and still pictures very liberally, and then started up the extremely steep southeastern slopes toward Mobutasa. Fortunately, the afternoon was overcast; a cold and strong wind whistled through the grass over the higher slopes, and, by the time we were nearing exhaustion, it cooled us well.

1 October 1957   Agamusei village, Awa

We arose at 6:00 a.m., and by 6:30 were on the trail heading for the forests which cap the range above our camp. We climbed steeply on *kunai* slopes for only a half-hour, and then entered the forest. Here we had to cut trail a bit, for the Mobutasa-Agamusei feud had apparently closed traffic. The walking was good, the elevation high, and the forest cool. In this rain forest, we crossed a high range running approximately east-west; and finally, after a steep forest descent, we emerged onto *kunai* slopes again and found ourselves high above the Aziana river, having crossed from the Lamari to the Aziana drainages. We had a wonderful view of the Aziana valley, and below us the Agamusei hamlets were visible. A half-dozen waterfalls gleamed as silvery bands from the hanging hillsides across the valley. One, located directly across from the Agamusei hamlets, was a really immense falls—but only 100 feet in height, I suspect. . . .

We dropped down the slopes—a descent of 2000 feet, I imagine, most of which I did sliding on the *kunai*—and arrived at the site of the camp of John Coleman's and Harry West's last Kukukuku patrol camp. Their shelters had been destroyed—except for one, which was filthy. Agamusei natives, who spoke the same language as our Mobutasa friends, were busily erecting frames and *kunai* roofs on new shelters for us. However, their crude construction, and their use of decaying, pig-dirtied old *kunai* on the low roofs, made the camp a most unwelcome sight. When our cargo-boys and police descended, we lined them and ordered all the roofs to be removed, the height of the buildings raised, and new *kunai* walls and roofs to be made. Now, after an enthusiastic afternoon of building, we are camped in six fairly large bamboo-and-*kunai* (some other wood was used, but not much) shelters, which should prove weatherproof in the impending rain. If it runs true to form, it will rain all night long, as it has for the past several nights. The Mobutasa men and boys with us were received well enough by the Agamusei people; but in the afternoon, the Mobutasa *luluai* demanded a showdown with Jack, for he claimed that the Agamusei men were arming and might attack them if the matter were not settled. Jack "held court," and managed to straighten out much of the feud and obtain what appears to be a reconciliation. However, fearing that all may not be as quiet as appears, guards will be posted and we shall not run any risks from now on.

. . . Our Agakamatasa boys, Tiu and Wanevi, keep looking at their home from this side of the river, which they have not previously visited. Agamusei people speak less Fore than did those of Mobutasa, but there are a few here who do speak Fore and it is with their help that we secure translation. Their Fore is obviously the Awarosa-Agakamatasa brand of South Fore which Tiu and most of our boys speak.

2 October 1957   Agamusei village, Awa

[65:30–66:1]   We spent today lazily, not venturing from our camp, and completely wasting any chance of learning much from the Agamusei natives. Laziness, lassitude, and ennui seemed to overcome us with the dullness of the day. We postponed, hour after hour, our climb down to the bridge over the Aziana which our police and cargo-boys were building all day; and finally, as the poor weather dragged on, our procrastination led to a change of plans and we abandoned the venture to our reading. The same applied to my plan to take the cine films in the Agamusei hamlet of Wanovi (Wanovibi), which I visited yesterday and in which I had hoped to do extensive cine work today. Dull light all day long was an adequate excuse, but the lack of ambition stifled all other possible enquiry and endeavor. Thus, in the midst of a place only thrice before visited by representatives of Western culture, neither of us could assemble enough determination to work at observation.

[66:27–44]   Our Mobutasa guests are still disgruntled. Although the discussion with Jack yesterday seems to have settled the war with Agamusei, they still wish to "bring court" over the case of the *meri* [woman] whose intrigues launched the struggle. She was apparently paid for by Mobutasa, to everyone's satisfaction, and hostilities were canceled by a bride-price truce. However, upon returning to visit her kinsmen in Agamusei, she was violated by resentful youths of her home village, who disapprove of "out-marriages," which deprive them of copulating partners. Now this "Helen" is back with her distraught spouse at Mobutasa, but he wants vengeance upon her "Paris" (or Parises). "Paris" and his cohorts have fled into the bush upon our arrival here with the Mobutasa crowd. The Mobutasa men want satisfaction, and Jack was forced to send three police to try to locate "Paris" and his confreres this afternoon. Resentful Agamusei warriors were arming and belligerent, and our police returned without the culprit and with the report that Agamusei was restless and might attack. We discount this but are being cautious again, with night guard posted, as last night.

26 SEPTEMBER–9 NOVEMBER 1957

[67:1–9]   We had a Kukukuku visitor from the nearest Kuk settlement of the Auroga group of hamlets. He says he will go back with us, but interpretation with him is very poor. He speaks only a few words with the local people, and his language appears to be totally different from that of my Kukukuku boy, Waiajeke (from Moraei). We are trying to bring Mobutasa and Agamusei headmen with us on our trip across the Aziana tomorrow. The Mobutasa leader is resistant to the idea—not being, I suspect, very keen about returning to his home via Agamusei without our protection.

[67:19–23]   Having finished *Madame Bovary*, I am now on Orwell's *1984*. Thus, civilization and cultures most remote to it approach each other here in our Agamusei camp. Worried lest the natives might have a lash at us with their arrows, I find Orwell's fiction remote indeed.

2 October 1957[10]   *Auroga village, Kukukuku*

[67:38–44]   Our policemen and cargo-boys have built an excellent rattan-and-vine suspension bridge across the raging stream [of the Aziana River]. The river is deep, descending rapidly and with such swift and strong current that swimming in it would be impossible. The span was well over fifty yards; and the bridge hung over twenty feet above the water at its center, suspended by vines hanging from two immense overhanging trees and by many strands of rattan "cable."

[68:2–42]   We crossed several steep gullies after leaving the few beautifully landscaped and fenced gardens of the Agamusei people which are found across from their hamlets on this side of the stream. Here they were using an extensive system of irrigation to their garden slopes, the water being carried in pipes made of bamboo. Such pipes, lashed to poles, conducted water many hundreds of yards horizontally across steep hillsides and bridged gullies to the gardens....

At 10:15, after four hours of walking, we stopped at a small stream and prepared tea. Then we continued on through the forest, and after about one hour emerged onto a short *kunai*-grass slope on the summit, where we met four Kukukuku men carrying excellent bows and arrows, their bowstrings tightly set. They said, through our Agamusei interpreter, that they were out hunting; but they were obviously sentries, and had already been heard shouting over the hillsides with a warning to the others of our approach, and had lighted a fire in the hot noonday sun on the *kunai* summit. They were most friendly, however, not a bit afraid, and embraced us smiling, laughing, and exchanging shoulder pats. They examined our clothing as we examined theirs, without fear or anger, and told us to pass. We asked them to come along to their villages with us, and they consented—voicing their independence, however, in a nonchalant statement that they would first have a smoke and sit down a while, and then later catch up with us. This they did; and soon they overtook us and passed us on the trail. They warned their villages ahead of our arrival by lighting large signal-fires behind us after we went on.

---

[10] Actually September 3; this error is in the printed version of the field journal.

[69:35–70:16]   While camp was being built, we were surrounded by Kukukuku boys of about five to twelve years of age. They were friendly, affectionate, and enthusiastic. Before long, it was evident that I was already severely lice-infested from bouncing them about on my knees and accepting their hugs and overtures. Very promptly, we rewarded a few boys with packs of matches, whereupon every other child insistently demanded the same reward. Adult Kukukuku men swarmed about our campsite, apparently pleased at our attentions to the children. Waiajeke tried translating for us; and in spite of his protestations that they spoke a different language, it was obvious that he was conversing well with them, that with a bit of difficulty they understood what he said (although he often had to repeat it once or twice) and that he followed all of their remarks. I soon started to compile a word-list. The children were among the most remarkable linguistic informants I have ever had, though I regularly use children. Thus, they understood very soon that I wanted to learn their language—without any translation to explain this. They promptly started to give me words, repeating them slowly and distinctly several times and—to my utter astonishment—waiting each time for me to write the word down before going to the next, often repeating it slowly as I wrote. Such cooperation is difficult to obtain from European-language informants. . . .

There must be complex name-taboos here. The children rarely, if ever, called their own name; but others were enthusiastic in calling out the names of their friends for us, often with a note of jest in the mien of the caller and a note of shy embarrassment in the person named. When asked a name, the person to whom the inquiry was made waited for a friend to call out his name.

[70:38–71:8]   Small boys had their abdomens covered with red mud when we entered the communal longhouse below our camp. Some had put this on in planned patterns, others simply were smeared therewith. Whether it is a matter of play or of ritual I do not know. Such smearing of the body with mud is not uncommon among other highlanders, as a sign of grief for a deceased relative or for protection during illness. I thought it might even represent protection against harm which might result from our invasion. . . .

The characteristic item of Kuk clothing is the bark cape, which hangs down covering the entire posterior from crown to below the knees. These heavy cloaks of pounded-bark cloth are worn by adults and even small children of both sexes. They are suspended from a ring of cord which is caught and supported by the hair. The hair is cut to leave a protruding tuft somewhat behind the vertex, which catches and holds the string of the rain-cape. In the large house, we saw rows of bark hanging to be processed into these capes.

4 October 1957   Auroga village, Kukukuku

[71:25–72:39]   We have had a most productive day of work. After arising at 7 a.m. and showering and arranging for our camp tent to be replaced by a roomier *kunai* house, we climbed some 1500 feet to the forest-line above the *kunai*-covered slopes and visited two villages. The first was a garden village situated over precarious slopes which themselves provided ample fortification. Here were four houses of the type usually used as women's houses. These have floors of bamboo matting raised one to two feet from the ground. Over these is pitched a thick dome-shaped *kunai* roof which hangs down below the floor-level. Here also were two other houses, one a mere shack and the other a more permanent floored house which the headman of Auroga claimed was his. All six houses were completely deserted, although fresh fires were smoking in four of them. I suspect that women and children rushed off to the gardens and bush to hide as we approached.

We spent a half-hour there and then rushed across to the larger village of Ureba, which is silhouetted against the skyline on the crest of a narrow ridge above our camp. Here we found not the open village of a dozen or so houses I had suspected, but almost two dozen; each of these was fully stockaded on all sides, the stockades rising from the slopes which dropped off steeply from each house. In addition, there were even stockades between individual houses. Stockades were usually about five feet high; but on one side of the village, the least steep, the stockade wall was 12 to 14 feet high. Thus, it was amply evident that our Kuks are prepared for attack at all times. Within the men's houses, large heavy wooden shields are always present, with numerous bows and arrows.

Gardens were neat and orderly; and especially interesting were the very ornamental flowering plants and shrubs which were placed among the food crops, obviously for decorative purpose. These gave them a flower-garden appearance. With neat fences, sometimes with a floral perimeter, and with interesting floral patterns among the large-leafed taro, and with a neat, checkered pattern of various crops—*pitpit,* a type of corn, taro, yams, and some sweet potato—they were among the finest native gardens I have seen.

Jack and I, with six others from our party, climbed to the villages, along with a large delegation of Kuks who had already come to our camp in the morning. The Kuks remained armed with axes and bows and arrows most of the time, although they always were friendly and ready to joke. In fact, after finding how harmless we apparently were, and how much we played with the children, they became a bit boisterous.

. . . When we arrived [at Ureba], the village was deserted. We came in through a high stockade at the highest point of the village; and looking down over the successive stockades and houses arranged linearly along the very narrow and steeply descending ridge, it was evident that no one was anywhere in the village—except at the far, lowest portion, where a horde of women and children were racing over the stockades to escape. We sent those Kukukuku men and boys who were with us down to find them and persuade them to return; and after a half-hour or so, a group of women and children did return and were amply friendly and seemed interested in our photographing. Women and children had come in large numbers to our camp last night; and although they had not shown up early, they showed little fear when they did come.

[73:23–45] Back at camp, hordes of Kukukukus appeared; they roamed about, snooping into all our cargo. They were mighty light-fingered and had to be watched every moment. However, this did not disturb us as much as the fact that every man and boy was armed with an axe (all were heavily sharpened down, as I described yesterday) or knife. Thus, surrounded and infiltrated with this axe-and-knife–armed band, we became a bit uneasy and Jack alerted his police. We had no incidents, however; and they remained most friendly as we took pictures and treated a half-dozen cases of severe yaws and ulcerations. Two women with advanced plantar lesions of yaws, whom we had found in the village of Ureba above, came down for treatment. It had required a good deal of persuasion and insistent requests before they were prepared to come down to our camp. When they came, however, they seemed to be pleased with the attention. The headman, who has already extracted a number of gifts from us, has two large chronic tropical ulcers on his leg; these we are treating.

... With increasing familiarity, the Kukukukus tend to become a bit cocky and boisterous. They are proud and self-confident, and in no way ready to be ordered about or "bossed"; and we have avoided doing so. However, even simple requests they at times seem to interpret as orders, and ignore them resentfully.

5 October 1957   Auroga village camp No. 2, Kukukuku

[74:27–45]   After passing beside another of the long, low, ridge-roofed huts made of *kunai*, which I called "communal houses" yesterday (and which the translators tell us are pig-houses where the Kuks live when caring for pigs, while others suggest they are ceremonial houses for initiation ceremonies for the boys—we really do not know), and passing further gardens and another hamlet (part of Chemogo), we made camp on a minor summit crowning a long ridge with forests on all sides. We were obviously in a center of thick population; and since we had walked for two to three hours, and since we were a bit tired and did not want to miss contact with any populations we could stay with, we made camp. Our Ureba guides and headman tried to talk us on, said we could get no food here, and tried every means to change our minds. We remained adamant; and after much bribing of the local headman with a mirror and small knife, and distributing matches and smokes, they said they would get food later in the day for us, once it was fully evident that we were staying.

[75:6–40]   After deciding upon the site for our second camp in the region of the Auroga hamlets, I climbed the steep hill to the hamlets of Igopiji and Kuanimugu. I found a half-dozen women's houses in each, and each obviously could hold a full population of from 50 to 100 people. . . .

Men's houses and women's houses are both raised 1–2 feet from the ground, and have floors of either crude wooden boards or of bamboo matting. The matting is of far poorer quality than among the Fore or Awa peoples. There is a fire built in the center of each floor; and we have often crowded into these tight huts through the small, low doorway to sit among the fleas and lice and Kukukukus. Their ectoparasites are causing us no end of suffering. We shower and completely change clothes—only to pick children up into our arms shortly thereafter, and soon find ourselves itching violently as a result. The pubic coverings are made of a type of grass, cultivated specifically for this use, and plaited or tied together at one end into small sheafs, the sheafs tied linearly into a fan-shaped covering. They pile one of these fan-like skirts or sporans upon another until a great mass protrudes anteriorly. In this mass of grass is a veritable menagerie of fauna. Some have new-looking outer "fans"; but the under ones are always old and filthy, and apparently decay off sooner than they are discarded. I find that most adolescent boys have about 10–14 such "pubic fans" piled one on top of the other and held by a cord about the waist. Men have many more—up to 20 or more.

[76:10–19]   Our Kuks are very self-possessed. They show less bashful withdrawal than most other highland natives; and once they become at all confident with us, they handle our persons and everything we possess with the utmost of ease and nonchalance. They tend to be light-fingered, and will pick up anything we do not keep an eye on. Furthermore, they know their mind well and speak it forcefully—with self-assurance. They do not hesitate to push the keys, move the carriage, and interfere—in any way that intrigues them—with the typewriter while they crowd around me as I type.

[76:29–77:25]   I brought out paper and pencil, and four Kuks made rather interesting pattern-designs, all producing the same type of pattern: a "weaving" across the page from left to right with parallel bands of broken lines, often a regularly broken line of acute angles. No human or animal figures were produced. Every one of the artists had to be shown that the pencils were drawing instruments. They were totally unfamiliar with the use of pencils. However, immediately after being shown, they grasped their purposes and attacked the drawing with pride and purpose—as individuals, paying little attention to any production but their own. I hope to collect many more Kukukuku drawings. They are more rugged individuals than many other New Guinea natives, and far more self-assertive and proud, confident, and even somewhat more arrogant than our Fore, Kimi, and Keiagana.

As traders they are keen, reminiscent of Latin Americans. Rather than accept whatever we offer, they bargain and haggle. Furthermore, they know how to bargain shrewdly and to set a price, to reject offers they deem unsuitable, to suggest better ones, and to insist upon prices we either cannot pay or have not the items to pay with.

A pig was brought to us this evening, and the Kukukuku owner wanted a small axe (not a large one—of which he already had two, he stated), a small knife (not a large bush-knife, he specifically stated), and a mirror. These three items were his price. We have no further hatchets, and thus had to bargain and bargain before setting an acceptable price. He refused a red belt, colored bandana, two small conch shells and a small knife and a mirror, but settled for two small knives, the conch shells, and a mirror—rejecting the bandana and belt.

After we had bought the pig, they wanted to see it shot. Jack shot it at close range with bird shot in his shotgun, and the blast excavated an impressive hole in the pig. However, although these Kuks are interested and intrigued by typewriter, radio, flashlight, cine camera, and all other manufactured devices, they soon digest their function and range of application and treat them with the nonchalance of old experienced hands. Thus, they hardly deign to look at the planes which fly overhead about twice a day (from Goroka to Moresby, we believe). We look at every one, even take out fieldglasses to study them. They do not look up to verify the direction or type of plane. Fieldglasses intrigue them for the first few minutes, then they demand to use them; and after a few minutes of use they are satisfied that they fully understand them and are no longer even interested in them. The same for camera, elaborate cine equipment, typewriter (which holds interest a bit longer), and radio.

6 October 1957   Tchaiorogoro hamlet, Auroga village, Kukukuku

[78:1–23]   As we enter villages, women and children pour out the far end and race for cover. However, some now come down to us, greet us, shake hands, and even bring small children to see us. . . .

Betel-nut–chewing is universal among the men; and the betel nut and the gourd filled with lime (ash) for chewing, made from the ash of a special bark, is found in every men's house. I am surprised, since betel-chewing is rare among the Fore, Kimi, or Keiagana.

Our camp was crowded all day long with men and boys from Tchaiorogoro, which is situated directly above our camp. On arrival, I raced up through the village, which extends for almost a quarter-mile along the rising ridge. The trail upwards winds between high wooden stockades on either side of the narrow track. It twists in and out through low gates in the stockades, and passes between stockades which separate one house or group of houses from another. At the top of the linearly arranged village was a large men's house, which I visited. Everyone was most enthusiastic in welcoming me; men and boys did not mind photos being taken, although many were quite surprised at the camera and most of them tried to look into its interior through the lens.

[79:10–13]   In late afternoon I wandered again about the village. I was invited into the men's house, and here I ate pandanus nuts offered me by the men and boys. In their own village, their slight reticence promptly subsided and they were ribald and forward.

[80:2–81:9]   Later in the day I returned to the men's house in the village, and again found myself besieged by those who wanted to palpate and examine my genitalia—and more insistent youths who wished to engage in fellatio or manual masturbation. Again, all ages appeared keenly interested in the possibility of talking their visitor into such practices; and there was much ribald gesticulation and suggestion. My *mankis,* even Kukukuku Waiajeke, were immediately similarly approached. In all of this, there has been no offer of women, no suggestion of their women or girls, and no reference to heterosexuality. The Kimi tend to be almost as ribald and forward, as suggestive and aggressive, and make even more insistent offers—but it is always of their young girls, from prepubertal ages to mid-teens. This aggressive, hospitable heterosexuality of the Kimi is in sharp contrast to the homosexual approach of these Kuks. Whether this is part of their culture—or rather an effect of early contact during which they still hide their women, avoid showing them, or speaking of them, etc.—is hard to determine. That homosexuality is very embarrassing to and very well hidden among the Kimi, Keiagana, and Fore is certain. That it is in the forefront of Kukukuku sexuality and mentality is very evident.

I should be most interested in fathoming the role of homosexuality and pederasty in this culture; but our brief contact with them will not permit this, I am certain.

We found an early case of gangosa today, and several minor yaws. These we will treat; but the former, an adult male with extensive nasal erosion, we failed to locate in the evening when we tried to find him for penicillin injections.

Late in the afternoon today, a Kukukuku arrived from Wenabi village to the north of here. He said he had been one day and one night en route. We gave him a mirror as a gift, and found that his language was the same (or very nearly so) as that of these Auroga Kuks. However, the headman of the Auroga group from Ureba—who is with us in our travels thus far, with several men and boys from his group—was deeply offended by this gift to the Wenabi man. He stated that they were old enemies, that these people were not free to walk about in this land, and that if we were going to reward them for so doing, he would himself see that his people took care of the intruder. Jack made a noble and apparently successful attempt at reconciliation. He pointed out that the Mobutasa and Agamusei leaders were here with us on our patrol, having settled their disputes and joined again in peaceful friendship. Why could not the Wenabis and the Aurogas be friendly?

The Kuks eyed the police flag-ceremony with skepticism and little pleasure at dusk. The Sergeant, with our interpreters, tried to tell them something of *kiap namba wan blong ol* [Territory Administration]; but this they seemed most skeptical about, certainly not offering to accept for themselves any such concept.

Using a long aerial, over 20 feet high and extending for some 100 feet or more from our *kunai* shack, we are getting fairly good reception on Jack's battery radio-set—everything from Radio Indonesia, Soviet and USA short-wave, New Zealand and Australia. Port Moresby comes through worst of all. The news informs us of the launching of the Soviet satellite [Sputnik] and its successful revolutions about the earth. Such astounding and wonderful news occupied all my thoughts this evening, although they wander back to the Kukukukus per necessity.

[81:19–35]   As I type, the Voice of America plays; and the childish announcing, the juvenile caliber of the broadcast, the infantile and intellectually insulting nature of the program from the United States is embarrassing for an American abroad. To make matters worse, an ostentatious, hyperdramatic announcement beams through at far-too-frequent intervals, stating—with fanfare—that this is a broadcast from the United States of America; the very tone of voice and inflection of the announcement is offensive and degrading. Certainly there is nothing in this short-wave broadcast, proudly announced as "beamed round the world," for an American to be proud of. As Jack hears it (just as when I have heard it in homes while abroad), I wish I could bury my head in the ground like an ostrich.

7 October 1957   Camp No. 3: Tchaiorogoro village, Auroga Kukukuku

Today we spent at [Anji hamlet of] Wantekia, and we have only returned to Auroga in the late afternoon.

[82:16–83:12]   Everyone in our party had some Kukukuku embracing him for at least part of the way up to Anji; and again, suggestions of fellatio were the only form of aggression besides genital-handling. This was all done in a most hospitable, friendly, ribald, and exuberant fashion; and on the high tide of Kukukuku enthusiasm, we were carried into their stockaded village. We spent some three or four hours in their settlement before starting back for Auroga, and during this time had a good chance of seeing much of our Kukukuku hosts. . . . A pilfering, avaricious, begging, pleading, bribing attempt to get every bit of trade-items we had from us, and a good deal of fawning over us, coupled with much genital-handling and even suggestions of fellatio, characterized their behavior—which was only slightly suspicious, and generally quite open and friendly the entire time. Women, girls, and small children were absent. However, a group of boys from four to seven turned up, and the impressive headman—the most impressive Kukukuku, from the point of view of leadership, whom we have thus far met—even brought his wife and small children up to meet us within an hour of our arrival. She brought with her a large load of taro and *kaukau,* which was given to us. This is the first food the Kukukukus have brought to us without our soliciting it and without long delay. . . .

A group of young men from 18 to 25 have been most aggressively friendly, and the younger (10 to 17) age-group are almost as aggressive. This younger group engaged in all the insistent begging and bargaining in all the attempts at wrangling mirrors, matches, razors, etc. from us and our police. They hauled out much of the yellow-colored [orchid-stem] braiding fibers used in weaving their headbands, diagonal breast-bands and belly-bands, as well as many bird-of-paradise and parrot feathers. These our police, Tarangau, and our cargo-boys all wanted desperately to buy. These are the main trade-items the Kukukukus trade with the Fore and Awa people. Feathers are in great demand in the highlands, and the yellow fiber for braiding is used by Fore and others in making armbands. The narrow armbands and the wider armbands of the Kuks are made of a dull tan or cream-color fiber; they use the yellow fiber only for headbands, breast-bands, and belly-bands.

[83:43–85:14]   While we were at Wantekia village today, I snooped about a good deal. When the Kuks, our retinue, and Jack happened to be off at another part of the village, I wandered through a fence just near the men's house in which we had been. Beyond this fence was a small house at the door of which a small boy—thin, emaciated, and perhaps 10 to 12 years of age—beckoned to me. Having heard, for all of an hour or two, the conversation associated with our visit, he must have been most curious. He was at the door, holding himself erect by the low door-frame, and with one knee bent in extensive contracture. There was a stench of old ulceration emitting from the house—the type associated with huge tropical ulcers or yaws ulcerations. I came to examine him and see what was the matter. He had a huge yaws ulcer on one foot, the entire lateral heel being involved in a massive yaws ulceration with secondary formation of a large tropical ulcer about five inches in diameter. He showed me his ulcer willingly, was timorous and anxious but pleased to see me, and—in spite of our being alone with no others present—consented to letting me carry him through the door out to the sun where I could view his difficulty better. Apparently he had been assigned this house alone (there was no evidence that anyone shared the stench with him), and he was now left alone in his misery in this slightly isolated abode. I called other villagers, and the Wantekia headman and our Wenabi visitor both understood my sign and gesticulations which indicated that I would give him penicillin injection and medicine if he could be carried back to our Auroga camp with us. They assured me by gesticulation that I could take him—implying, I felt, that I could carry him there myself if I wished.

When we were finally ready to depart, I had to carry the lad out from the house myself; and our police and cargo-boys made a good stretcher for him. Jack succeeded in impressing the reluctant father and another Wantekia man into carrying the child. . . .

At Wantekia we found another man with extensive yaws—he appeared of his own volition after we had arranged to take the boy with us for treatment—and Jack found yet another severe case of yaws. These three came with us, and at our camp we treated them with six ml of procaine penicillin (1,800,000 units) and an additional 1,000,000 units of aqueous penicillin intramuscularly. We shall give all three patients further injections of procaine penicillin in the morning before we leave, and I hope that may be sufficient. The little lad has extensive hamstring contractures and perhaps some knee ankylosis; and thus, even if his ulceration and yaws should now heal well, he will remain moderately incapacitated. However, any necessary reparative surgery can be done at a later date once a government station opens in this region, and one is now planned for this very Wantekia valley.

[85:27–31] At Wantekia, the Kukukukus were quick to see the favor and attention I paid to Wanevi, and Jack to Asoi; and they promptly sought to drape themselves about the lads and later to tease them, push them about, and even to hit them. I had to withdraw the boys several times from their excessive jealous attention.

[85:40–86:3] The headman at Anji is an impressive, cooperative, friendly, and rather pleasant man who obviously swings weight in the village—perhaps also in all the Wantekia hamlets and villages, of which there are many we could sight from Anji. This headman promptly brought us food, introduced his wife and small children to me (the first Kuk to do either), and agreed promptly to bring sick yaws patients to our camp at Auroga for treatment. He did not beg and cajole for trade-items (perhaps because he received enough reward from us spontaneously), and only upon seeing us and our police-boys use soap and wash in a stream did he ask specifically for anything—that being soap, which we later gave to him.

8 October 1957   Camp No. 4: Iwane hamlet, Simbari Kukukuku

[86:42–46] We are camped in a deserted hamlet of six women's houses and one men's house on a high ridge overlooking the valley in which all the Simbari hamlets lie. This is the first set of houses and gardens we have reached after nine-and-a-half hours on the trail from Auroga.

[89:3–19] As evening drew nigh, we came down a long, easy descent to the hamlet in which we have camped. On our arrival we found only one young woman and a boy of about five years—huddled together outside their house, with their backs to us, trembling with fear. The fire in the house was still hot; we had obviously surprised them. There was no evidence of recent occupancy of the other five women's houses, nor the men's house, although we did not open the barricaded doors of most to make certain. From the far side of the hamlet we could look across the valley at many gardens where smoky fires were lighted; these appeared to be used as smoke signals. We finally spotted several men and women on the far slopes and made repeated efforts to call them over to our side, Waiajeke shouting the messages but with no success. A woman and girl far down the slopes below us, who apparently departed from our village just in advance of our arrival, ran away when we called out—as did the girl and young child we saw in the hamlet—as soon as we started to build our camp directly in the center of the small village.

[90:7–35] Thus, we camp tonight in a deserted Kukukuku hamlet of the Simbari group, and are on our precious rations without any food available from the natives. Tomorrow we shall move our camp down the valley to settlements in the center of the Simbari group, a trip estimated to require about 12 hours.

We have seen very little wildlife other than birds. A few cassowaries have been heard, tracked, but not caught. One has been sighted. Cockatoos and great hornbills have been seen and we have shot two pigeons, one of which we ate (the other we gave to the police) today. Only one snake has been seen, a poisonous death-adder of small size in the grass which was being cleared for our Agamusei camp. No mammals of any sort have appeared, other than the pigs and small dogs in the villages—and all the village dogs I have seen are miniature pups, no large ones at all. Our cargo-boys have taken down trees in which there were holes made by the *cuscus*, or New Guinea possum, but none have been found. Today, on the trail, we spotted a beautiful green frog, colored almost identically to a leaf. Also today, we finally ran into dense leech country. Leeches infested the trails on the high ridge which we followed for much of the day. Their density did not compare with that which we found on our trek south toward Papua last month; but there were plenty, and the feet of most of the cargo-boys were bleeding from leech bites. Animal-life in New Guinea is poor. I constantly marvel that one can traverse such vast stretches of virgin bush without seeing more fauna. Even insects have been few and unimpressive; and we have seen only a very rare, small but beautifully colored butterfly. The forests are so silent, and they tower along the trail so majestically; I feel as though in a Gothic cathedral!

8 October 1957[11]   Camp No. 5: below Kaiguanbi hamlet, Simbari Kukukuku

[90:48–91:16]   In the pouring rain, and with some disturbing leaks in our new *kunai* roof, we huddle on a ridge above a wonderful roaring stream. The Simbari Kuks have been among us—or rather, we among them—all day today; and they have been friendly, proud, arrogant, and at times somewhat over-friendly in an attempt to win trade-items for nothing in return. . . . However, there is a pleasant spontaneity about them, a naive curiosity—perhaps not so naive as it appears—and a distrustful desire to examine everything we have in great detail.

[92:17–93:17]   After a most refreshing swim [en route to Tchetchai village from Changai village], we proceeded down along the western bank of the main stream—which flows generally south—and passed numerous adequate campsites wonderfully situated for bathing and enjoyment of the stream. We feared to take them because of their remoteness to population, however, and the possibility that, should we take them, we should see little of the Simbari Kuks and get very little food from them. Finally, after some two dozen Kuk boys and men had joined us from villages high on the slopes and ridges above the stream, we ascended to the village of Tchetchai. All the way up, as when the boys from these villages first met us down at the stream, they were most jubilant, draped themselves about us, vied with one another for our hands and tried to hold them—either pulling us up or down (thus hampering ascent greatly) all the way up to Tchetchai. At Tchetchai, I raced up through the village to see if the level area we had spotted through the fieldglasses from Changai was adequate for camp. The village was totally deserted. It contained about 18–20 houses and was arranged, like most Kuk villages, linearly

---

[11]Journal entries for two successive days are dated 8 October 1957. This error is in the original journal; in his entry for 23 October 1957 (page 153 in *Kuru Epidemiological Patrols*), DCG notes that he had lost track of the date, but the error has been perpetuated in the published journal. In order to facilitate comparison of these excerpts with the full journal, we have left the dates unchanged.

along a rising ridge. Only some 100 feet above the village was a wonderful flat camp area which I thought was perfectly suited for our contact with the Kuks, and from a tactical consideration also excellent. I climbed further up the hill for 5–10 minutes and found no sign of further houses, although many garden houses were scattered on the nearby slopes in pairs or singly; it was to these huts, rather obviously, that many Kuks of Tchetchai had evacuated on our approach.

For a quarter-mile or so up the ridge, there was only muddy slope, densely encroached upon by shrubs and *kunai;* and only in one place was there a site adequate for a good camp—which would involve rather extensive clearing of secondary growth of shrub. When I descended back through the village, I suddenly came upon a crowd of several dozen Kuk men and boys, most of whom fled down the trail as I approached. Jack had moved on with our patrol, and the Kuks had apparently not been aware that I was not with him but in their deserted village. With his departure, many had slipped out of the bush to watch the patrol climb the next slope; and I, running, descended upon them. A few came back to my pleas of *"kanje"* (which means "come here," we believe). These were enthusiastic in their embrace, holding my hands and, as usual, pulling me too rapidly along the trail covered with ankle-deep mud. After crossing two small ridges, about 15 minutes' walk beyond Tchetchai, we came to a *kunai*-covered slope in the hollow of which, near the ridge peak, was a level, nearly horizontal plot large enough for our camp. Jack suggested this as our camp, rather in desperation after long dispute, for I tried unsuccessfully to bring him further up to a more exposed and higher portion of the ridge—always preferring as much height, wind, and view as we could possibly achieve, in spite of adverse conditions such as greater distance from wood and water in the valleys. In this, at least, I am not very thoughtful of our carriers.

[94:29–95:45]   Our repeated requests for food for our troop were passed by without being heeded. When pressed, they routinely stated that there was none to be had, in spite of the plentifully laden gardens all about. As afternoon wore on, the situation became critical; and Jack, with Kosinto and Sergeant Malekor, paid a visit to Tchetchai. Here he found a rapid exodus at his approach, and not the deserted village I had previously encountered. A man remaining behind conversed with him through Waiajeke's interpretation, and finally made it fully evident that they had no intention of supplying us with food. Jack distributed trade-presents to him and to the others who had wandered up, and—being arrogantly refused the food—made it clear that he expected no trouble when he ordered the police to take food for themselves from the gardens. At the sight of the rifles, and in response to this threat, the Simbari Kuks reconsidered and decided they had better sell us some food. Gardens were hurriedly visited; and shortly after Jack's return to our camp, a good supply of sugar cane, *kaukau,* green cooking bananas, and some—but very little—taro and yam were brought in and paid for with beads, matches, mirrors, salt, and a bit of paint and *girigiri.* All these items are desired, but the mirrors and matches go best of all.

During the afternoon, I wandered up our ridge—about a 15–20-minute walk—to the hamlet of Kaiguanbi above us. Here are some seven or eight women's houses and one large men's house. I spent an hour or so in the men's house, with a few men I found in the village and with the host of boys who brought me up to the village. Nothing unusual was in this house except for a package of trade-salt, the major item of trade which the Barua Kukukukus to the northeast of here sell to all other Kuks. It is a hard, firm, stone-like salt, packed in bark with a tight band of vine which is firmly sealed by plaiting and braiding. . . . Jack has obtained information that here at Simbari there is salt-manufacture as there is at Barua, and the Simbari Kuks themselves trade salt with their neighbors. . . .

Clouds descended in early afternoon on our camp, obscuring the fine view of the valley below, where, far down, the mountains drop off steeply as one leaves the highlands for the Papuan flatlands. We hope to head southwest for Muniri, then west for Moraei and finally, completing the Kukukuku patrol at Moraei, launch southward to the Yar people and downstream to the Purari River to the Papuan coast, resupplied with new cargo which should already be awaiting us at Moraei. The Simbaris have been the most difficult Kukukukus we have yet seen, these natives being much like the other Kukukukus but more apprehensive in their dealings with us and a bit more anxious for us to move on. They can hardly be enticed to enter our *kunai* house; they rushed out when we demonstrated the radio this evening, and nothing could induce them to return.

As dusk approached, our camp was deserted, and the Kuks have all returned to their villages to crouch about the fires discussing the intruders. We hope they decide to tolerate us, to supply us with food on a trade basis and, especially, to find us more and more acceptable. I like these Kukukukus greatly. In fact, I suspect that I prefer them to most of the natives of the Okapa region, who are very fine people. Their independence, their acquisitiveness, their curiosity, their arrogance all tend to make them more fascinating people. Their vanity and aesthetic sensitivity, and their complex and prudish lasciviousness, make them an intriguing people to study.

9 October 1957   Camp No. 5: below Kaiguanbi hamlet, Simbari Kukukuku

[97:11–28]   A swarm of eager boys accompanied me back to our camp, dragging me by the hand most of the way through the mud and over the steep ridges. Back at camp, I organized the word-list and then hiked with Wanevi up to the village of Kaiguanbi, where we spent about two hours—mostly in the men's house, examining every item found therein. . . . I took several photographs of the packaged salt-stick I described yesterday, either a trade package from Barua (where there is salt-manufacture from a reed of some sort) or locally manufactured. Inside the house, I took flash pictures of the boys assembled with Wanevi and another picture of the salt-stick; but the flash scared even the adults, and they would not approach the men's house with me again, nor enter it when I urged them to. The flash really worried them, and I could not assuage their fears. In contrast, at Wantekia and Auroga the same flash was treated as an interesting novelty and there was much clamor for the flash bulbs.

[97:46–98:28]   In the afternoon, I moved the table and collapsible chair out of camp to a small grove, where I worked at catching up with this journal and assembling the stack of unsorted papers I am still carrying about with us. As a journal, this can only serve to reflect the disorganization of my work on the patrol. In general, the Kuks are too unused to European contact to meet us with much confidence. Our smattering of language and poor interpreting make it impossible to obtain much reliable information in the short time of one to two days with each group, and thus I have been very unsuccessful in obtaining any significant data about the medical lore and thought of these people. We have asked about kuru several times, and they deny knowing anything about it. I have found their body-parts and body-functions vocabulary as rich as is usually the case in languages of New Britain and New Guinea. Medically, we have seen a fair amount of severe yaws and secondary tropical ulceration, which certainly indicates a great deal more in hiding. A fair amount of scabies, with secondary impetiginous ulceration and superficial skin infection has also appeared. Coughs—approaching bronchitis—are not uncommon, and URIs in children are fairly frequent.

Aside from this I can say naught, for we see too little of the people, and that little too fleetingly. Further sojourn would be desirable, but limitations in food which the Kuks can supply—or will supply—forces our onward movement. We could remain, using the food we carry, but that would limit completely the scope of our patrol and force a retreat to Moke soon. As it is, we shall get to Moraei just in adequate time, if we conserve our food cautiously and try hard to get local supplies. There at Moraei we hope to find a large cargo awaiting us for our trip south to Papua. We shall make our patrol crew much more compact and dismiss many police, *mankis,* and cargo-boys, as well as dispense with luxuries.

10 October 1957   Camp No. 6: Maiguanga village, Muniri Kukukuku

[99:11–40]   Not knowing whether Muniri would prove a feasible day's walk, whether we should have to make a bush camp, or what time we might arrive at a camp at Muniri, we made an effort to get off early and managed to start on the trail up the ridge above our camp toward Kaiguanbi hamlet by 6:00 a.m. Just below Kaiguanbi, we turned and descended into a narrow valley along a garden fence, climbed, then crossed a second small valley, and climbed steeply up to the large village of Gorogwanga with over a dozen houses. . . . We spent 90 minutes here, attending first to a cargo-boy who had fallen with heat exhaustion on the *kunai* slope up to the village, even at this early hour. He had constricted pupils, weak and rapid pulse, no fever, and much perspiration. I suspect that all of our group are low on salt (we have not been rationing it), and thus we stopped to issue salt to everyone; and our cargo-boy rallied on salt and water—in addition to being given the "works" of native medicine, struck with *kusai* (the ferocious stinging nettle along the track), doused with water, given ginger to chew raw, and beaten lightly with a bamboo rod. We removed him from the too-heavy box he was carrying, and assigned another lad to help him along the track.

[100:44–102:15]   We arrived at the first hamlet of Maiguanga (containing only three houses) after passing through a fine garden with yams, taro, sweet potato, sugar cane, and bananas, and some small swampy enclosures fenced in (as they were at Kaiguanbi) for growing the reed-like grass used in making *kanila* (the louse-laden pubic-covers of the Kukukukus). Here old men embraced us, and through Waiajeke's interpretation we learned that the Muniri people had just built a *haus kiap* on the next crest for our arrival. We were astounded at such news and such enthusiastic welcome, after the cold reception our party had received at Simbari. We took compass-bearings on all visible points, enjoyed the panorama southwards out of the highlands as the ranges drop off toward Papua, and then started down the steep slope to a river separating this hamlet from the ridge on which the high conically roofed *haus kiap* was built in typical Kukukuku fashion. At the river we found a really large stream—bridged well, and with immense pools adequate for swimming, and with tier upon tier of exciting waterfalls. Finally, several hundred yards above us was a mighty cataract thundering down 80–100 feet in a gigantic roar from cliff-sides high above.

We promptly undressed and started a wonderful two hours of swimming and rock-scrambling with our 80 cargo-boys looking on, a few also stripping off their *kais* and *watis* and swimming too—or rather bathing, for few trust themselves in deep water, and, as is usual with highlanders, few can swim. (Our Papuans from Yar are a notable exception.) Many of our police also washed. We climbed up the waterfalls and rapids, avoiding the impossible main currents (which were far too deep and strong to brave) until we reached the major falls. Here, the entire large river plunged far out into space and fell about 100 feet in a tremendous falls. To one side, a small portion of the stream dropped from still higher up the cliff to an even lower point, forming a secondary falls about 200 feet high, but very narrow. The mist and turbulence of the great basin into which the stream fell made it a more forbidding place for swimming than the somewhat quieter pools with less staggering falls entering into them further downstream. All the Kuks—about six from Simbari, five from Muniri who came out to meet us, and another several dozen from Muniri hamlets hereabouts—scampered over the rocks with us, but carefully avoided the waters and did not attempt to undress and enter the water. They appeared afraid, and I suspect few can swim. Our nudity attracted considerable curiosity and embarrassed attention.

In late afternoon, we started up the slope from this river playground to our camp prepared for us by the Muniri Kuks. Here several dozen men met us—including two in trousers and undershirts and wearing blue plastic rings which John Coleman had distributed to them on his patrol last year, thus appointing them village officials. These *tultuls* had already provided a supply of *kaukau*, taro, sugar cane, and bananas adequate for our large party. Muniri, like Simbari, had been visited twice previously by Europeans: the 1956 patrol of West and Coleman, and the 1956/57 patrol of Coleman. The two "officials" were the only really spontaneously cooperative Kuks we have met on all of our travels through the Kukukuku region. The Wantekia headman did supply us with food without request, and the headman of Auroga did appear and offer us a modicum of assistance. However, at Simbari no one helped us at all; and what contact we had with the natives, we had to search out. The *haus kiap* here at Muniri was of the style of a Kukukuku men's house without the raised floor. It had just been built. A few nearby rectangular *kunai* huts were adequate for housing the cargo-boys, and all that we had to erect were walls and a level floor in our house and a small cook-house.

While this work was in progress, a crowd of about 50 Kukukuku men and boys—only a few armed with knives and axes, and very few with bows and arrows (a real change!)—swarmed about Jack and myself, trying the cameras, fieldglasses, our clothing, and examining everything they could. Their headbands and much of their feathers and fine decorative ornamentation was traded to our cargo-boys and police shortly after our arrival. I set up the typewriter to type up yesterday's journal account and then started the two headmen, and two adolescent boys to whom I took a fancy, drawing....

[103:29–42]   As imitators of each other, the Kuk artists do remarkably well. As soon as one starts, hesitant others, who have been fingering pencils for minutes and unable to draw a stroke, start right in. When a curved line is first hit upon by one, others watching him and also trying to draw, promptly introduce a curved line. Similarly, straight lines, broken lines, closed polyhedral circles and spirals all require but one of three or four drawers to introduce them, and promptly most others follow course and use the new form in their drawing. All try to fill the paper, but all do so systematically, working slowly across it from one edge where they start their work. They very rarely skip to a wide blank space of paper, and almost never attempt a small individual figure or pattern on a blank part of the sheet.

11 October 1957   Camp No. 6: near Maiguanga hamlet, Muniri Kukukuku

[106:15–24]   Two of our police and several cargo-boys have developed sore, cracked feet. All the others are well. Kosinto has epididymitis and inguinal adenitis and tenderness along the spermatic cord and dysuria. I suspect that he has flared up GC en route, but must keep an eye on him for other possibilities. Tarangau and many of the others complain of their knees. We have been fortunate thus far, however, in the absence of any serious illness. The acute *limbun* poisoning of a half-dozen boys and Anua's acute meningitis were far more troublesome and serious matters on the last patrol than anything we have yet encountered. I hope our luck holds out.

12 October 1957   Camp No. 7: above Kataramapinti hamlet, Moraei Kukukuku

[106:38–107:11]   We broke camp at Muniri at 5:30 a.m., and by 6 a.m. were all on the trail with one of the Muniri headmen and two boys—both Simbaris, who are still with us—as guides. We climbed steeply up the ridge on which Maiguanga hamlet and our camp were located, and finally reached the stream which rushes down to form the great waterfalls, but reached it above the falls. We followed this stream, sometimes along its slippery, steeply pitched, rocky bed but occasionally heading into the forest to parallel the stream on a ladder of roots. We climbed steeply for hours. I suspect we climbed to well over 8,000 feet. At noon we were still climbing and stopped for tea, a tin of crab meat, and bread (baked by Andi in the camp oven) and butter and jam. By 1:00 p.m. we had reached the ridge of the Lamari-Vailala watershed and had a fine view northwestward across to Pintogori, the Okapa pine forest, and the Fore regions. Then began the seemingly endless, steep descent. Without spikes in my shoes, the number of long, steep log bridges over ravines and down slopes made walking very difficult.

[107:28–108:8]   On the walk from noon to 2:00 p.m.—Tiu leading—Wanevi, Waiajeke, and I raced together ahead of the rest of the patrol. All three boys sang and chanted numbers in English from one to 375 whereupon we interrupted this venture by stopping to admire the great view to the west and northwest, which encompassed cloud-free Mt. Michael and the entire mountain horizon from the south of Mt. Michael to far north and east. Ivaki, the Purosa hills, and the Kaza River (which flows down the Purosa-Ivaki valley into the Lamari) stood clearly before us. I took several photos and we engaged in Fore banter—much of which I could follow, much they helped me with. After an hour or so, I was again handling complete Fore sentences well with correct grammatical structure. This raced descent, after a rapid ascent to the pass, was perhaps the most pleasant walking I have done on the trip. The real virtue of the boys and my reason for keeping the "swarm" was amply evident. Here in an unrestrained atmosphere of free interchange and complete irresponsibility, I came closer to the Fore and their culture than I had managed to for many a week. . . .

During those two hours a sharpness of wit and repartee, a frankness of expression of opinion, of likes and dislikes, criticism of Europeans seldom uttered to our faces, and even a glance at the boys' ability at self-mockery and ridicule were the rewards I received. Tiu is masterly along the trail, swinging a bush-knife more expertly than any of our police, in his exuberance taking down numerous trees which need not be taken from the way—just for the boyish pleasure therein.

[108:40–109:37]   On the way up the mountain divide today, we passed through two bush camps of the Muniri Kukukukus. One had five houses, another had four. They were substantial hunting villages, obviously without garden areas and obviously temporary. A great section of our walk was through forest, which was largely of pandanus, the edible nut-bearing variety. . . . Our cargo-boys have often stopped and cut down the huge pandanus palms to get the great nut-mass they bear. The Kuks, however, climb up the trees for these—allowing for further fruit-bearing, I imagine. At these Kuk forest-camps—situated in the center of huge stands of pandanus—our boys started to take down the trees, and Jack was forced to *tambu* [forbid] this, for we thought it was hardly playing fair with the Kuks' forests.

Soon after we arrived at Kataramapinti, I accompanied the crew of Moraei boys down through their village and sat with them for about two-and-a-half hours in one of the *haus bois*. Here I rapidly became completely flea- and louse-infested (the usual result), but the sojourn was well worth the subsequent pruritis. On our way back up to our camp, which had been erected during my session down in the village, we were passed on the trail by a girl of about 16 who, as she descended the trail, scattered the eight boys who were with me off the trail and into the bush, where they turned their backs on her and hid their heads in their *nambai* (rain capes). A few minutes later, we came to two women's houses, facing in one direction and the front screened partially by a fence. Here the road branched—one track passing behind the houses, the other in front. I took the track in front; the boys took the track behind. I called to them, and one explained (through Waiajeke as interpreter) that they could not look at the women in front of the houses. These then are the only such examples I have yet seen of female avoidance. What age, sex, clan, and family relationship determine it—or if it is as valid as it appears—I do not know.

13 October 1957   *Camp No. 8: near Anjapte and Watcheramapinti hamlets, Moraei Kukukuku*

[110:29–43]   We arrived here at the *haus kiap* of Moraei—just built this month for our arrival, in response to notification of the Moraei Kuks who visited our Pintogori post that we planned to make this trip—a good hour before the patrol, and found all the replenishment supplies from Pintogori for which we had arranged. Medical supplies, trade-items, two full bags of rice, salt, sugar, two cases of tinned meat, trade-tobacco, *laplap* cloth, and a case of tinned foods for ourselves, in addition to rum and cordial; all, in other words, that we had requested, had arrived, accompanied by two native police from the Patrol Post. The police we had sent on ahead the day before yesterday had gone on to Agakamatasa (which we can view clearly in a sweeping view up the Lamari from here) to try to arrange for an added supply of food to be brought down here on our patrol, for by now we have had ample experience with the Kuks as inadequate providers of food.

[111:14–49] Our cargo-boys and Jack came in an hour or so after our rushed arrival, and we were soon surrounded by a cordial and happy group of Moraei Kukukukus. No food had yet been brought; but in mid-afternoon men and women went out to the gardens, and in late afternoon a rather good supply arrived. Just about this time, a large group of men and boys from Agakamatasa arrived, carrying a very large additional supply of food. The trip here must have taken them six hours or more of difficult walking, including crossing the Lamari on the new bridge which we have had erected for the purposes of this patrol. Thus, by late afternoon, we were assured of adequate food for our hungry group. Jack went through our 70–80 cargo-boys, seeking volunteers to continue the trip south to Papua and the coast. Few were anxious to go; most of them were completely unwilling. He finally induced 33 to come along, and the others will return to their homelands via Fore Agakamatasa tomorrow.

In spite of the hardships of the trail, we have been living in utter luxury on the patrol thus far. Thus, two nights ago we lay in bed at Muniri, listening to the symphony orchestra from Sydney, with an American guest violinist, playing Beethoven's "D Major Concerto." We had just finished a four-course meal of soup, excellent curried chicken (canned), rice, *kaukau*, bread and jam, and cheese—and along with the meal a full bottle of Bordeaux Superieur, a French wine I had picked up in Goroka last month. Reading Orwell's *1984* in bed, listening to the symphony with excellent reception, sipping the last of the fine wine—we could not possibly hope for greater luxury (all this with a night guard of watch-fires burning, posted in the event of Kukukuku attack, in a region only twice previously visited by Caucasians). Last night we had tinned stew, rice, *kaukau,* and barbecued wild pigeon. The split bird, rubbed in salt and cooked over an open fire, is a real treat. I had never tasted the New Guinea wild *balus* before. These then are the hardships of such patrolling. Before leaving Muniri yesterday, we had canned kidneys in wine for breakfast!

14 October 1957   Camp No. 8: near Anjapte and Watcheramapinti hamlets, Moraei Kukukuku

[114:13–34] We arose to a fine day, and quickly got the police off hunting for a possible route south—since all Kuks and all visiting Agakamatasa Fore do nothing but deny the possibility of the trip, deny any knowledge of a route, and (when forced to admit a trade-route) insist that it is south from Muniri, starting from our camp a hard and long two days back. We have told them we hope and plan to follow the Lamari south; and they insist the river is dangerous, that there is no population and food supply, that there are canyons, side-streams, and cliffs that make passage almost impossible. However, we have sent out police to find and start cutting a trail, and hope to start the day after tomorrow.

I went down to the hamlet of Watcheramapinti and remained there all morning taking pictures. About five women, six small children, and eight older boys were there. I visited their gardens, which are excellently and securely fenced in with aesthetically pleasing fences; the cleanliness and decorative flower-arrangements make them unusually pretty for food-gardens. Flowering plants with varicolored leaves and shrubs are planted among the corn, sweet potato, banana, cucumbers, and *pitpit*. Occasional dead trees and logs left in the garden add to its interest, as do rocks in a rock garden.

[115:13–27]   The headman of Kataramapinti appeared and received Jack's wrath for having insisted that all the Kuks had moved to this northern region of the Moraei complex, when it is evident that most have moved south to his Kataramapinti region. After seeing Jack's performance of rage and anger and reiteration of his threat to return and "have his hide" if we find, on our way south, any trails he had refused to tell us of, he reconsidered and told us that he would come with us down the route as far as he knew it. Thus, today it now appears that we have a route south which is not too difficult—for the first day, at least. Our police have returned, along with a huge adult cassowary they have shot, and report finding a route down and across the river which we crossed on our way north from Kataramapinti. Thus, we are a bit less worried than we were yesterday about the route south.

[116:17–29]   Yesterday our police shot two large black birds with vulture-like heads. They claim the bird does not eat carrion but survives on nuts and fruits and is related to the white, sulfa-crested cockatoo which is found everywhere about this portion of the highlands. These birds had brilliant red stripes on the wings and chest, and a few pink terminal wing feathers. The boys say they are good eating; but we did not try them, and left the feast to the police and boys.

We took a muscle steak of the cassowary, which—after much hammering with the back of an axe and bottles to tenderize it—turned out to be edible and tasty but tough. We also took cassowary heart and liver, both of which are good but for which I have little taste.

15 October 1957   Camp No. 8: Moraei Kukukuku

[118:15–119:14]   On the radio this evening, we hear further kuru news: that kuru is probably hereditary, and that it seems restricted to the Fore-Okapa region. Since all kuru data must stem from Vin or myself, we are mighty astounded at all these radio reports. The film team, returning to Moresby, must have given a story, for we heard in the Mobutasa region a Territorial radio account of the kuru research film, in which the film-team members quoted rather well the very words I used in describing the purpose for which I had requested the film and what the film should attempt to portray. I am grateful to them for being so scrupulously careful to repeat only what we told them.

To be left out of all the extensive publicity accounts of kuru which keep appearing—while the Administration blows its horn about kuru studies, kuru films, kuru projects—is a bit ego-deflating. However, in spite of the initial wound to pride and vanity, I would in all honesty rather have it so. I know full well that kuru study—whatever the Film Unit does or the Administration's aides provide (in entomological, botanical, etc. assistance)—is all part of what Vin and I have planned and for which we have laid the foundation. If it contributes, in any way, to solving the problem, fine! If not, it is wasted effort and only embarrassing to have to talk about it. Thus, I am most pleased to be on the "black list" as an American, with the comfortable knowledge that the interest and stimulus and plan of attack behind all kuru work stems from Vin and myself alone thus far. If we meet with no further success, the less publicity the better. If we can get more out of our kuru survey and study, then I shall be all the more pleased to have remained in anonymity until the proper time for professional disclosure of any significant findings. I can only hope, however, that I shall not find Dr. Gunther and the offices in Moresby so hostile as to interfere with our further study of kuru, or our plans for future work. In general, recent communications have been so favorable that I have been encouraged. The radio statement tonight came directly from the second kuru paper, a manuscript Gunther has just had in his hands and is quoting from; or it came from our discussions with the Film Unit people, at which time we often reiterated the problem kuru caused in pay-back murders and assaults and other reprisals for kuru magic.

It is fortunate that the Administration in Goroka and Moresby do not know of this Kuru Research Patrol through uncontrolled Kukukuku territory and down an unexplored route to Papua. If they did, we certainly should be hearing great tales of our progress on the National radio.

Two Kukukukus have—as a result of Jack's threatening, cajoling, bribing, and display of displeasure at their having lied to us at Kataramapinti (that everyone had moved to here) and here (that everyone had moved back to Kataramapinti)—agreed to accompany us south, for several days at least; and we are very pleased at this. Today we approached them for further help; and after bribing them again with a considerable distribution of trade-items, they agreed to supply six additional carriers to our line to come south with us for one day, carrying food for the first day.

[120:27–121:6]   In one house [at Watcheramapinti hamlet], I found a package of native trade-salt from Simbari. It was a half-consumed package, about one foot long (one foot or so having been already consumed). We offered to buy it for a *laplap* and twice the salt quantity it contained, but were refused the bargain. We upped the price to two shells (large *kumus*), one large knife, twice the quantity in salt, and a *laplap*. Our offer was determinedly refused. No full, unopened trade-packages of this rock salt were available. We could only get them in Simbari, we were told. I had seen these in the men's houses in Simbari, but had not been able to buy one. Now we really wanted even this half-consumed item, for it is certainly an old traditional item of intertribal trade which is of very considerable anthropological and historical interest, and of immediate interest is the fine braided and laced pattern of the pandanus leaf and vine packaging. We had to settle for a small piece of the salt, which I took for eventual chemical analysis.

We spent about three hours in this hamlet. Before leaving, I examined a sick man seen earlier and found him to have a low-grade fever, excessive perspiration, no chills, but moderate scleral icterus and mucosal anemia, with a yellowish and pale facies. He had no lymphadenopathy, but marked splenomegaly, with a notched spleen extending three finger-breadths below the umbilicus in the mid-clavicular line. The spleen tip reached medially to the umbilicus. This smooth, sharp-edged spleen was accompanied by a slight, non-tender hepatomegaly (only two fingers-breadth below the right costal margin). I suspected indigenous malaria with a recent prolonged attack of clinical malaria.

[121:23–36]   Before leaving this Watcheramapinti hamlet, we told the sick man (who was too ill to walk back the tough half-hour trail to our camp) that we would send medicine to him, and prepared to leave. He then—having argued and discussed matters at length with the women who appeared to own the trade-salt and who seemed to have the final say in the matter of its disposal—told us that he was willing to sell it for *laplap*, knife, and salt. We gladly accepted, and one of the women agreed to bring it to our camp for trade. We asked that she also bring food for our patrol, and she complied by collecting a *bilum* full of corn and cucumbers for us. We took this opportunity to urge everyone to bring our patrol more food, and the women promised they would.

16 October 1957   Bush camp No. 1 on the shores of the Kataramunga River, en route to the Yar people

[123:13–26]   We made camp here at 1 p.m., after only four hours of trail progress—and that at a slow pace. We left our Moraei camp at 7 a.m., having arisen at 5 a.m. and packed and lined the cargo by 5:45. Delay was occasioned by an insufficient number of cargo-boys to carry our cargo. We had counted on 26 two-man loads, including the three bags of *kaukau* we are bringing along for the first day. Six Kuks from Moraei had promised to carry these for one day for us, but they failed to materialize. Those Kuks hanging about our camp in the morning were completely unwilling to carry anything. Thus, we were treated by them—as Kuks have a habit of treating us—with quiet curiosity, a lack of comprehension of any requests for service or assistance, and (when adequate translation makes lack of communication no longer any excuse) a bland but definitive refusal.

[125:18–22]   We have distributed to the carriers the bags of *kaukau* and one-half bag of rice, and thus some six cargo-boy loads are released and we should be better off. As it is, however, many—in fact most—of our loads are above the regulation maximum of 35 pounds per cargo-boy. Many are carrying 45–50 pounds or more.

[126:1–15]   We lined the entire patrol in the camp this evening, and gave 0.5 gm of chloroquine phosphate to each person. Thus, malarial prophylaxis is taken care of for the next week. Mosquitoes are now biting. Upon descending to the Kataram, we encountered the so-called "sweat bees" in great swarms. I did not know any common name for these little bees when they plagued us at our lowland camps on our previous Yar patrol. Jack knows the black variety, which is here; but the yellow-brown species (which predominates by far, and which is a great nuisance although it does not bite or sting) is new to him, and this lighter-colored insect was the one which was so troublesome on our last patrol south. When bush is freshly cleared, as now in our campsite, the bees are particularly dense. They swarm thicker and thicker as the day wears on, and—just as it happened on the Yani River side during our patrol south—with sunset they disappear.

[126:31–37] On the way down, the boys picked a highly aromatic yellow-to-purple–colored leaf, ovate in shape and some eight-to-twelve inches long, which they roll up and insert through their nasal septal hole, which is almost one cm in diameter (large enough to pass the colored drawing pencils into it when they are drawing for me—a practice they gleefully discovered, and one that demands surveillance if I am not to lose colored pencils).

[127:2–43] The traditional fighting enemy of the Moraei peoples appears to be the Auroga group, Auroga being really closer to Moraei than our indirect approach via Simbari and Muniri (after leaving Auroga) would indicate. Tiu tells me that the fighting stockade we saw at Agakamatasa was only recently erected, because the Auroga people threatened to attack them—the Agakamatasa people being allies of the Moraei Kuks. No fight occurred, however. In Moraei, they tell us a rather difficult-to-believe yarn of a recent attack by Auroga people and the killing of many Moraei people....

17 October 1957　　*Bush camp No. 2 en route to the Yar people, at the site of deserted Kuraripinti village*

We broke camp at 6 a.m. and climbed rapidly up the steep ridge to the deserted site of Laye (Lalye), which we visited yesterday in the afternoon as we investigated the trail ahead for today. The climb was made in about one hour, and then we descended a steep gorge, across a small stream, up another much lower ridge, down and across a small stream—repeating this process a number of times and never descending very far nor climbing very high until we descended to the Kuata River, a stream big enough to have several water-holes a few feet deep, adequate for bathing and a modicum of swimming, and obviously much larger when in flood. Here we stopped at 11:30 for tea, lunch, and a swim; and then we climbed a steep hill of 500–1000 feet, and from the top had a fine view south and west over to the deserted site of Kuraripinti.

At about 2 p.m., as we approached Kuraripinti ground, we descended into a stream gorge and followed a long way down this shallow stream-bed. While in the stream-bed, a violent torrential rain broke upon us. Our police and carriers rushed under ledges for shelter—all grabbed leaves to cover their heads, and very few made efforts to protect their loads from the rain. Foolishly, we had neglected to make all rice-loads waterproof; and soon I was forced to hold the patrol up and build shelter for the rice-loads, until Jack and the rest of the patrol caught up with us. Then, since we were standing in the downpour, in too muddy and low a forest to make camp there, we sent Sergeant Malekor ahead to find an adequate campsite, which he soon did. Rain ceased, we cleared forest and built a camp with sago all about and plenty of sago leaves for roofing nearby. This was our first sago stand on our way south. The Lamari is off to the west, a bit below us; and although we cannot see it, we can hear it.

[128:30–32]   The Yar people with us [whom I had brought back to Okapa from our previous patrol] are still in territory that is strange to them, and are unprepared to rush ahead to their home village of Weme as yet.

18 October 1957   Beside the Tu River, below Kuagatnunga hamlets

[129:6–10]   We are in a comfortable sago-roofed lean-to shelter beside the roaring Tu River; the Yar people have finally recognized their home-country, and at this river they first knew where they were. They even have a name for this large stream, which the Kukukukus call Kuagatnunga.

[129:35–131:11]   We walked and cut trail from 6 a.m. through about 1:30 p.m., without even stopping for a rest, thus making good time. . . . We crossed numerous shallow valleys, each with a small stream, and finally came to this Kuagatnunga stream, in which I swam in a water-hole for a half-hour before crossing it and finding, just above its banks, a fine level campsite covered with a very beautiful spike-flower with huge leaves like lily pads. Here we waited about one hour for Jack and the cargo-boys—who had started after us and had taken their time en route—to begin to arrive. Jack informed me that hungry cargo-boys were exhausted along the route, straggling for over a mile back; and many were exhausted to the point of being "flaked out." We decided to make camp here, and I sent off two police with our Kuk guide and three other Kuks to climb up to the houses and gardens of Kuagatnunga (which are reputedly near here) to persuade the local Kukukuku to visit us and to come with food to sell. Our group began to clear bush and to make camp; and, in spite of threatening rain, at about 3–4 p.m. we had a clear afternoon and a clear, starry early evening and night. . . .

Our cargo-boys—roaming the surrounding forest in search of poles, sago and other large leaves for roofing, and vine for "rope"—came upon a human skeleton only some 75–100 yards from our camp. Investigation reveals that this is a burial place in a neat, highly landscaped, small grove of *mareta* pandanus. A stretcher on which the corpse was undoubtedly placed, well above the ground in this pandanus clump, remained shoved to one side. A decaying *nambai* (rain-cape) and faded *munyiri* (chest-band) all suggest a man; teeth number 32 erupted, and pelvis— although small—is that of a man, I believe. It appears that the body, once decayed, was tipped off the stretcher and the bones left on the ground below their former raised position in this small sheltered burial-site.

Shortly after our return to our camp from this burial-site, we were told that another fresh burial had been found nearby. Wanevi was too afraid of the skeleton and burial-site to bring the camera there for me, and handed it to me and ran off; with two cargo-boys and Asoi, Jack and I traveled about 10 minutes upstream, crossing and recrossing it knee-deep at wide fords, and finally passed several groves of pandanus on the river shore. From one, a hideous stench of decaying flesh was wafted to us with every breeze. The boys clung tightly to their noses, ran in quickly to see the burial for a moment, and ran out, refusing to remain near it any longer. Jack and I crossed the stream again, and went into a very charming grove of pandanus with the jungle cleared about it. In an overpowering stench of a bloated, black, oozing and swollen corpse (seated in a hunched-up position, and tied with neck slightly bent, knees against thorax, and the right hand up over eyes and forehead), we studied this burial. It was tied about six to eight feet off the ground into the center of the pandanus grove; the site was cleared to make it a very pretty one and aesthetically planted around the periphery. Along the river, from where the pandanus tops were seen looming above the surrounding thicket, the burial-ground gave a very pleasing appearance. At the site, however, the nauseating stench, the swarms of flies and beetles crawling over the black, oozing, and swollen flesh, and the drone of these insects together with the roar of the passing stream, made the burial-site a most unwelcome place to loiter for long. We took pictures quickly and rushed out. The head of the corpse (which, from the state of decay and preservation, must have been about one week dead at the most—and perhaps only a few days) was shaven, and I could not make out whether it was male or female.

On the way across the stream, Asoi, carrying my camera and lenses, tried to save Kuru, Jack's dog, from being washed downstream, and himself fell into the stream with my camera. He promptly pulled the camera out; but water had gotten inside and has ruined the film, including the pictures of both burials and of many flowers here in the bush, and interesting shots of *cuscus* [opossum] traps along the trail yesterday.

[131:16–49] Upon our return to camp, our sleeping-shelter and a cook-shelter were both finished. A group of Kukukuku from Kuagatnunga were sitting in camp with our Moraei Kuks.... They have houses and gardens up the ridge north of the Kuagatnunga River, very near to where we cut our trail; and we have thus "passed them by" to move down to a site which they use principally for burials. The group included two typically attired older Kuks, three young adults (all dressed in *laplaps* instead of in the traditional garb of the older two; one of these also wears a shirt). There were two boys, of about 10 and 12 years respectively, with them also. They had brought down from their gardens a small stack of green cooking bananas and handful of *kaukau*, all of which we bought. Our old guide, who today was obviously somewhat lost en route, claims to be unable to direct us further to Mononi, but he says that these Kuagatnunga men will help him find a direct track. Our Weme Yar man asked to be permitted to rush on alone, having finally "found himself" with our arrival at this stream, and we have let him go on ahead. He says he can reach Mononi tonight, sleep there, and arrive at Weme in the morning. There he will prepare his people for our coming, collect plenty of sago for us (which he claims is abundant enough and processed in sufficient quantity by Weme people to feed our entire group), and see that a shelter is ready for us. If all this works out as planned and hoped for, we shall be mighty lucky.

[133:31–46] Our three Yar people, speaking no Kukukuku at all, are very friendly and solicitous with the Kuks. Thus they—more than our cargo-boys, police, or *mankis*—spend a great deal of time with the Kuks and constantly have the Kuk boys with them. Paretai sports an absolutely new T-shirt this evening, and I learn that it is a gift from one of the Yars. At the insignificant wage which natives get per annum of hard labor, this is a mighty sizeable gift. Such spontaneous giving in friendship, or in an attempt to cement bonds between natives, I have seen often now; and its generosity and spontaneity and goodwill often mocks the stingy, cautious doling-out of trade-items in which we engage—buying huge *bilums* full of sweet potato for a handful of salt or a teaspoonful of colored beads or a few *girigiri*, buying a pig for a cheap mirror or a cheap knife or a *kina* shell, and paying for a backbreaking day of lugging heavy cargo over rough trail on precipitous slopes with a few sticks of tobacco or a shilling.

[134:19–28]   I suspect marriage contact exists between the Mononi Kuks and the Yar of Weme as does Yar-Kimi and Yar-Fore marriage contact, as I found in my last visit south to the Yars. Two Fore women from Mugaiamuti have married into Moraei. We have not, however, seen a trace of kuru among the Kuks, but the general hiding of women and sick could easily account for this even if it were present. I shall try to interrogate our Kuks cautiously on this matter tomorrow, for we have the closest rapport and confidence with these few Moraei people that we have yet achieved with any Kukukukus.

[134:39–45]   Our walking today was to the limit of the endurance of the cargo-boys and police. At least thus far, Jack and I are not retarding our progress. However, it is hardly conceivable that a European could struggle over these trails with such loads as our boys are carrying, and I admire them greatly. They are in good spirits and, even with bruised and sore shoulders, still happy and cooperative.

19 October 1957   Weme village (Papua), Yar Pawaiian people

[136:6–20]   Early this morning, from 6 to 8 a.m., we worked on medical treatments: dressing the many cargo-boys who have peeling, bruised, and even ulcerating shoulder injuries from the cargo-loads; medicating the numerous complaints, including two PUOs whom I have on SDM; and treating two cases of yaws in Kuagatnunga Kukukukus. Everyone of the Yar and Kuagatnunga and Watch-eramapinti groups assured us that at Mononi (near the northeast of Weme) we should find the last Kukukuku settlement, and these "Yar-Kuk half-castes" would provide us with ample food. At 8 a.m. we broke camp, rapidly ascended a short way along the Kuagatnunga River (which is formed by two confluent streams just above our camp), and then began a steep ascent above the river-shore burial-ground which has been our camp. Fortunately the terrific stench of the decaying platform-chair burial in the grove of pandanus about a quarter-mile above our camp did not reach us during the night.

[137:43–139:19]   Mononi turned out to be a large area of extensive lush gardens on steep riverside slopes, with a diffuse hamlet of four or five deserted houses. We were surprised to find the village deserted. Our police searched the hamlet and finally pieced the story together from our Kuagatnunga guide: that recent deaths in this site had left them with very few adult men, and the group had deserted the village and moved into a new bush camp.... We bathed in the Mononi River, had tea and a light lunch (including a ripe pineapple salvaged from the gardens), and then set out on a rapid walk which brought us here to Weme by 4:30 p.m. As we approached Weme, the trail became excellent, and we made at least three miles per hour on it instead of the usual slow one-to-two miles per hour....

The Mononi group is a Kukukuku group, we are assured, not a Yar group.... I was filled with joy as I waded down the Weme, rounded a bend, and suddenly saw the fine houses of Weme village—raised high on poles, situated beside the river on a flat grass-covered plot, with a neat row of five large dwellings behind a compound of three large houses, the front one of which was a *haus kiap* for us and the two next houses for our police and cargo-boys. We had suddenly emerged from the stone-age savagery of arrogant and unpredictable Kukukukus to a quiet, sophisticated Papuan village surrounded by lush gardens filled with pawpaw, banana, and pumpkin, and bordering a wide, quiet, almost-sleeping stream. The rough and wild mountainous bush had disappeared in the space of fifteen minutes' walking, and a bend in the river brought us from the rugged country of the Kuks to the flat, lazy, and luxuriant country of the Yar Papuans.

Waiting for us were Pe'i and Urahau, the village councilor and village constable whom I knew well from my sojourn with these people at their Wi'ir site on the Yani River in August. They saluted sharply and handed me their village book—issued from Beara but never containing a census or a government inspection of this site, government patrols from the Papuan side having stopped at So'o or a bit above it. . . .

I soon found that the village had all the people I had seen at Wi'ir and few, if any, new faces; thus it appears that they were all at Wi'ir while we visited them there. Jack soon arrived, greeted them, and found that all the men could speak fluent Motu with him. Thus, we are immediately in a much different and better position than upon our last visit, when interpretation had to proceed through Fore and Kimi, spoken only by a few from this Yar settlement. The people are most happy to find Jack speaking Motu with them, and they have welcomed us with excellent hospitality. Huge quantities of food were already piled; and, after learning that they would prefer salt above all our other trade-goods, we bought the food they offered with salt, primarily. They unhesitatingly demanded a fair and just price, knowing the proper trade-value of the food they had; but they did not stint in quantities at all, and loaded us with immense piles of pumpkins, bananas, sugar cane, etc. so that we already have enough for our entire component of about 60 men for two or three days. The women have not run away. Although they did not show up, a walk through the village revealed them peeking out of the houses at us; and we found them coyly shy when we climbed up and entered the houses, shook hands with women and girls, and examined what they were doing.

[139:31–140:25] The yaws cases we treated in August have all recovered completely; and, in particular, Urahau, village constable, has no more difficulty in walking. The severe yaws ulceration of his toe is healed. In the evening, while we were treating sick and injured in our party (all minor complaints; fortunately, nothing serious has appeared), the Yar people brought in a girl of about 15 with a severe but healing burn of her right leg and thigh. An old gentleman with obvious chronic weight-loss, and arthralgic and left-flank and lateral chest-pain complaints, presented nothing on brief physical examination on which I can make a diagnosis; and thus I am really at a loss as to what he has. His lungs are clear, and there is no GU complaint. We shall have a thorough medical inspection of all our group tomorrow, and of the villagers who desire it.

Discussion with Pe'i (the councilor) and Urahau (the village constable) reveals that the trip south to So'o involves two days of walking to the shore of the Subu-Lamari, where it becomes navigable. Then they make rafts (usually three-man rafts with three large logs lashed together) and shoot down a very fast stream which they claim is navigable in low or high water and which has many sharp bends and rocks protruding from the water, and, in places, flows through a canyon. This, of course, is the portion of the trip south that worries us. The Yar people deny any alternative route overland, and insist that rafting is safe—but admit they do not try to carry more than three on a raft, and that cargo would be made wet en route. They agree that canoes are the answer; and in fact when they return from coastal work, they come upstream to this place to "take off," using canoes from So'o village (which afterwards return, usually with others from Weme who are going to the coast to work). We have planned to go to this riverside camp, camp there, and build canoes for the trip down. We have little canoe-building knowledge or experience in our police; Jack has done more of it than anyone in our party. I fear that the Yars themselves, using and making rafts only, may be none too expert either. So'o and down should be no problem at all; but now, as on my last visit to these people, the jump down to So'o looms as a real problem. . . .

Fortunately, food here appears plentiful enough for us to remain for a while. However, our party more than doubles the population of Weme, and thus will, if we stay too long, tax heavily the food supplies, I fear.

20 October 1957   Weme village, Yar Pawaiian

[140:43–141:17]   This evening our plans are changing—nay, changed. We spent a day of rest and leisurely activity, coupled with feasting by our crew of cargo-boys, *mankis*, and police. We now contemplate a group of ten to go south to Beara, having added (to Jack and myself, the four police, Andi, Asoi, and Wanevi) the *luluai* of Kasokana, who is among our cargo-boys and one of the best workers. The Yar people, however, are backing out a bit and giving us problems. Thus it now appears, as I suspected, that we shall be most lucky if we can get as many of them as six to come with us. The track to the Lamari-shore camp is good and clear, and food supply still appears to be no problem; but now they have pointed out that they are themselves inexperienced canoe-builders, and canoe-building may be a time-consuming problem requiring more than the two-to-three days maximum we have planned to throw into it. Now they suggest rushing off at once to So'o with a party of Weme men and coming back with two or four So'o canoes to the Lamari-shore camp. They can be there with the canoes six days after leaving here—i.e., two days to the shore, two days on raft downstream, and two days up in canoes. We shall thus stay here for four days, and then spend two days traveling to the Lamari shore to meet the canoes.

[141:28–142:44]   This morning we assembled all patrol personnel, and (with the *mankis* and Jack helping) dressed all major and minor sores and cuts and bruises and attended to all complaints. Practically every one of our party of some 60 has complaints of one sort or another. Two are on sulfadimidine, and two have sufficient secondary infection to be given procaine penicillin shots. No one is seriously ill. The Yar people brought up the girl with severe, healing burns. Jack dressed her burn, and I injected further penicillin and gave sulfadimidine as well. The man who is undiagnosable here in the bush, whom I have on one gram of streptomycin daily, was also brought up again. Three cases of rather extensive yaws were presented (two adult women and one boy of about seven). These patients we dressed and treated with penicillin.

My constant fear—iatrogenic illness—is with me, sitting here beside me as I typewrite by hurricane-lamp light on a bark table erected under our house (which is raised some six or seven feet from the ground). This is in the Kuagatnunga Kukukuku boy of about eight or nine years who has come to Weme with us. He had several infected lacerations, some mild impetiginous infection of a scabies-like dermatitis, and also a few incompletely healed furuncles. These I treated locally, and then gave him a single 5 ml injection of procaine penicillin (1,500,000 units), since I feared he would leave today and never receive further therapy. . . . This afternoon the boy came to me crying with pain in his knee, and a minimal but detectable edema of the entire right lower extremity. He indicates clearly that his pain is along sciatic distribution, and thus I am rather certain that his difficulty is a sciatica induced by the prolonged pressure of a large (5 ml) injection of very slowly absorbed material (i.e., the oily procaine penicillin). There is no local tenderness or palpable mass at the site of injection, and no sign that the injection-site is infected or locally involved at all. I suspect the trouble is as I have stated above: sciatic pressure which will be totally relieved as the penicillin is absorbed. . . .

Since any trouble from such "shoots" may prejudice the people against them— and they now accept them readily—it is very unfortunate to have any reaction at all. Undoubtedly we have far more septic abscesses, local tumors, periosteal traumas, and other injection complications all the time in our treatment of the native population, in view of the poor injection-technique of our assistants; but the patients are seldom observed long enough after the "shoots," and the complaint often would not be attributed—either by the patient or the medical assistant—to the injection. By now I have a group of some 20 Fore and Keiagana cargo-boys, our Kimi interpreter, and several Yars crowded around my lantern-lit table; and, as I have been writing, all my thoughts and attention have been directed to the fine little patient beside me to such an extent that I have only now become aware of the crowd admiring the function of this horrible machine.

[143:11–31]   Our component south is now set at the ten I listed above. I am most embarrassed and distraught at not being able to take Tiu, Taka, and Masasa; and I am stiff and perhaps a bit abrupt in my behavior toward them as a result. Tiu has noticed it to the extent of asking me if *yu got kros long mi* ["Are you angry with me?"] today. If there is any further canoe-space and we decide to take any others, I shall certainly try to make it one of the boys, for they—at their young receptive ages—are the most important people to introduce to the coast and what small civilized centers there are there.

A huge katydid-like insect (a long-antennaed green orthopteran, at least three inches long) has just landed on this machine and interrupted my typing, to everyone's amusement—including the little Kukukuku whose lice I am, at this moment, sharing and which are producing a most intense inguinal, thigh, and genital pruritus. His clothing is most sparse, only the bark rain-cape and a small *kanila* (pubic covering) of the non-voluminous type worn only by prepubertal children. Even in this scanty dress, he manages to hide a horde of ectoparasites. His smile at my katydid-engendered plight has refreshed me greatly and eased considerably my apprehension about his sciatic irritation, which I now believe is subsiding.

21 October 1957   Weme village, Yar Pawaiian

[144:27–146:40]   This morning a delegation of Yars came and told Jack that the group had made a final, definitive decision to move to Wi'ir and make their permanent village there. Thus, our suspicion that the problem was still brewing, that dissension and disagreement flourished in the group, and that they were divided and uncertain as to whether Wi'ir would or would not be made permanent, seems to have been just the case; and last night, in conference, the group made the decision. This brings them only four or five days walking from Moke-Okapa (or Lufa, for that matter), if the trail we cut from Paiti is kept open. However, Jack is forced to tell them that employment opportunities with the Australian Petroleum Company and other Papuan coastal labor are better on their traditional but longer route south than on the trip through Moke to the highlands, where government employment is poorly paid and scarce and where a highland labor-pool assignment with a trip to the coast as plantation labor could never provide them with either the pay-scale or the type of work they are accustomed to. It is most unfortunate that one cannot encourage their highland contacts more, for their country offers probably the best New Guinea-to-Papua road that exists.

My anxiety has abated. The little Kuk boy comes to me early this morning and I ask, by sign, whether his leg hurts; he indicates that it is all better and causes him no difficulty at all. No edema is present. The sun has not yet risen above the trees; our small patrol is on the trail, morning tea has been served and we await breakfast....

4:00 p.m.

 ... After getting our small avant-garde patrol off to the south, Jack and I went out to the Weme gardens and found there the women of Weme—with small infants and girls—in a sago-making picnic. Jack had asked yesterday for the opportunity to take pictures of sago-processing. Here they were starting work on a newly felled large sago-palm. It was about 20–24 inches in diameter at the base, and the leaves (or fronds) must have risen over 40–50 feet into the air. Urahau indicated that from the base, cut about four feet above the ground, a section of trunk some 20 feet long or so could be used for sago production. Three women were busy pounding or cutting out the sago-containing pith from a segment about eight feet long, located about 8–10 feet from the base. Here the bark had been split and peeled back, but kept as retaining sidewalls; and three women, standing beside each other on one side of the horizontal trunk, pounded the pith with machine-like regularity using a cutting tool made entirely of wood. The hard wood cutting-end was shaped like a cone concavely indented, and this was regularly brought down with a precise advance with each blow so that a regular, smooth, progressive indentation into the pith was produced; the bark retaining-walls, close to the women's laps, filled slowly with the shredded pith. As large masses of this pith were cut free, they were scooped up by hand into *bilums* and into the more finely woven sago bags which I had first seen in my previous trip to the Yars....

The sago-palm stand was in the center of immense, flat riverside garden plots bordered by forest on all sides but the river shore. The women carried sago-pulp–filled bags out to the stream and there, at the stream's edge, set up their sago-washing. This was done by using the funnel-shaped frond-base of an immense sago leaf. The narrow end of this open, narrowing trough (about 6–8 feet long) was placed over a large tub made of an immense sheet of bark (some 5 × 9 feet) which had been placed over a fire and heated, then folded into a non-leaking tub. Sago pith was emptied from the *bilums* and bags into the broad end of the frond trough. In its center portion, a sago-bag—finely woven with some native reed or fiber I am not familiar with—was fastened inside the trough, with its narrow top held neatly open by supple bits of wood and vine so that any flow through the trough passed through this open-mouthed bag. Above the bag, water was poured upon the piles of crude sago-pith, and the woman doing the processing then squeezed the shredded sago-pith firmly with her hands. This squeezing released a sago-containing suspension which streamed down the frond trough through the sago-bag. The bag retained all the pulp and fiber except the finely suspended edible sago, which passed through and ran out the distal portion of the trough into the bark tub, where the cream-colored effluent collected; settling, it left an orange-brown supernatant into which the effluent continued to pour. After a few dozen firm wringing squeezes of the pulp, the processor moved down to the bag and squeezed it out firmly, emptying thereafter the pulp back into the proximal portion of the trough for further washing. When washed several times, the squeezed-out pulp was discarded in handsful to the ground, where it slowly piled up. After the bark tub became about half-filled with washings containing the suspended edible sago, a conical secondary receptacle was placed within it under the spout-end of the frond trough, so that into this funnel the turbid, opaque, milky effluent could continue to pour without disturbing the sedimentation in the remaining portion of the tub. Thus, as the two eventually overflowed, only orange-brown supernatant spilled out; and sedimented sago particles remained in the bottom of the bark tub. In all, the process was very neat and nice. All the work was done with native materials set up quickly and expertly for the purpose; and although an arduous and exhausting process, it was done with smooth practiced art and no unnecessary expenditure of muscular work.

22 October 1957   Weme village, Yar Pawaiian

[148:34–44]   Problems are always looming; and at the moment the worst one is a swollen, painful, tender right foot. I misstepped in dusk while descending the ten wooden-ladder steps of our house and tumbled, twisting my right foot in the metatarsals rather severely. I have no fracture, but the swelling is increasing and I have a great deal of pain on weight-bearing. We shall start for the Subu canoeing site on the morning after tomorrow, and I am very worried at my ability to get along on the rough trails with the many log bridges, etc. Fortunately, I have tomorrow to rest. I can think of few more fool accidents that might happen, and the time is most inopportune.

23 October 1957   Weme village, Yar Pawaiian

[150:45–151:24]   The Mononi Kukukukus are here at Weme, visiting us and the Yars; and although they and the Yars at first claimed little mutual contact, it is fairly evident that they are quite at home here, that many speak a bit of the Yar language, that they are friendly with all the Yars and well acquainted with all of them. Today, Urahau (V.C. of Weme) says he plans to adopt two of the eight Mononi Kukukuku boys here with us, having promised their father before he died that he would do so. Thus, extensive contact is certain, and the Mononi Kuks seem quite agreed to the adoption; and the two 12-year-old Kuk boys seem quite pleased to stay here. . . .

Kukukuku villages were all stockaded, isolated on ridges and well-protected prominences, and often hidden in the bush. There is ample evidence of current warfare and continued fear thereof. The Yars have apparently had no warfare for a long while. The note in the Papuan village-book written at Beara, where Urahau came to complain about the "Borae" murder of a Yar woman, I thought referred to the Fore when I first saw it in August at Wi'ir. Now it appears that it was Moraei Kukukukus who were responsible, not the Fore.

[151:44–152:30]   My foot is much better this evening, but I am still crippled, limping precariously about; and I am not sure how well I shall manage the long walk tomorrow. Because of this, I hope to get a very early start. Sergeant Malekor, weak from acute gastroenteritis, is slowly recovering on sulfa therapy; and he too may leave slowly in the dawn hours with me. Bo'oru, my Yar guide from Kasarai to Wi'ir, will guide us.

Jack censused the village today; and during census I checked all for yaws and other complaints, and made a very cursory medical check. About half of all inhabitants have cutaneous tinea imbricata, a half-dozen with extensive tinea over most of the body. . . . All children under adolescence have large spleens, the great majority with spleen tips reaching umbilicus and extending down into the left lower quadrant of the abdomen. Thus, in a severely malarial population with moderate yaws (severe yaws when I was here in August) and a good deal of secondary skin infections, one should not be surprised at a poor life-expectancy; and I am more surprised by the general good health of the young men.

24 October 1957   Bush camp en route to So'o from Weme, Yar Pawaiian

[155:3–33]   As I type, I am febrile, with markedly tender left inguinal adenopathy, secondary to the numerous small waxing and waning tropical ulcerations on my leg and foot. I have peripheral vasoconstriction starting with pale nail-beds and the beginning, I believe, of a chill. I have mild diarrhea and moderate malaise and generalized hyperesthesia. In addition, the sprain of my right foot (two lateral metatarsals) reveals, on removal of the pressure bandage, far more residual edema than I thought would be present. Thus, with innumerable complaints, hypochondriacal concentration thereupon, and a bit of real illness, I have started myself on 3.0 gms sulfadimidine and shall shortly take another 1.0 gms.

Thunder, still far off, rolls over the forested hills about our camp, and it rumbles threateningly. I am "too crook," to borrow a Bakerism I have begun to accept, to write further.

25 October 1957   Bush Camp No. 2 between Weme and So'o, Pawaiian

We had no appreciable radio reception on Jack's battery-set last night, our aerial being blocked and snared by branches and soaring timber. The news came through in low intensity, and we soon gave it up. I was too ill to write or read. Late in the night, the lightning and thunder which had been flashing and rumbling all evening reached us, and torrential rain poured down upon our camp. Our goru-palm roof withstood the first ten minutes of downpour and then began to leak. I had to roll up my blankets and get up to sit in one of the few dry places in our lean-to, awaiting termination of the shower. Our police, *mankis*, and cargo-boys were all sitting up around fires in their leaking shelters. A quarter of an hour later, Jack too had to abandon his bed and sit up with me. We prepared a hot chocolate with rum, which took some of the wet chill out of us; and in about one-and-a-half hours, the rain stopped and we crawled back into damp blankets.

[156:4–34]   Today we started on the trail at about 7:00, and reached our riverside camp (in which we are now comfortably installed) by 11:00 a.m. . . . On arrival here we found our seven cargo-boys and two police whom we had sent ahead, along with three men and a boy from So'o and the four Weme men. Three Weme men and Corporal Asarumba had made a raft, gone downstream, and already returned with two very large dugout canoes from So'o. They could not get from the So'o people the four canoes we had hoped for, so they settled for two very good ones. They report that one full day of canoeing can bring us to So'o, whereas it takes two days poling and pushing upstream to get here from there.

[157:3–23]   Shortly after our arrival, all our police left camp to hunt. Herebebi, our "second in command" among the police, came in with a very large wild pig he shot near our camp; and thus our group is well fed this evening.

It has been pouring without stop since mid-afternoon, and our preparations for embarking in the canoes have been interrupted and much slowed by the rains. However, while the weather was still fine, Jack and I inspected the two canoes. One is wide enough to take our patrol boxes and the other is somewhat narrower; both are about 25 feet long. We took one downstream to the junction of this Tobosa River with the wide, swift, impressive Subu (Lamari) which we now see flowing almost smoothly for the first time. Just below the junction, it shoots some mild rapids which we must negotiate tomorrow; and our Yar people say they would rather not join the two canoes with a platform until we have passed these rapids. Also, in descending the Tobosa, a platform-joined canoe-pair would be difficult to manage on the rapids and in dodging the stones. Thus, we have decided to try this trip in separate canoes. We can take the 10 we have planned on. Three or four from Weme wish to come with us, and the three men and a lad of about 12 from So'o will naturally be with us.

[157:39–158:11]   The So'o people speak a language identical to that of our Weme Yars, but they claim they are not Yars. They say they are true Pawaiians, implying that Weme Yars are not. Both groups insist that the languages are identical. Weme people have not only married So'os, but two So'o boys here with the canoes that have been brought up were adopted by the Wemes as small boys. Thus, this pattern of Wemes adopting boys from far afield looks like a longstanding practice. Kukukuku and Fore boys are now with them.

Our Papuans and a few police were swimming in the rushing stream today; but none of our highland cargo-boys, who are either non-swimmers or weak swimmers, entered the water. I would love to have been swimming, but the fear of crocodiles deters me—for they are reported here by everyone, and there must be some truth to the reports.

We are now set for the trip south. Perhaps at the last minute I can find room in a canoe for Masasa; but thus far I plan only to take Wanevi, who has been so helpful and cooperative on the entire patrol. He is enthusiastic about going, although a bit afraid of the water and awed by the size of the streams. However, during the last week of stream-swimming, he has learned to swim a dozen yards or more.

26 October 1957   So'o village, Pawaiian

[158:26–159:9]   We have arrived in civilized and sophisticated So'o; and I write in a fine, airy, somewhat-leaking Papuan rest house in a torrential downpour on the very banks of the Purari River. So'o, about which we have heard for so long, was more easily reached than we had anticipated. Our trip was as follows:

By 9:00 a.m. we had given slips of guarantee of payment to our carrier line, packed our cargo for the canoe trip downstream, and loaded the canoes ready for departure.... After we had packed, our three Weme Yars who are coming south with us and the So'o canoemen informed us that four should handle the larger canoe loaded with most of our cargo and two the smaller canoe, leaving no one to handle the third small canoe (which we had found hidden along the Tobosa River and which had been laboriously repaired for use). Thus, all possibility of taking Masasa (whom I had really counted on taking) or Tiu and Taka (had there been room) dissolved. Until the moment of departure, they sensed my hope and hung about pleadingly, dejectedly, and despondently. Finally, as we took off, the three older lads sang good-bye happily to Wanevi, warned him not to stand up in a canoe—in which none of them had ever ridden before (with the exception of Tiu, who went down to the Lamari-Tobosa junction with us yesterday)—and waved an enthusiastic, if sunken-hearted, final good-bye as we sped down the strong current and mild rapids of the Tobosa toward the Lamari.

[159:34–161:2]   The trip down the Tobosa seemed to take only a few minutes; the trip was familiar and the canoes traveled rapidly. The river was high, but no higher than yesterday. It had rained moderately last evening and during the night. At the Tobosa-Subu junction, we brought the canoes to the bank to survey the Subu rapids immediately downstream. The water was a bit low for a fully safe passage; and the canoemen decided, on studying the stream, to ask all (excepting themselves) to leave the canoes and walk a quarter-mile or so along a forest track and down a small stream tributary to rejoin the Subu shore downstream from the rapids which the canoes would pass without passengers. They traversed the rapids without difficulty, but down a rather impressive stretch of whitewater in which inexperienced canoemen could easily have caused catastrophe by improper swinging of their weight. We reembarked and started downstream on a trip which was not interrupted again until we reached this village. The Subu remained rapid; it was usually over 100 yards wide, and the canoemen took the canoes down the deepest, fastest currents with precision. Many rapids had to be negotiated; we shipped water on every one. At times, rather large amounts came in; and our canoeman in the bow kept using his feet to bail out the canoe as we sped along....

On some of the small sandbanks where the stream was a bit slower at the river-edge than usual, we saw tracks which were probably the marks of crocodiles sliding back into the water or coming ashore. However, we saw no very interesting wildlife along the entire stream, and the forest was unusually silent. Only an occasional loud shriek of a bird was heard. New Guinea bush tends to be silent in comparison to rain forests of Central or South America.

Both sides of the river—but especially the eastern—were bounded by ridges and ranges rising behind a lowland belt perhaps a quarter-of-a-mile to a mile or more in width, like small mountains rising at least 1000 feet above the river level. The Lamari winds through this hill-and-small-mountain region through virgin forest until just a short way above the confluence with the Purari coming from the west. Here, on both shores, we spotted gardens with banana trees and moderate-sized clearings, but no houses in either. Then a few hundred yards further down, exactly at the confluence of the two streams, the Lamari (Subu) widens to several hundred yards and the Purari is at this point about 300 yards or more wide—so that where the two join, one can look across an expanse of at least a third of a mile of water. On the eastern bank of the Subu, facing the Purari's mouth and overlooking this vast expanse of water, is a small hamlet with a very large house. Another smaller house stood in a grove of coconut and banana trees, and was surrounded by a small garden. This uninhabited site is an overnight resting-site for Weme rafters coming downstream to So'o. It is wonderfully situated, and I hated to pass it by without spending a night at this fine location. We cut across the "bay" to the swift Purari current (which is moving at the rate of about seven knots) and headed downstream. Obviously, large launches can go much further up the Purari beyond its junction with the Subu. Upstream on the Purari are Namaina and Uri, the two places Urahau immediately named for me.

[164:15–165:8]   We asked the village boys [here at So'o] about crocodiles in the river, and they say there are many but they do not come up the bank into the village. However, the shore is less than 100 yards from our rest house and the village, and we are not over 20 feet above the stream level. The boys tell us that very recently a village youth was taken by a crocodile just here at the canoe landing, as he was bent over the stream washing. The croc got his foot, then crushed his thigh and dragged him in and under; his body was not recovered.

Immediately upon our arrival, fresh, slightly green coconuts were bought; and Jack and I feasted on the fine meat, very similar to the meat of pandanus nuts, from which the Kukukuku boys laboriously extracted slivers of meat with their teeth, cracking each individual spindle-shaped nut for a small sliver—a most laborious job for little reward. Pawpaw, *kaukau*, taro, bananas, and cucumbers have all been brought; and our now-small group is overfed. While we were standing on the shore marveling at the ease of the trip from our last camp, and regretting that we had not brought others in our party on rafts—as we would have, had we known more about the trip—we heard an outboard motor approaching from upstream at a fast rate. An immense canoe, over 25 yards long and twice as wide as our widest canoe on today's trip, came speeding down the river with a powerful U.S. outboard motor. It was the Australian Petroleum Corporation workboat coming down from the large A.P.C. camp far up the Purari River. The boat must easily make 15 knots with this outboard. It had left the camp at 7:00 a.m. and was here at 4:00 p.m. We flagged it down, and the native boatman brought it in to shore. There were no Europeans on the boat. The boat was bringing down mail to the Papuan coast and Kikori. We were told that they would gladly stay overnight here and take our party downstream tomorrow morning. Since So'o people would gladly relinquish the two-to-three-day downstream canoe trip to Beara and the long pull back, we have accepted—although I greatly regret rushing so quickly off from So'o, where I should like to have had a better chance to look about.

27 October 1957   Kapuna, London Mission Society Hospital site on the Wame River

The Wame, the westernmost mouth of the Purari in the immense Purari delta, brings us to our first night with Europeans once again. We arrived in the late afternoon, after about eight hours of river travel in the powerful A.P.C. outboard-driven canoe which carried easily all of our cargo and patrol members from So'o.

[166:21–167:20]   Just after entering the Baroi River, our pilot pointed to some trees near shore, filled with huge colonies of flying foxes. We pulled in to the bank and cut our way through most inhospitable river-shore country—muddy, densely overgrown, mosquito- and sandfly-infested—to a plot near the trees from where we had a close view of the dozens of black hanging forms of flying foxes and heard the high-pitched chirping of the beasts. Jack and our police blasted at the colonies (which filled two trees) and brought down four of the animals, hitting several others which did not fall but remained hanging in the high branches of the trees. They shot these injured-but-clinging animals several times, and they still did not fall. When the boys ran in to collect the killed animals, one of the flying foxes had a screaming small cub in the marsupial pouch. Wanevi wanted to keep this little one as a pet—the natives often want to keep wild animals as pets—but Jack tabooed this, for they are extensively infested with large biting ticks. Wanevi's disappointment was real, much like that of an Australian youngster denied a stray puppy he has brought home for a pet. I have none of the enthusiasm of all the others in our party for hunting, and am bored by it. I also dislike the gun-play. I could not stand the biting flies, and went back to the canoe while the butchery was on and sat in the hot sun in preference to the muddy, insect-ridden, and thorny vine-blanketed shore. I have seen few pieces of land on which I would less rather camp or cut trail than this.

... We did not stop at Kairimai village, in our hurry to get to Beara before dark; but only a few minutes further downstream we came to another shore-clearing. This was not a native village but a European homestead which, we were told, now belongs to Burt Council, a colorful Australian figure in Papua about whom I had already heard a good many "tales" in the usual line of Territory gossip, the major commodity of exchange in European communities.

[168:12–47]   Before we reached Wa'abowa we spotted four crocodiles, about five to seven feet long, basking in the sun on the narrow silt sandbanks or on logs at the shore. Wanevi spotted the first two before anybody else; and having never before seen a *pukpuk*, he was filled with astonishment and glee. We turned to shore and took shots at each of these crocodiles, but only one was hit. This "croc" was instantly killed and fell into the water and sank, thus making it impossible to retrieve him. Jack tells me his skin is worth about £5. White ibis-like birds, some floating stately downstream on log launches, were frequent along the river; but we saw little other animal life. D'Albertus creeper decorated the shore in only one place—and there we saw only two plants—but Jack told me of regions up the Fly River where long stretches of shore are crimson on both sides with this decorative vine climbing over the trees with long chains of sparkling red flowers. However, I am told that the flowers and vine usually harbor great colonies of ants, which make it highly inadvisable to collect them into one's canoe—another affirmation of the danger associated with the pursuit of beauty.

As we approached Kapuna, sago-stands along the Purari were immense; Jack says they are larger than any he had seen in his five years up the Fly River in the Western District. These immense sago patches often extended for several hundred yards along the river-front and far back deep into the jungle. Thus it is amply evident that the sparse local population for whom sago is the main staple is in no need of concern lest food run short, although there is a different yield and different quality of sago from various varieties of the palm. There is no really controlled replanting of sago. However, the natives will take shoots from a good tree and plant them near where they have felled it; they use suckers for this replanting. Sago is used for all walling and roofing among the Pawaiians; but the Koriki people (whom we are now with) use the nipa-palm leaf—which will last for up to five years, while one of the sago leaf will become brittle and start to leak in two to three years.

[169:31–170:27] We passed by the Kapuna Hospital, not knowing what it was until our native pilot told us it was a hospital. On hearing this, we turned about and landed and were met at the pier by the missionary doctor, Dr. Calvert, and his wife, also a physician; both were New Zealanders. They are running this L.M.S. [London Mission Society] Hospital, which is the only hospital (they are the only doctors) in the Beara subdistrict; this is the only L.M.S. Hospital in the Territory. The government, we were told, has a Native Medical Assistant and an Aid Post at Beara, but no hospital and no European staff at all. Thus it was immediately evident that this hospital and our hosts were of more interest to us than Beara itself; and however far Beara might lie downstream, we had to stop here, look about, and learn what we could of the medical situation here.

I asked if they could accept our sick patient from the So'o village who was with us in the canoe; and after a brief discussion of the case with Dr. Calvert, we brought him from our canoe to the hospital. Calvert immediately revealed that he was a cautious and conscientious and well-qualified medical man by his interest in the case, his decision to observe and study before rushing into therapy, his cautious evaluation of our patient in cardiac failure, and his musing that perhaps he might have a beriberi heart disease—a distinct diagnostic possibility of which I had never thought. Thus I am well satisfied to have the man in Calvert's hands. We soon made the decision to unload our cargo and party here and let the A.P.C. boat go on to Beara, for we could be taken down to Beara in the doctor's canoe on the morrow, he told us. A short while thereafter, two huge canoes tied side-by-side and powered by one four-horsepowered Archimedes outboard motor came slowly down the river and to the pier. It was packed with women and children and men, all dressed in good clothing and laden with extensive manufactured goods such as suitcases, mirrors, tinned food, flashlights, lanterns, even pressure kerosene lanterns. We learned that it was the famous Koriki-Iai leader, Tommy Kavu, who a half dozen years ago was in strong conflict with the government because he was leading a movement said to be a powerful delta region cargo cult. He was very commanding in bearing and general appearance, and he spoke a fine English. I was told he has served in the Australian Navy, where he was an able seaman, and had often visited Australia. He was bringing a load of Mission teachers back to the Mission, and all his passengers stopped here for the night. He agreed to take us and our cargo to Beara in the morning, thus relieving the doctor of the time-consuming trip. We were most pleased to accept.

[170:42–171:13]   A building for our police, Weme boys, and *mankis* was found; food obtained; and our patrol well housed for the night. Jack and I stayed with the doctor and his family, ate well, and slept between two white sheets—the latter for the first time in nine months, except for a few nights at Vin and Gloria Zigas's in Kainantu.

31 October 1957   Beara Government Station, Gulf District

We have lost the chance of seeing records in Kikori and of getting to Daru. On the radio this morning I spoke to Scragg, who has already arranged for Qantas to fly all of our patrol into Port Moresby from Baimuru tomorrow. The Catalina will put down near here at Baimuru and take our cargo and men in; and if we cannot get them all on, we shall send the remaining men by the steamship *Rui*. I should much rather have got on the *Rui* and made the slow uncomfortable coastal trip to Kikori and Daru, and I now realize that the way to have done so was simply to have gone ahead with this plan and not to have gotten in touch with Scragg until we were leaving Daru on the *Rui* for Moresby. Now we must concede to the Public Health Department's desire not to have a foreign investigator transported by them on the accommodationless *Rui*, and to get our Kuru Meeting over with. Thus, we shall be in Moresby tomorrow.

[172:7–19]   It did not take long to understand [Beara Govt. Station A.D.O.] Clary Healy's designation of this Port Romilly-Baimuru-Beara region as "the rectum of Papua," and it certainly proved to be one of the most uninspiring places I have yet seen in the Territory. There is not a foot of path leading out of the small Beara station, and the shores of the river offer not a square yard of open space nor even the possibility of beating one's way through the bush. Swamp, nipa-palm stands, and muddy jungle are all that are to be found; and flies and mosquitoes abound. There are no nearby villages except for one a short way downstream and across from Beara. The natives of this village keep away from the post, and none of their children attend school at the post.

8 November 1957   Lae

[174:26–176:43]   Vin, Jack, Lucy, and I flew on the Skymaster, arriving in Moresby from the South (Australia), nonstop to Lae this morning—our "Kuru Conference" completed. Bill Smythe remained at Hotel Papuan, unable to fly to Rabaul until Tuesday; and it looks as though we have lost him to the kuru project. In spite of initial plans for him to join us now in Okapa for the final windup of studies, there has been a reevaluation of staff and Rabaul cannot easily be left without laboratory or pathological facilities and without what little help he gives to the clinical workers (as would be the case should he come to Okapa with us). I am sorry not to have Bill, but the decision is realistic and sensible in spite of the disadvantage it places on us.

In spite of big promises, high plans, ambitious support, and great encouragement given to kuru study when I arrived in Port Moresby a week ago, particularly with the impetus given to interest in kuru and in our work by the enthusiasm for the study evidenced by Professor Eccles (neurophysiologist from Canberra and president of the Australian Academy of Science), Professor Sunderland (neuroanatomist from Melbourne, and secretary of the Australian Academy of Science), and Professor Robson from Adelaide (medicine), all three of whom had made the trip to Okapa to see kuru and had been shown the patients and our records and charts and papers in press by Vin—in spite of all this initial support, we have ended up our Kuru Conference and week of sojourn in Port Moresby with only an hour meeting with Dr. Gunther and two one-hour meetings with Dr. Scragg. The rest was limited exactly to what we accomplished on our own and what decisions we made. Secretarial aid, in the form of a full-time secretary at Okapa, has fallen through; so has medical assistance by addition of a member to our clinical team. Material aid has dwindled from high promises to meager reality. What we have netted are:

1. this new Imperial portable typewriter for my use in the kuru work;
2. this trip to Lae to see Mr. Kevin White, botanist who may help us in our ethnobotanical and toxic-plant studies;
3. part-time employment of Gloria Zigas to do typing for the kuru project;
4. shipment to the USA of all blood, urine, CSF, and autopsy tissues collected for trace-metal studies and stored in Port Moresby;
5. complete editing, re-cutting, and arranging of the 30-minute kuru film; and full preparation (writing, tape-recording, synchronizing) of an appropriate soundtrack;
6. promise of further venules and trace-metal–free bottles from Dr. A. V. G. Price, and of calcium disodium versenate and of BAL for treating kuru patients;
7. promise of £150 additional kuru funds;
8. authorization to buy 500 feet of Kodachrome 16mm cine film for further kuru documentation;
9. tacit acceptance of the accomplished fact of my present and continued participation in kuru study, and of participation of U.S. investigators in the project; and
10. assurance that Dr. Robson would bring up a clinical team from Adelaide to continue clinical follow-up of our patients after my departure.

Lucy and I were guests at Joe Szent-Ivany's home last night for supper. I had worked all day at the Iduabada Technical Training School, where I had projecting and tape-recording equipment for preparing the soundtrack for the kuru film. After a 4:00–5:00 p.m. meeting with Scragg at the Public Health Department, at which we settled final plans, I rushed back to Iduabada and worked on until 6:30 when I caught a cab into town to meet Szent-Ivany and Lucy. At 10:00 p.m. we left Szent-Ivany's home and I returned to dark, deserted, tomb-like Iduabada School and resumed my work. By 2:00 a.m. the soundtrack was written and prepared in synchronization with the film on tape.

Lucy has received an atrocious clipping from the Sydney *Sunday Herald* on the "Laughing Death," a name with which tabloid journalists have dubbed kuru. Dr. Gunther assures us that a several-page article on kuru is about to appear in the tabloid *People* in Australia, and *Time* is shortly to carry two columns of story on the disease. Vin assures me that the *Time* story will be OK, for *Time*'s representative visited him, came to Okapa and saw patients, and wrote the story from Vin's information. I hope he is right. Thus far, newspapers have made a farce of the pathetic plight of the Fore, and have usually presented a much-garbled account of the kuru investigations. I mistrust all journalism, and find the publicity and misquotings of facts we have supplied to others a most embarrassing matter. If only we might pursue our studies unpested. Research cannot be documented blow-by-blow for the lay press; the very thought of an effort in that direction could stifle successful inquiry.

Having returned after 2:00 a.m. to town from the Technical Training School, I had some difficulty getting to work at 6:00 a.m.; but we did get moving. Jack arranged—by quick trips to the Konebada natives workers' compound to see our boys, and to the police barracks to see our three policemen—that there would be no mixup tomorrow in getting them on the special chartered flight direct to Kainantu, by which we are returning our party to the highlands. I am furious that I have had to leave Wanevi out at Konebada. Natives are not permitted in the Papuan Hotel or any other hotel in Port Moresby. There are no transient accommodations for natives. There are no restaurants or even milk bars, etc. at which natives can be served. Slowly, Papua and New Guinea may progress in the next few decades to the despicable advanced stage of racial segregation with which we in the U.S. suffer in a few southern states. Actually, such segregation and suppression is complete throughout the Territory; but in the general primitive state of development, it is not so evident yet as it will be. Thus, unable to entertain Wanevi appropriately, to provide for him well, or to keep him with me in order to show him about, I am frustrated at every side. Now again our charter from Lae to Kainantu will be filled, and I cannot take him to Lae with me as I should like to. He is still stunned and bewildered by the ocean, by the coast, by boats, planes, and this sudden descent into "civilization." Asoi has taken the experience more in stride.

9 November 1957   Lae

[179:29–42]   Drs. Holland and Munroe, two Canadian entomologists with the Ottawa Department of Agriculture (Science Service Department) who are working here on fleas and other insects, were introduced to me at the hotel today. They know Bob Traub well. While we were away on the Kukukuku patrol, they visited Moke with John Barrett, entomologist from Aiyura, to make the entomological studies pertaining to kuru I had requested. Nothing much has come of these, but they did collect the two species of *Rattus* which infest the *kunai* roofs of the Fore dwellings, and on them they found many fleas and, especially, many mites. No taxonomy is yet completed. They tell me that the pig flea is also the human flea. This is of interest; for the women and children, the principal kuru victims, have more pig contact than do adult men. Munroe and Holland are interested in kuru, but obviously not enough so to race into it as a problem.

VII.
Bush Correspondence
28 September–16 October 1957

HAMILTON to J. BAKER and DCG [Okapa Patrol Post]
28 September 1957

Dear Jack and Carleton,
  Your pass has just arrived. I have given the medicine-list to Liklik, and he is sending out for the patients immediately. The lists of bloods are here, and will send them on with this.
  John is not back yet. I missed you all terribly last night, but I wasn't all alone though. After all the boys had gone and I was reading by the fire, I heard thuds and bangs coming from the kitchen—just as well I was reading a love story and not a murder—I went out to investigate and found the cat and all the kittens from the office locked in the store and having a high old time knocking the tins off the shelves. Did you bring them up, Jack? Or was it the mother cat's own idea? They have taken possession of the house now and I am gradually taming them, but the training will be worse. It's not a nice business cleaning up the puddles after them.
  It is a beautiful day here this morning. Hope you are having just as good weather. Looking forward to hearing more of your progress.

Cheers,
Lucy

DCG to HAMILTON, BERKIN, and ZIGAS Yakeia
28 September 1957

Dear Lucy, John, and Vin—or whichever of you is at Moke,
  We made Yakeia after 8½ hours on the trail—a steep trail, up and down most of the way. This includes seven hours of walking, one half-hour rest, and one hour for lunch. Here we can look back to Moke; and thus when you look down the valley south, you should see far on the horizon, a bit to the left. Tomorrow we cross the Lamari to Mobutasa, which you can also see directly down the valley if it is really clear.
  Yakeia natives use a new *tok-ples* [vernacular] and the people dress differently, although something like a cross between Moraei Kukukuku and the Fore. They are really "bushy" but have been moderately friendly, have fed us well and supplied our immense line with ample food.
  [95]
  All the kuru case-records I am sending should be stored in our Kuru House with the other records. I have brought them up to date with notes, etc.

I am also sending a new temperature-list, starting with September 30. It contains a list of medications, including appropriate medications for the new children with kuru whom I have asked to have brought in. In addition, I am sending to Lucy, as she requested, one copy of Smythe's Fore grammar analysis.

Sincerely,
Carleton

HAMILTON to DCG and J. BAKER   30 September 1957
Dear Carleton and Jack,

Am still here on my own and am getting a bit worried about John [Berkin], though I expect he has just stayed in Kainantu to celebrate his birthday. I tried to get on the radio on Saturday and this morning, but had no success. I have sent word down to Ray Nicholson to come up tomorrow and try to get through if John doesn't turn up today.

Just as the *luluai* from Okapa arrived with your pass, we were having a very exciting time. I got a volunteer to eat some of the plants that cause the *skin guria* [kuru-like tremors] and dream man [hallucinations]. He sat down at a table here and munched away at some old bark I had of a tree called *agara,* and ate a few of the *erereba* leaves. Most of it he seemed to spit out, but he said he swallowed some. It was obviously not very pleasant to eat, and he was in pain towards the end. Then the shakes started, only in his arms and body. Then suddenly the shakes stopped, and he swept all the things off the table and would have wrecked the place if I hadn't had Homeguei standing by with the handcuffs. His eyes did not become glazed, but he was very violent in his actions; and I don't think they were put on. He tried to get a knife from one man who was in the garden. He is still wandering about somewhere with the handcuffs on. Homeguei didn't think it was necessary to lock him up, but they have warned all the children on the station to stay inside. I am glad the old calaboose [prisoner] refused to eat it, as he is a much bigger man and would probably have done more damage. You might have had another murder on your hands, Jack, if he had.

I have collected a lot of the *erereba* leaves and have them drying. Yesterday was very wet and foggy all day, so rigged up a dryer here with three pressure-lamps under a piece of corrugated iron sitting on two chairs. Today is a nice day, and the leaves are outside again. I hope somebody comes in today to take the stuff out to catch the plane tomorrow.

Thank you, Carleton, for Bill Smythe's paper. The man who ate the leaves is in again, attached to Homeguei, thank goodness. He is still very distressed, panting, and the pupils of his eyes are only tiny pin-points. He does not talk, but he appears now to understand what is said to him.

I have plenty of the leaves of the *erereba* now; but Dr. Smadel did ask for a pound or so of seeds, of which I have found only two among all that have been brought in to me. If you could get some seeds or fruit and send them in, it would be helpful.

Had a visit from [the missionary] Mr. James on Saturday. He wanted to know something about saws and timber. He had a chat with the carpenter-boy and seemed to be satisfied. He wouldn't stay the night, but set out at about 10:00 p.m. in the fog to walk back to Purosa.

About this stuff that is to go to Moraei: apart from what is on your list, Carleton, what else is there and when is it to be sent? John no doubt knows, but as he is not here, neither Homeguei nor I know what to do about it—that is, when to send it or where the rest of the stuff is.

Ogia, the fellow who ate the leaves, is now in the kitchen. He is not violent now, but is kind of happy. He is doing dances, but still not uttering a sound; and he is much more rational and will do what he is told. The pupils of his eyes are enlarging.

All the best,

Cheers,
Lucy

HAMILTON to DCG   1 October 1957
Dear Carleton,
Did you get my note telling of the adventure I had with the fellow who ate the "ereba" leaves—he went berserk. I have sent the plants away to America as requested by Dr. Smadel. The dried leaves are ready to go as soon as a vehicle comes in. I also sent specimens of the other plant which has the same effect: *agara,* a tree.

The entomologists were in last week, but I don't think they accomplished much as far as kuru was concerned. John Barrett, the young fellow from Aiyura, was more helpful. I took them all down to [kuru patient] Sama's house one day. I doubt whether they have recovered from the walk yet. John [Barrett] and I got quite a few bugs and took samples of pieces of *kunai* roof to go through back at Aiyura. It is a pity he couldn't have stayed longer. He may have been able to accomplish more.

Have got the names of quite a few other medicinal plants and am busy collecting them.

All the best for your trip.
See you in Port Moresby.

Cheers,
Lucy

DCG to BERKIN, ZIGAS, and HAMILTON   Moraei village, Kukukukus
                                                                14 October 1957
Dear John, Vin, and Lucy,
All the cargo has arrived, and it is more than we anticipate needing. We are in a good camp at Moraei and have had a most successful Kukukuku patrol. Now we hope to head south, but the story is the usual: no route, no guides, no people, no carriers, water's too high, bush too dense, slopes too steep, no food, etc., etc., etc. We are trying to get food supplies adequate for the trek and hope that we do not encounter gorges, ridges or slopes, and streams which block our way. If at all possible, we shall get to Weme. If our food holds out and we can make it on to So'o and south, we certainly intend to try. If not, we shall return via Wi'ir and the route John and I took.

We are sending back herewith much cargo which is too much a "luxury" for the trip south. Among it are some mail and films and other items which require prompt attention.

[96]

We shall get in touch with you as soon as we again reach civilization. Of course, Moraei is hardly "civilized," but at least we can get a message back from here. Waiajeke is going on to Agakamatasa and will stay there a week or two, then show up to work with his beloved Masta John, for he really is set on going to Moresby with you, and not with us. Tiu, Taka, Masasa, and Wanevi are with us; and Jack is taking Asoi and Andi with him. Andi is cooking well, and all is still in good order. Tarangau has been having knee trouble; he has a bit of back pain, and we would be worried about him on the trek south. Thus he is returning, on his own decision, with the two "buggered-up" police. Kosinto developed inguinal adenitis, dysuria, and numerous other symptoms which I thought best to treat with penicillin and SDM. He has had three days of this before leaving us. He may show up in a few days at Moke. Look him over, treat him further if he still has swelling and tenderness. He has SDM for the road.

We cannot think of anything else to add. There will be plenty to tell you back at Moke, or in Moresby before that, if we ever get there. The cine films and undeveloped still negatives to Smadel, and the surface-mail registered letter home to my mother (with a manuscript in it), and the tin to be further wrapped and mailed surface-mail as a package are the only really important matters. The three folders with kuru papers to be brought to Moresby are also important; bring this to Vin's attention.

Good luck and many thanks.

Love,
Carleton

P.S. Keep any kuru brains from autopsies standing about in 5–10% formol-saline. Any that are trace-metal–free, fine! All others, in just plain formol-saline. Get as many other tissues as you can on them in the same formol-saline. If you get any in Zenker's wash them 24 hours in water after 24 hours in Zenker's. Then wash, for dehydration, in 40% ethanol and then to 70% ethanol.

BERKIN to J. BAKER and DCG   15 October 1957
Dear Jack and Carl,

I am sending this pass on to you, not knowing where you are but hoping it will reach you eventually. I received this radiogram this morning; and with a conversation I had with Kosinto, both Vin and Mick Foley suggested I send it on to you posthaste. Vin also got one of the same wording. Mick was asking about Kosinto, "that horrible *luluai* you have with you." He wanted him to go to the coast with a mob from Goroka.

Good luck. Cheers.

John

[enclosure]
14 October 1957. Dr. Gajdusek. Kuru film now in Port Moresby. Your immediate presence desired collate film prepare sound track. Advise arrival date immediately. Scragg

HAMILTON to DCG and J. BAKER   15 October 1957
Dear Carleton and Jack,

Looks as though we might see you a bit sooner than expected. Hope you got the camera, Jack. We have been expecting to hear from you before this.

Have received a list of the botanical names of the specimens sent to the botanist in Lae. I don't think it is very accurate, as he has given a different name for "ereba" than that given by the botanist in America and he has given different botanical names for plants which I have mistakenly duplicated in my collection but given the same native name. However, I suspect many of the names are correct, and I thought it best to send off the list with the native names to Dr. Smadel and he can pass it on to the toxicologist. They might get a lead as to something.

Have now moved to Moke and am starting on the weighed diets. Unfortunately, the *tultul* has sent Sama off to Okapa to die and won't bring her back under any circumstances.

Cheers
Lucy

DCG to BERKIN, ZIGAS, and HAMILTON   Moraei   4:30 a.m.
16 October 1957

Dear John, Vin, and Lucy,

We got your last-minute notes and thank you; they came up yesterday on the eve of our departure. We leave in a half-hour on a newly cut track south and have investigated the route enough to know that without mishap we should get at least one day south of Moraei without difficulty. Food-supply and equipment are adequate, leaving nothing at all to be desired. All we now await, with misgiving, is the terrain and forests and rivers. We certainly hope that no insurmountable barrier appears. After today, it will be virtually impossible for notes of any kind to reach us until you hear from us again. This note, when it arrives in your hands, may be arriving when we are already in Papua, if we have any luck.

The telegram about the film is only evidence of what I told the film-unit people. I told them that we might get to Moresby in less than one month and that, if possible, they should rush the printing and telegraph readiness so that we might be sure of seeing the film at the time of the Kuru Meeting in Moresby; this is the result. They really have worked fast. There is, however, no real hurry. We are not certain to have anyone from the unit there even now to edit it with us, and we can manage that even without anyone present but ourselves. Thus, I suggest telegraphing back the following message:

> Director Health, Moresby. Ref your cable of fourteenth. Completing Kuru epidemiology field patrols in two weeks. Arriving Moresby first days of November prepared for Kuru Meeting and film editing at your convenience thereafter. Gajdusek

This should solve all problems. From Beara, if we get there instead of back to Moke before that, I shall telegraph Director of Health Scragg, asking him to send authorization to you, Vin, and to Lucy for prompt transport to Moresby at his designated date in early November; I will be able to give him a definite date of arrival in Moresby once we reach Beara.

Thank you, Lucy; you have done wonderfully in getting the stuff off to America, and the sending off of the tentative identifications of plants is a most excellent idea. Tell Vin not to forget to bring the three folders containing the manuscripts of the three completed kuru papers. Other kuru records we will not need with us in Moresby, and we can return and work through them for a final report once the meeting is over and we wind up the project.

I am very glad that you are getting the food-balance work done, and sorry that Sama has left. Do not push the matter with her at all, and do your best on any other patients who turn up. Have any new cases started to appear about Moke? This would be most important to know. I have written again to Smadel and shall mail the letter from Moresby, explaining that everything possible that he can arrange to have done is desired on the specimens you send in; and I think that from previous communications, they already know this.
[97]

Sincerely,
Carleton

VIII.
General Correspondence
4 October 1957 – 28 January 1958

SCRAGG to DCG 4 October 1957

Dear Dr. Gajdusek,

Thank you for your letter of the 24th September and the attached copies.

I regret that Dr. Graeme Robertson's report on the brain sections has not yet reached you, and I enclose a copy.

We are interested in ordering reprints of the kuru papers, but would like to know the details of the second paper. To date, I have only sighted the one for publication in the *Medical Journal of Australia*. We would like to purchase 100 copies of this paper. On receipt of the second paper, I will advise further.

Please send me a telegram as soon as you have set a definite date for the Kuru Meeting in Port Moresby, so that we can arrange for the necessary persons to be present. The necessary approval is herewith given for Dr. Zigas to come to Port Moresby. Mr. Baker should take the matter up with his own Department.

Sir Macfarlane Burnet has written that he will no longer process any material received from you, as you are channelling things more and more into American centres. I feel this is possibly so, but I am sorry that this has occurred so close to the completion of your current studies.

He states that the CSIRO Chemical Industry Section and Dr. Graeme Robertson will still continue to deal with the material sent to them.

Yours sincerely,
R. F. R. Scragg

BERNDT to BURNET 8 October 1957

Dear Sir,

May I take the liberty of commenting on press reports (*The Sunday Times*, Oct. 6, '57, Perth; and *The West Australian*, Oct. 8, '57) concerning kuru (designated a "laughing disease"), localised in the Eastern Highlands of New Guinea? The reports noted that this was associated with hysterical outbursts of laughter, as its main symptom, and that the disease itself was "discovered" by Dr. Carleton Gajdusek on one hand and Dr. Vincent Zigas on the other. While realising that press reports are notoriously inaccurate when setting out material of this kind, I would like to mention the following points:

    a. The first social anthropological work in this region was carried out by myself and my wife in 1951–52, 1952–53, when we came across "kuru" and made a preliminary sociological study of it, drawing both on our own observation and on verbal accounts.

b. We tried to interest both medical officials and district officers in this disease at that time, recommending that detailed medical work be done and that it should take into account sociological data.

c. References describing this disease appeared in *Oceania* vol. XXIII, No. 1, 1952, and vol. XXIV, Nos. 3 and 4, 1954.

d. Dr. R. F. R. Scragg of Port Moresby approached me by letter (June 21, 1957) asking for comments; and I subsequently corresponded with Dr. Carleton Gajdusek on this topic. We are particularly interested in the sociological implications of kuru.

e. In the above correspondence (d), no mention was made of hysterical laughter as being a main or even a noticeable feature of this disease. In the detailed material obtained during our research, this aspect was certainly not predominant; and few examples of hysterical laughter can be associated with it. It should be emphasised that our anthropological-sociological study was made under what could be termed relatively traditional conditions, and not in an artificial situation of an organised clinic or hospital—which I understand was set up at Moke as part of the Okapa Kuru Project.

f. Some sociological consideration of this topic (kuru) has been included in a volume on the Eastern Highlanders by myself, to be submitted for publication within the next month or so. A shorter study of kuru is planned for publication in *Sociologus* (Berlin) early next year. It is hoped that in this latter article we can take into account the medical implications (if we can have access to the reports published by Dr. Gajdusek) in their sociological context.

Thanking you, I am

Yours sincerely,
Ronald M. Berndt

BURNET to BERNDT 14 October 1957
Dear Mr. Berndt,

I am sorry that newspaper reports on kuru were brief and confused. I did my best to refer enquiries to authorities in New Guinea, as my own association with the problem is of the slightest.

If any further occasion arises, I shall mention your prior studies of the condition.

Yours sincerely,
F. M. Burnet

[98]

HAMILTON to SMADEL 15 October 1957
Dear Dr. Smadel,

Lists of identifications of many of the foods eaten by Fore people have come to hand. I shall forward you these lists, and perhaps from them a lead may be obtained as to foods which contain toxins which could cause kuru.

These lists are not yet complete. I still have specimens which have not as yet been identified.

Most of the insects are eaten only by women and children, but non-kuru from Fore and other regions eat them as well as kuru-sufferers. It could be that those who develop kuru eat more of a certain species than others; but they do not obtain these foods when in hospital, and appear to eat very few insects after the onset of kuru.

It has been pointed out by the entomologist that some of the insects grouped by the Fore people under the one name belong to several different species. It is perhaps possible that those people who develop kuru eat a species which may contain toxins, and other people eat a species which do not contain those toxins, but both species are known by the same name by the Fore people.

There are many varieties of fungus which are eaten in considerable amounts by everybody—that is, kuru-sufferers, non-kuru Fore people, and people from other areas where kuru does not exist. These fungi have not as yet been identified.

I am collecting specimens of medicinal plants. There are very few which are commonly used even throughout the Fore area. Most are very localised—that is, known of only in one village or by a few old men in that village. I have been told of some, but few people seem to have used them.

I have made lists of all foods eaten, and am questioning kuru patients, non-kuru Fore people, and people from other areas as to the foods commonly eaten. From this I should be able to discover whether there are certain foods that kuru-sufferers have been in the habit of eating more than other people. When completed, I shall send you results of these investigations.

Yours faithfully,
Lucy Hamilton

KURLAND to SMADEL   18 October 1957

To: Dr. Joseph E. Smadel, Associate Director, NIH
Thru: Acting Director, NINDB [H. A. Imus]
From: Chief, Epidemiology Branch, NINDB [Leonard Kurland]
Subject: Kuru

Thank you for the copies of Carleton Gajdusek's letters of September 22 and 27, which were reviewed. In his previous letter, Dr. Gajdusek mentioned that the Fore people are well nourished and that food was not in short supply. Perhaps cannibalism has some hidden meaning; but when I learned that grandpa or someone else was occasionally served for dinner, and that spiders and other insects were eaten by women and children, I wondered if a large segment of the population isn't barely surviving and if the best food available doesn't go to the man of the house.

I note that Dr. Gajdusek is going to tremendous trouble to obtain blood specimens; if these are for genetic typing, I seriously doubt that they will tell him very much except the blood groups of the Fore people and little if anything insofar as distinguishing the affected from the unaffected in this inbred population.

Speculation of a diffuse neuroectodermal hypersensitivity developing because of the ingestion of human tissue is interesting, but I believe only a remote possibility.

The isoallergic phenomenon can be induced only with the greatest of difficulty unless adjuvants are used and the material given by injection. Furthermore, I would suppose the men rather than women and children would be most likely to partake of such a feast.

The study of spiders (and perhaps other insects and odd plants) sounds like a good lead since, for the first time, he has found some selectivity in diet for the group principally affected. I haven't had any experience with the toxic qualities of spiders; but Dr. R. E. Crabill of the Smithsonian Institute informs me that a few of the species mentioned are said to be used for food by natives in South America and Thailand, but that several of the species mentioned by Gajdusek are unknown to him. Feeding experiments of freshly ground spiders and other local insects in monkeys or other animals were suggested. In addition to the possibility of ingested toxins, I suppose one could be bitten occasionally when trying to capture spiders; and the species in New Guinea may be different from those which the authorities on the subject have heard about before. However, it is unlikely that spider-eating is a newly developed practice here, and it would be difficult to explain a new disease unless a new practice was introduced or a genetic mutation had occurred in the rapidly reproducing agent (spiders, etc.) rather than in the Fore host.

I hope that Dr. Gajdusek will have the opportunity of meeting with us when he arrives at the NIH, for we would like very much to have him tell us of his fascinating experiences.

SMADEL to DCG and HAMILTON
[telegram]   21 October 1957.   Investigations on plant materials received from Hamilton indicate pit-pit samples numbered 6, 7, 11, 13, 14, and 15 and Agara strongly positive for alkaloids. Other pit-pit samples slightly positive. Kegeta 38 and Erereba negative for alkaloids. Arranging for bioassay and toxicology. Need five pounds each Agara bark and leaves. Need ten pounds Kegeta 38. Need pit-pit botanical specimen with fruit and five pounds each of pit-pit samples fourteen and fifteen. Can you repeat native feeding experiment with erereba and agara given separately at controlled doses? USDA [U.S. Dept. of Agriculture] tags follow airmail.

Smadel

SMADEL to DCG   21 October 1957
Dear Carleton,
[99]
The identification of the specimen sent earlier, family Araceae, genus *Homalomena*, has been confirmed for one of the specimens sent by Lucy Hamilton. The reaction of the man who had eaten Erereba and Agara, as described by Miss Hamilton, has stimulated high interest among some members of the staff of the National Institute of Mental Health. It is this group, headed by Dr. Seymour Kety, who will attempt to establish a suitable bioassay method to be used in conjunction with Dr. Horning's chemical isolation and fractionation studies. Furthermore, it is possible that the staff at the Addiction Treatment Center in Lexington, Kentucky,

might wish to test some of the crude material in human subjects if toxicology studies indicate this is feasible.

As indicated in the telegram, we are requesting 10 additional pounds of Kegeta 38. Dr. Horning has requested this, even though it is negative for alkaloids, because according to the information sent with the plants this material is eaten by the population group suffering from kuru, but not by other groups in the area. He proposes that extensive workup be made in an attempt to obtain clues in relation to kuru.

I hope these requests do not overtax your group with the limited facilities available. Some of these materials look very interesting, and we hope to obtain information that will make your efforts required for sending the materials worthwhile.

Sincerely yours,
Joseph E. Smadel, M.D.

SMADEL to HAMILTON   31 October 1957
Dear Miss Hamilton:
Thank you very much for sending us the list of foods eaten by the Fore people. I have circulated copies to those here at the National Institutes of Health who have been working on materials sent us by you and Dr. Gajdusek. Dr. Horning in particular may have some helpful comments on some of the foods listed.

Sincerely yours,
Joseph E. Smadel, M.D.

REES to SOUTHERN   31 October 1957
Dear Mr. Southern,
I am forwarding herewith the report on the estimations for copper in the ashed blood samples. The differences between controls and samples from kuru-sufferers that were suggested by the exploratory spectrographic analyses do not appear to be substantiated by the quantitative atomic absorption results. I suspect that this arises from the fact that the concentrations of copper are very close to the limits of detection.

I am forwarding a copy of this report to Dr. Anderson.

Yours sincerely,
A. L. G. Rees

ROBERTSON to GREENFIELD   31 October 1957

National Institute of Neurological Diseases and Blindness
National Institutes of Health
Bethesda, Maryland

Dear Dr. Greenfield,
Many thanks for your letter. I can only make one response—that science must go on! I am sure that you will do it infinitely better, and with far more authority, than I could. I was asked to examine the specimens by the Walter and Eliza Hall Institute,

and it was hoped that I should go to New Guinea to study the patients neurologically. My only interest was in trying to solve problems. I suppose that I would have been able to do it as well as the Murray Valley Encephalitis paper.

The sections I showed you were from a brain which had already been cut up in New Guinea, and we now have many sections from two intact specimens which were sent later.

I sent reports on the first specimen to the Director of Health (Acting) in New Guinea, with a copy for Gajdusek (which was not sent on), and then a copy direct to Gajdusek when a telegram from him asking for the findings arrived—but I have had no reply from either. It does seem reprehensible to send specimens to two places without informing the other, if only because of the work involved, for we haven't unlimited technical help and I have to sweat over the sections, owing to lack of continuous experience.

The only intimation to the Walter and Eliza Hall Institute that Gajdusek had sent specimens elsewhere was quite recent. I am baffled by it all, and obviously do not understand all the facets—therefore the less said the better. I have mentioned my reaction to Sir Macfarlane Burnet and he agrees about it.

Thank you so much for sending me the informative proof—it was something which many others would have forgotten to do.

It was so nice seeing you and Mrs. Greenfield again. With kindest regards,

Yours sincerely,
Graeme Robertson

CURTAIN to DCG   4 November 1957

Dear Carleton,

I am enclosing more information on your sera, with particular reference to the normals.

First, I think I have discovered a new serum globulin in a number of your normals. It is detectable only by moving-boundary electrophoresis (best with interferometer optics) where it moves in pH 8.60 sodium diethylbarbiturate buffer (I = 0.10) with a mobility of 2.01 (cm^2 volt. sec. × 10^5). Its concentration ranges from .6–.9% of the total protein (absolutely about 0.7g%), and I have found it in twelve of the fifteen normals I have studied in the moving-boundary apparatus. The normals are otherwise notable for their high γ and low albumin. This finding is common amongst primitive peoples. The patterns are in no way like your kurus.

I have two hypotheses with regard to the new globulin. One is that it is an artifact, due perhaps to fibrinolysis of the clot (I cannot find any information on this possibility). The best experiment I can devise would be for you to send me sera from yourself and European colleagues under the same conditions as the normals. If not an artifact, the absence of the new globulin in three samples would prompt me to look for an hereditary influence. Would it be possible to find out the relationships of the donors of the 100-or-so normals?

Most of the electrophoresis in the literature of sera from New Guinea and Pacific peoples have been done on paper, where my new component would not be detected. I would like very much to secure samples from coastal natives. Could you arrange this with the Department of Health officials?

As to the last batch of kurus, I am puzzled to find these essentially normal. I attach a list of these.

In addition to the moving-boundary runs, I have screened most of the normals by paper electrophoresis. I am interested to find that one of these, F107, has a pattern similar to kuru (raised β).

Regards,
C. Curtain

DCG to SMADEL

Port Moresby
(in the throes of our Kuru Conference)
4 November 1957

Dear Joe,

This letter contains many things which require fairly prompt attention; and thus I apologize at the outset for the numerous requests, and the delay in getting them to you and in supplying you with much information.

Only a few days ago I came through to the sea—the Papuan Coast—after descending the Lamari River into the Purari and thus "crossing the island" with a group of natives who had before this trip never rafted, canoed, nor seen salt water. Before we knew it, we had "made history"; and National and Territorial radios were reporting an exploit—often highly dramatized and exaggerated—which in the patrol seemed only a pleasant excursion. Having thus "made history," I am only now discovering the appalling journalism that broke over our heads while I was out of all contact with civilization for over six weeks with the wonderful Kukukukus; finally descending southward from the highlands through sago country and the Pawaiian peoples of the Upper Purari; then down the mighty waters on raft, canoe, and larger canoes to the coast.

Enclosed are a set of Australian newspaper clippings that have been sent up to me—all nonsense and appalling but only the beginning of a real barrage of sensational journalism which is about to hit us, I am told. Furthermore, during my absence, the Australasian *Time* and *Life* representative camped with Vin Zigas, waited and waited for my return from the wilderness, and finally gave up when word reached them—which I had sent from my patrol—that I was heading deeper and deeper and further south rather than returning to Okapa. Thus, I am told, any day a U.S. "story" may appear; and I cannot vouch for a word of accuracy, and I am most disturbed. However, a Catalina [plane] was sent to pick up our patrol, telegrams went out to Vin Zigas and Lucy Hamilton and Smythe (who has worked with us medically and on Fore linguistics), and I managed to have a "Kuru Congress" organized here only a few days after emerging from the bush.

Patrolling has fully established the geographic extent and limits of kuru and the cultural affiliations, cultural spread, native trade-routes, and marriage contacts of the kuru-afflicted peoples; and this immense task of 1500–2000 miles of some of the hardest walking in the world is now done!

Proofs from the *NEJM* article have arrived, but obviously it will be out before I can reach them by mail; and since I trust you picked up the misspelled *Fore* (spelled For*ei* in the first paragraph of the paper) and other minor corrections, I am most satisfied. The sooner it gets out the better, now that the press has garbled up the story from misinformants.

The Australian article is all ready for press, but the arrival of glossy prints for appropriate illustration has failed to materialize. I asked you some time ago to send such retouched glossy prints to them as soon as possible; and the matter now becomes an emergency, for the press there is making shambles of the matter and a definitive medical report must immediately appear in the Australian journal. Editors of *The Medical Journal of Australia* (Ronald R. Winton, Editor, Seamer and Arunde Streets, Glebe, Sydney) telegraphed me today that they are still anxiously awaiting these illustrations from you. If, by any chance, you did not get the letter in which I requested them, listed the specific shots, and asked that these be prepared and sent to the *Journal*, please let me know immediately by cable. If you did get the request and can hurry up the dispatch of these photos to Sydney, please do so. [100]

This, then, will complete the first two preliminary reports, satisfy U.S. and Australian crowds, and leave us time to prepare definitive studies. *Klinische Wochenschrift* has, by the way, accepted for publication the third—much more definitive—paper; and I am now revising that and bringing it up to date for a really good German article, which will be the most complete report on kuru (the third in medical literature). The final definitive studies I shall write from the U.S.

We have just had visits by Prof. Eccles (Australian National University Canberra), who is one of the world's foremost neurophysiologists, Prof. of Medicine H. N. Robson of Adelaide, and Prof. Sunderland (neuroanatomist from Melbourne); and the three, after spending only $1\frac{1}{2}$ days at Okapa and seeing only nine patients—for in my patrolling in the bush, we have let the clinical side and case-collecting lapse—of our series, are beside themselves with enthusiasm, and they state that without any doubt this is the most important neurological discovery of decades. They apostrophize and hyperbolize at length, but the gist of the matter is that they are rushing back to Australia with the story; and Eccles says that he would have dominated the International Congress of Neurology in Brussels with the kuru story, and that the entire new gamma-fiber story of motor-neuron function would have been dominated by kuru had he only known about it. He sees the possibility that the new gamma-fiber concept and theories may line up with the lesions in kuru very closely. It may well be that this absolutely unique type of "tremor" and incoordination is primarily a disturbance of the gamma-fiber radiations. At any rate, the professorial delegation thought Pearce Bailey should be in on kuru immediately; I told them that, through you, he was. They thought Denny-Brown should fly out at once! They thought that the entire neurological world had in kuru its choicest and most promising problem.

Finally, with us, they believe that some toxic factor must underlie the process; and our continued suspicion of a trace-metal imbalance, toxicity, or deficiency looms most likely—especially in view of the Wilson's disease story, manganese poisoning (Moroccan miners), and mercury (chronic) neurological lesions. We mused about vanadium, molybdenum, zirconium, tungsten, and a host of others—and must admit that slow chronic poisoning, or sudden massive insult with progressive degeneration even after contact was stopped, were possibilities we had not eliminated or even investigated, and that continued study through the periodic table of all tissues and fluids on hand was the most urgent matter. Thus, I am now writing to beg you to conserve aliquots of whatever is on hand, to not exhaust them on any one metal-study, and to do all possible to get wide screening for trace-element deficiency, or, more likely, poisoning. It is true that kuru occurs in a mineral-rich part of the world, near massive heavy-metal deposits (gold, nickel, etc.), and between two great river drainages (Yani and Lamari) and not beyond these two!!! Although soil or water mineral contamination (or deficiency) would not explain age and sex distribution, there may be some other subtle environmental or cultural factor that does—even, perhaps, endocrine—and the trace-metal possibility still seems most probable of all the toxic suggestions.

We are still pursuing ethnobotany, tracking down botanical and entomologic species of all the specimens eaten. *Erereba*, the kuru-tremor plant, does not produce tremors or marked reactions such as Lucy described to you in a few other natives and Europeans who have tried modest doses. However, the natives remain adamant in their story of its effects, and further experimental trial will be done when I get back to Okapa. It remains an interesting plant. We are wondering about a mycotoxicosis and other possible toxicities resulting from food spoilage or storage, but cannot yet pin any such event upon kuru diets, or even on the age- and sex-group most affected. Shortly, the zoological and botanical taxonomic lists of Fore and kuru-patients' diets, pica, skin exposure, etc., etc. will be off to you for survey by any and every "expert" who might be able to spot in the huge list of genera and species interesting toxic possibilities.

Trace-copper studies on kuru- and control-bloods done at CSIRO in Melbourne seem to rule out the possibility of gross copper-poisoning. Other metals have not been looked for extensively as yet, and I doubt if the CSIRO group will ever. Trace-metal–free brain—sent to you—was unique, and we certainly hope every conceivable cytochemical and histochemical and trace-metal study will be done on it. Two other "trace-metal–free brains" are fixing now and will shortly be off to you. We gave Prof. Sunderland of Melbourne one kuru brain. Others will shortly be off to you.

I hope to send out in the next few days the immense file of trace-metal–free blood, tissues, and especially CSF which has been stored in iceboxes here in Moresby, unstudied. This is a completely unreplaceable collection representing a half-year of fieldwork, and it cannot easily be repeated. Since it looks as though nothing can or will be done on it in the near future here, I am sending it all to you, refrigerated; and I hope that the chemists and NINDB groups will do justice to the immense effort which went into collecting this series of specimens.

[101]

My plans: The 100-odd currently active cases (our series is now close to 200 patients, well over 50 of whom have died) must now be rapidly relocated, reevaluated, and restudied. Those who can be brought in will be given further clinical therapeutic trials, especially of calcium versenate or other chelating agents, and of any other drugs we can get in the next few weeks. We shall do everything possible to get further blood, CSF, blood films, autopsy specimens, and ethnobotanical collections and food, soil, fire-ash, etc. specimens together and off during the next few weeks. Then I shall leave and wend my way back to the U.S., with the firm intention of continuing on the kuru problem. First, in analyzing the records, the films, the epidemiological files and writing the appropriate reports; secondly, in taking some part in the laboratory studies; and finally, in planning a return attack in a year or so on the kuru problem, which must be cracked. I like the possibility of being placed for a period of months as "Visiting Scientist" at the NIH when I return, for as you know I am fully without income now [102]. Any prompt and immediate source of income upon my return may help get me out of the hole that I am sinking into as I continue to work on kuru. The "job" will let me get the paper-writing done, the data analysis complete, and give me time to "look around" for something more permanent—that it must permit a continued interest in kuru is sure.

I am told by Vin that further cine film was sent and is awaiting me at Okapa. The last six rolls of 16mm Kodachrome are "travelogue" and background anthropology from the beginning of this last venturesome patrol. We bought and paid for that film, including for all developing, which therefore you in the U.S. should not again pay for since Kodak in Australia has already collected for the full processing anywhere; I felt that this patrolling into "new country" fully justified the use of the film as we used it. Here in Moresby I am now cutting and editing and preparing a soundtrack on 25–30 minutes of fairly good black-and-white film on the clinical picture of kuru, and this "professional" film will be off in a 16mm copy to you very shortly. Actually, I have full permission to use any portion of it we wish in a final, definitive, longer kuru picture covering every aspect of the illness including serial follow-ups on individual patients in time as the disease progressed. We shall try to use cine film liberally in our final spurt of kuru work which we are about to launch. It appears, furthermore, that there will not necessarily be a lapse in clinical follow-up. The Australians are very impressed and are trying to prepare a clinical field-team to take over my records and follow all of our cases upon my departure. I have secretarial help now in preparing copies of all active-case records to leave behind for them; and thus, with luck, there may be a chance of continued and uninterrupted clinical follow-up of kuru. Professor Robson himself wants to bring up a team for the further clinical follow-up and therapeutic trial study of the disease.

Since our remote patrol post cannot sustain any larger numbers, the Government will have to, it appears, limit investigators interested to those it invites. I am ensuring our return by arranging for it now, and all are in complete agreement here. Our visiting professors were most frantic lest fire wipe out our newly constructed, thatched kuru-house with its completely unreplaceable stacks of records; and taking this warning, I shall soon start shipping them off to you for file pending my return. Copies of active-case records will be made and left here for further follow-up study.

This then should bring you up to date on kuru, Joe. I cannot make Bangkok, for kuru is more important. I shall, however, be off very shortly; and although my route home must be a long one, I am heading directly for NIH, and I shall stick with the paper-writing until we have milked the records dry and with kuru until we have solved it. AICF studies are too big a thing, I realize. I shall be interested in seeing what others are doing with them—perhaps they may even help in kuru—but kuru comes first. If that passes, AICF leaves plenty to do. Furthermore, I am brimming with new projects which I shall keep bottled for the moment. If, by any chance, you or anyone else can fly in to see kuru during the next few weeks (en route to Bangkok or some such venture), let me know immediately by cable. After my departure, there will be another lapse for a while and few cases on hand for demonstration to a passing visitor.

The Australians are, understandably, most anxious to get our kuru article out. I beg you to do everything possible to get the glossy prints for illustration off to them; otherwise we shall have a hefty "international incident" on our hands.

Sincerely,
Carleton

P.S. Please hang on to all the enclosed clippings in the "Kuru files."

ECCLES to BAILEY 5 November 1957

Dear Dr. Bailey,

I have just returned from visiting New Guinea, where I met Dr. Gajdusek and had the opportunity of visiting the area where he has been working on this extraordinary disease, kuru. I am enclosing the official report [see Appendix 3] that we made as a result of our visit, so that you can see how highly we regard the work that he and Dr. Zigas are doing. I feel that this disease is the most interesting neurological disorder that has been found in recent times, and it is worth the most intensive possible investigation. I know that your Institute has been giving him very good support and I would like to let you know how much he and Dr. Zigas are appreciating it, because up to date (for reasons that we need not go into) Australian help has been negligible. I have come back from New Guinea fascinated by the problem and full of admiration for the heroic efforts of Drs. Gajdusek and Zigas. What they have done there is simply magnificent, and to me and to my colleagues it was a thrilling experience to go there and see, at first hand, what had been done. You will probably have Dr. Gajdusek back in Bethesda in the near future for a brief visit; but I think you will find that he will return there at the earliest possible moment, as his emotional attachment to New Guinea and this problem of kuru is so great.

It was very nice meeting you in the summer, and seeing all your interesting exhibits relating to Sir Charles Sherrington. I have been meaning for some time to write and ask you if you had any spare copies of your photographs of his letters. I would be very glad to have them. I have many of his letters here—mostly about personal things, but there are several that you might like to have copies of. If so, I would be happy to have them copied here and send them to you.

My very best wishes.

Yours sincerely,
J. C. Eccles
Professor of Physiology
Australian National University

SMADEL to DCG 7 November 1957

[telegram]   Illustrations sent to *Medical Journal of Australia* November 6. Extremely sorry for delay.

ANDERSON to REES 7 November 1957
Dear Dr. Rees,

Thank you very much for your report on the ashed samples. I have not heard directly from New Guinea for a while, but I understand there is a conference up there at present to discuss implications of all the investigations.

I am

Sincerely yours,
S. G. Anderson

BURNET to WINTON 8 November 1957
Dear Dr. Winton,

I hope these notes will be useful. I do not think it would do any harm to hint that it is a pity this investigation was diverted to Americans instead of being done by our own people.

With kind regards,

Yours sincerely,
F. M. Burnet

[Enclosure]

Kuru

Kuru has been known to be present for many years in the Eastern Highlands. It was observed by Dr. R.M. Berndt in 1951–1952 in the course of social anthropological work, and some description of its social aspects given by him in *Oceania 23, 24,* 1952, 1954. Dr. Berndt has other work on the topic in preparation for publication.

A study of the "sorcery" angle of this and other diseases in the area was made by the Government Anthropologist, Mr. C. Julius, in 1956; and tentative arrangements were made at the beginning of this year for the Hall Institute to cooperate with officers of the Administration Medical Service in the investigation, particularly since there seemed some possibility that it was a post-encephalitic manifestation. These plans were superseded when Dr. Gajdusek took over the investigation.

Most people concerned are aware that Gajdusek had some special qualifications for the investigation but feel that it could have been done at least as effectively by Australian investigators, and that it is an unhappy circumstance that the work has been diverted to American laboratories.

Julius's report [see Appendix 1] on Fore sorcery in relation to various diseases is a fascinating document, not only in regard to kuru. His description of the way in which dysentery, which has been known in the area for only ten years, is in process of having an explanation in terms of sorcery built up, is particularly interesting. The elaborate traditions in regard to kuru point to its being a long-established disease.

DCG to SMADEL 10 November 1957
Dear Joe,

Vin, Lucy Hamilton, Jack Baker, and myself arrived in at Kainantu from Lae on a specially chartered plane this morning, flying up the long Markham Valley and climbing then into the more interesting highlands once again, the highlands in which I feel more "at home." We shall put out by jeep caravan for Okapa in a short while; and I shall launch a brief final clinical follow-up study with some further evaluation of drugs (more BAL, testosterone, and versenate, I hope), and laboratory studies on as many patients as we can revisit and persuade to reenter our hospital. In 2–3 weeks I shall pull out, and on the 11th of December fly to Hollandia from Lae, bound homeward at long last.

[103] From Hollandia and Biak I shall fly to Kuala Lumpur and remain there only a few days, instead of the months I had previously planned to spend there.

Kuru study must continue; and, as I told you, an Adelaide group may (??) be able to keep up some sort of clinical follow-up pending my return to the scene in a year or so for a follow-up of my own.

I shall shortly get the records and files off for you to store with the Kuru file to which I hope to turn immediately upon my return to the U.S. The two trace-metal–free brains now awaiting shipment to you are about hard enough, and I shall get them off with Vin in the next few days after returning to Okapa. These are the last such specimens which will be forthcoming. There is some long bone and plenty of spinal cord with them! Thus, this may be the final neuropathological shipment on which you can count for a while, and I beg that everything that can possibly be tried, be done.

Reports from CSIRO in Melbourne show traces of Cu and Al in most of the kuru bloods and tissues studied and in some autopsy tissues. Mn was present in a few bloods and tissues, and Zn in a few others. No other trace metals were identified by spectroscopic survey of ash for all elements; and Mo, W, Zi, rare earths, and other rare possibilities failed to turn up. Of course it must be borne in mind that the special technique used might not be sensitive enough; there might be very little of an offending heavy-metal atom in blood studied. Since Prof. Eccles agrees with us that only trace metals are thus far known to give a picture anything like kuru, we must still pursue this line. Cu levels on bloods have been completed at CSIRO by quantitative techniques; and they, like the previously done Cu estimations at Port Moresby by the good agricultural trace-metal lab, have not revealed any elevations above expected normal. Blood alone was studied. Manganese, aluminum, zinc, etc. have not yet been studied; and these did show up in some specimens (Al in all!). However, aluminum-capped McCartney bottles with rubber washers were used, and this may explain the finding; then again, Al is present in the highest levels of all "trace metals" in soils, and in the field it is most difficult to assure total absence of any dust. We may be able to check back to see if any of the Al was found in specimens in vacuumatic venules as well as in McCartney bottles, but I am not sure of being able to determine this. The cap supposedly does not come in contact with the contents, but—

From Moresby I shipped off to you the vast file of CSF, bloods, autopsy tissues, etc. collected in specially prepared "trace-metal–free" containers or in standard venules. These should be used as whole blood, obviously, for long storage has hemolyzed them a lot. They have been stored under refrigeration and should still be fine. Even if some are contaminated, they should still be OK for trace-metal work.

From Kainantu we shall now mail off a large supply of toxic plants, or plants suspected of toxicity which have been of particular interest in the kuru region. Furthermore, we have a botanist coming to join us, who will help make complete the huge taxonomic identification task—which is already nearly complete—of all the foods and medicinal plants of our kuru-afflicted populations. Many already appear to be from genera known to contain toxic species. Just how we shall sleuth out which, if any, might cause kuru is hard to say. All efforts thus far have been to little avail. Kuru patients do not admit to using anything special, or to being excessive consumers of any of the suspicious plants, as far as we have been able to learn.

Kuru still "looks" like a solvable problem. As Eccles says, it is undoubtedly the most exciting thing in neurology in his whole long career in the field. That its "solution" may help with other neurological problems seems obvious. But kuru remains thus far unsolved.

The peculiar age/sex pattern leans me still toward genetic predisposition at least, for—in spite of criticism to the contrary—it is the type of thing genetic predetermination can do. What of all the age-specific heredofamilial neurological degenerative disorders? And endocrine disorders? What of all the sex-specificity seen in such presumably genetically predisposed afflictions? This may be a different age-curve and a more striking sex-curve. But kuru is an undeniably unusual affair. However, we are still pushing the possibility of inapparent antecedent infection or thus-far-undetected deficiency or toxicity. No luck, however, as yet.

The botanical specimens going out are:

1. Agara (Fore) (also called Pinto)

    Botanist identifies this as *Himantardraceae galbulimima* Belgraveana. Big tree, growing wild in bush. Leaf and bark eaten to induce "dream man." Sometimes eaten with erereba (another Fore "drug"). Eaten after the evening meal. Eaten raw. Sometimes some is kept in the mouth during sleep. Eaten by women as well as by men, when they want insight into the future. Induces "skin guria" (Pidgin for kuru-like tremors), as does erereba. Not eaten by children. Fruit also is eaten raw to induce "dream man." Lucy believes that bark and fruit are eaten more than leaves, but leaves are said also to be effective. We are sending both bark and leaves of this plant.

2. Kegeta (Fore)

    Botanist identifies this as *Desmodium* sp. (probably *Desmodium sequax*). Shrub six to eight feet high. Grows wild in bush. Leaves eaten with pig and human flesh cooked in bamboo and *mumu*. (Cannibalism is almost gone, but not kuru.) Found at edge of a Moke garden on 7/7/57. Bark chewed with bark of Obakoo (a tree) and Mapaia (a variety of sweet potato), and it is given to victims thought to be suffering from *tukabu*. Some is rubbed on the skin, and the site of supposed *shutim nil* is then supposed to show up on the patient's skin. Many kuru patients (but not all, and not exclusively kuru victims either) claim to have used this plant. The leaves are eaten generally—by many kuru cases—and bark only used as a *"tukabu"* medicine."

In Webb, L.J., *Guide to the Medicinal and Poisonous Plants of Queensland*, Council for Scientific and Industrial Research, Commonwealth of Australia, 1948, we find:

Desmodium umbellatum DC. "Supposed by some to cause 'Chillagoe disease' of horses (Bailey 1909, 139)."

Desmodium nemorosum F. Muell. Reputed poisonous to stock (Bailey, 1909, 140).

Desmodium brachypodium A. Gray. A tick trefoil. This and *Glycine tabicina* were suspected of causing "string-halt" in horses near Stanthrope, April 1940 (Qld. Herb. Rec.).

Further *pitpit* samples have spoiled, and Lucy will now start collecting new specimens. *Erereba* leaves were sent off along with some rhizomes about 5–6 weeks ago and you should now have them.

A new possibility: *Oenanthe javanica*—consumed regularly by most Fore—is related to *Oenanthe crocata*, poisonous with spinal cord action on hind legs with paralysis. Oenanthotoxin resembles cicutoxin (from Webb).

Also: Iga (Igami)

> A type of yam identified as *Dioscoraceae discorea* sp. Yam bearing aerial tubers. Aerial tubers and underground tubers eaten widely.
>
> Webb states: several species of this genus with alkaloid dioscorine produces CNS action like pictrotoxin with paralyses.
>
> Lucy finds that men use this more than women(?). Or at least they plant and care for it, but perhaps they feed it to women(?).

Vin has received your reply to his letter and is very pleased to be so promptly answered. He and Gloria, his wife, are taking six months leave from March onward, and in June/July will cross the U.S. from Europe en route back to New Guinea. If any way to invite them to visit NIH etc. can be arranged, it would be fine. I shall be there long before that and discuss this further with you.

Time stopped in to get a kuru story while I was on the last "cross-the-island" patrol. Furthermore, in Moresby I was besieged by reporters—even equipped with tape recorders—and I am most disturbed and annoyed. "Laughing Death," which is a hideous misnomer, fills most Australian papers and magazines with highly distorted and well-padded accounts. Vin saw the Australian *Time* representative and seemed impressed by him, and Vin expects a more balanced account from them. I am, however, skeptical and wish to hell that kuru were less a thing to fire popular imagination—for it will, I fear, be played for all it is worth by the press.

[104]

We shall now make further 16mm cine films on kuru. The 30-minute professionally photographed kuru film (black-and-white, and nothing spectacular in photography) was what I wanted it to be: a fairly detailed and all-inclusive account of the clinical picture of kuru. I edited, spliced, and rearranged it and then made, tape-recorded, and typed out a full soundtrack for it while in Moresby. It is off now to Canberra for a professional commentator and the Commonwealth Film Unit to work up the map, etc. and my text into a final product. Copies of the film (done in 35mm but to be printed in 16mm) will be sent off to you, as I have requested; and I hope to be carrying one with me, also, when I leave.

Eccles, Robson, and Sunderland all pointed out my old Prof. Denny-Brown of Boston as one of the world authorities on this type of illness, and strongly urged that we let him know all about kuru soon. He may already have heard from you or others. If not however, please, Joe, will you inform him and see what he may have to offer? I hope to see him shortly, when I am back home.

We have a host of world-repute entomologists at work in New Guinea at the moment, and three have already been out to Moke briefly to see kuru. They have identified or are identifying the "food" insects and larvae, and we talked them into collecting the *kunai*-thatch–infesting rodents (two species of the genus *Rattus*, I am told) and into combing the rats for mites (which are very plentiful), fleas, etc. and in hunting out other ectoparasites. I do not have their taxonomic results as yet. It will be worth having, however, although thus far I cannot get excited about infectious possibilities.

The Fore houses, as all highland native houses, are smoke-filled. However, it has recently come to my attention that Fore women and children, especially, may be particularly exposed to a perpetual smoke-filled atmosphere (women's houses are smaller than men's and have no vent-hole). Thus I shall soon get more ash specimens off to you from kuru-victims' houses, along with controls. What to do with them is a problem, but it is certainly worth knowing whether they are unusually rich in Mn, W, Pb, Zn or some such metal.
[105]

Sincerely,
Carleton

BAILEY to SMADEL
Memorandum.   Subject: Dr. Gajdusek's project.   12 November 1957.

Dr. J. G. Greenfield, Visiting Scientist, has expressed concern about lines of communication between Dr. Gajdusek and the Australian group in New Guinea. Attached is a copy of correspondence from Dr. E. Graeme Robertson, neurologist and neuropathologist from Melbourne, to Dr. Greenfield (October 31, 1957), and to Dr. R. F. R. Scragg, Acting Director of Health, Port Moresby (August 16, 1957).

Dr. Robertson received the first sections of brains from Dr. Gajdusek; attached is a copy of the report which he forwarded to Dr. Scragg with a copy to Dr. Gajdusek. According to Dr. Robertson, no acknowledgment was made by Dr. Gajdusek and he was not advised of the brains being sent here until he heard from Dr. Greenfield.

Dr. Greenfield states that we have sufficient brains (12), which have been studied to date, to establish the disease from a pathological standpoint. What remains to be done is essentially epidemiological, and more precise clinical studies.

Do you have any suggestions to better relations between Dr. Gajdusek and the Australians?

Pearce Bailey, M.D.
Director
National Institute of Neurological Diseases and Blindness

DCG to SIMMONS   12 November 1957
Dear Roy,

It takes a lot of courage to write to you, having involved you in one of the all-time classical blunders of scientific investigation. I set out on the six-weeks Kukukuku patrol carrying ample venules and your glucose tubes ready to bleed Kuks galore. However, the rugged country and time-consuming trip to reach them proved that there was no possibility of getting out of the Kuk regions into which we had penetrated any specimens short of long unrefrigerated carriage. I abandoned the idea.

However, before leaving on the patrol I had alerted our very good European Medical Assistant [John Berkin], who has been assisting in our program here at Okapa, to my plan to rush back nonstop overland the specimens; and he was fully briefed on how to pack them into plastic bags, into the thermoses, and to wire immediately for Jeep transport to Kainantu to be rushed in—and he and Dr. Zigas were ready to keep all refrigerated until it was finally placed in ice on the plane bound for Melbourne. All just as planned. However, I changed my plans; and after ending up our major Kukukuku work, I came to the Lamari shores poised ready for a gamble at returning to the Yar people to the south and crossing the entire island down to the Purari and descending it to the Papuan coast. This plan demanded lightening our cargo. Thus, a large portion of our supplies and many of our cargo-boys were sent slowly back to Moke-Pintogori (i.e., the Okapa Patrol Post), with letters explaining to John Berkin and Vin Zigas my plans and asking them to attend to numerous things. However, I made no mention of the unused glucose tubes I was returning for eventual uses later. Instead, I had packed them into the bottom of one cargo-box, unopened, empty; and since all was going by a leisurely overland route without any note from me explaining who had been bled or where, I never dreamed of the possibility that—in zealous following of directions—the empty returned tubes could have been found and interpreted as collected specimens, as was done. Thus, John did exactly as I had instructed him, and packed them all off and rushed them to you. It is only now that I am back and looking for them that I have become aware of what has happened. In Moresby (for we did succeed in reaching the coast) Vin appeared at the Kuru Meeting we had there, and I told him of my plans and arrangements to have a large group of our Kukukuku friends come in to Okapa and be bled directly adjacent to our iceboxes—thus avoiding all the dangers of long delay in overland, unrefrigerated transport. Then he told me that the specimens had been sent out in good order, and I almost collapsed.

If, Roy, you have convalesced from your hospital sojourn after the heart attack I assume you suffered on opening 300 empty tubes, I hope you will be willing to read on.

To your letter of 14-10-57:

With these reports, all specimens we have sent to you and of which we have records are now recorded; and it appears that we are very close to being able to sit down and make really valid analyses and evaluations of the data. In spite of the overwhelming conviction that some hidden toxin is lurking behind the kuru story, the genetic predisposition or full determination probability remains unshaken and thus far the most likely—although even I myself am somewhat skeptical. Our patients are dying on every hand and the entire problem grows more spectacular as time progresses. Unfortunately, newspapers have gotten hold of our story; and all data on kuru—which ultimately must stem from Vin or myself, since no one has seen anything of it other than what we have revealed or shown to the few who have visited (and a brief view thereof by anthropologists Berndt and wife in 1953)—is being printed in a most distorted and completely misleading fashion. Even direct quotes from our letters and reports appear in press in much-distorted versions, and our own statements are misquoted. However, the November 14 issue of the *New England Journal of Medicine* should have our first preliminary report in it.

At the moment I am in a telegraph battle with the editor of the *Medical Journal of Australia,* who is trying to rush into print with our second paper on November 28th before making the essential alterations and corrections in the original manuscript—without which the Australian article will be inaccurate, incorrect, not up-to-date, and very misleading. I hope I can hold his presses long enough for him to make the changes, which are already posted in the mails to him. If not, I shall certainly refuse to publish another one of the many further kuru studies which we already have data for in the Australian journal. There is absolutely no difficulty in getting our kuru papers printed anywhere we wish, and in a great hurry. That being the case, I am very annoyed with the rush precipitated upon us for hurried publication by editors and politicos involved. If they will wait long enough to get the corrections and additions—the Australian paper as is does not even have the neuropathology report which we have fully described in the *NEJM* paper, and that at least must be added if it is to be any better than the newspaper accounts—I shall be most happy. If they jump the gun on me and publish before I have seen galley proofs, I shall be most disturbed. After all, it is our paper; and if we do not care about delay in publication and if we insist on seeing proofs before publication, I think we should have that right. Otherwise I shall be forced to retract much that appears in Australian accounts on kuru. Sorry to inflict all this upon you, Roy, but the latest telegrams have me worked up.

To get back to your blood-group studies. I am enclosing sheets which should straighten all this out and bring all specimens you have studied to full documentation and proper assignment to groups.

I am sending by refrigerated air-shipment several further specimens. These include several on Kukukukus who have walked into our Patrol Post, as I arranged that they should; and they have just been bled here, and their blood specimens refrigerated promptly. Also, other interesting specimens to be included in the Kuru and also the Non-kuru files are included. All are fully documented on the enclosed sheets. [106]

In a short while, I hope to get off to you the final series of kuru bleedings. Lois will take the sterile sera on these and ship them off to the Kuru file in the USA, where I shall work on them, if you will be willing to get these to her. Aliquots should go to Cyril Curtain for his work and, if Dr. G. N. Cooper (Chief Bacteriologist of Public Health Laboratory of the University of Melbourne) could be given specimens of the kuru-blood sera for streptococcal-agglutination studies he is doing—Lois will get these off from the specimens you give her—that will take care of all we have in mind. Cyril Curtain may be willing to take clots for further hemoglobin search. In addition, during these final 2–3 weeks of our field project—after that Roy, there will be no further specimens—we shall try to bring the number of adult normal subjects of racially pure stock from each of the kuru-area and surrounding racial groups ("cultural-linguistic groups" would be better) in your file to about 100. I shall not collect indiscriminately, and everything collected will be aimed at bringing to final statistically adequate sampling each group you are now studying; and as I send specimens, I shall document them fully.

I must pass across Asia and Europe en route home; but in about one–two months from the time I finally let you know I am leaving the territory, I should be there at work on kuru once again—with all the files, notes, and data in my hands. I shall then be doing lab investigative work on kuru specimens, and planning for a new 1958 field-attack on the disease, and making full analysis of the immense stacks of data for definitive publications on kuru. From there, I can do all the correlations and writing necessary in collaboration with you to get the kuru and kuru-area blood-group genetic survey worked up. I shall also have all the cartographic material on hand for a series of extensive ethno-cultural maps to go along with the kuru-region large-scale maps we shall be preparing. Thus, Roy, please accept my assurance that you will not be left "holding the bag" as much as you were with the Cape York and the New Britain data.

My apologies again for the "impossible" error in the shipping of the glucose tubes. Let me know how things work out. You can still reach me here until the first days of December.

Sincerely,
Carleton Gajdusek

DCG to CURTAIN   13 November 1957
Dear Cyril,

I am late in answering your letter of September 27th, and I have much to say. Roy Simmons continues to get shipments of blood specimens from kuru cases and non-kurus, and he can continue to supply you with every sort of serum you may desire. However, it is, in your work—as in his—essential that your specimens be accurately and exactly correlated with the phase and stage of illness (in kuru sera) and at least with the individual and his ethnic location (for the controls). To do this I must check the names on the lists of results you have sent me against my records. We have extensive charts on all patients, and complete records on all specimens we sent out. Controls are also documented as to age, sex, state of health, date of bleeding, and ethnic group, village of residence, etc. It is obviously high time to make certain that those specimens you are studying are correctly grouped. It is in an attempt to do this for you that I now write. Also, Cyril, I must express my appreciation and thanks to you once again for your help and cooperation in this most intriguing problem. As you have no doubt seen, we have been plagued by reporters and newsmen and radio men on every side; and I am furious, helplessly at their mercy and most disturbed. It would not be so bad if radio and news reports were inaccurate, misleading, far-fetched and sensational and ridiculous—for what else does one expect of them? But from them it is obvious that much of our correspondence and, what is worse, manuscripts of our preliminary reports now in press have leaked out to newsmen. Finally, in an attempt to preserve the *Medical Journal of Australia* from the embarrassment of publishing an outdated, already-superseded, and somewhat inaccurate account, I am now having a running telegraphic battle with the editor—trying to convince him of the need for withholding publication until the much-amplified, revised, and corrected manuscript now in the mail en route to him is received. Politics, national pride, personalities—all of which have nothing to do with the study and investigation of this dread affliction of these fine native people—are forcing the editor's hand; and if I lose out in getting a well-proofread, edited, and corrected paper in the Australian press simply because of their political/national pride panic, I shall never publish a word further of this kuru story as it unfolds in the Australian journals. Now, to the work of this letter, if I can cool off enough.

First, the list of corrected names of the first series of sera are:
[107]

These listings should serve to help you arrange the data into appropriate groups. When I list "Kuru," rather than "Died of Kuru," I mean to imply a typical case which is still surviving and probably will die shortly. All problems such as those I cannot identify require checking against original labels, Simmons's or Lois's records, etc. If you rush this information to me, I shall probably be able to straighten everything out before early December (when I plan to leave for the U.S., to go on with kuru lab work there and the preparation for another intensified field/laboratory approach next year). I am most interested in learning what develops in this study of serum proteins. However, I must reiterate the need, which you fully appreciate, of further and further controls. This is so because a great deal of study already has indicated most unusual serum-protein patterns in New Guinea natives. We did a series of New Britain sera last year with Zara Wheaton at the Hall Institute; and everyone showed enormous gamma-globulin elevation, and paper electrophoresis was most remarkable. I know that Bill Smythe in Rabaul has been finding greatly elevated globulins by paper electrophoresis and precipitation tests (crude as they are) in most New Britain natives. The Western Highlands group is rather dormant now, I believe; but I do know that Walsh did have some highland globulin data, and I was told that there again "abnormal" globulin patterns did appear. Now, there is nothing I should like better than a good lab-lead on kuru; and the possibility that your results are on the track is the most exciting possibility yet turned up in kuru, Cyril. However, please complete tables based on the case-identification I am herewith providing, and let me document carefully all of your "controls." Thereafter, if you get to me the revised, regrouped data when you can eventually get it compiled, I shall very promptly make correlations and breakdowns within each group by age, sex, stage of disease, etc.; and we can even spot such things as malaria-affected populations with spleens, etc. (some portions of this region are malarious), and malarial-free populations. The supply of "normals" should be fairly adequate, for I have been sending rather large Fore, Kimi, and other local tribal-group blood collections to Roy Simmons.

It should be borne in mind that whenever the names of kuru cases are identical, the specimens are different bleedings on the same patient; and these should be grouped together in chronological sequence in making up your tables. If you can find the time, I would very much like to see what further results have turned up; to get further data on the specimens you have already reported which I have been unable to identify; and to see your control data with all the information on names, numbers, lots, etc. of these "controls" so that I can document them properly for you. It is possible that age, sex, invalidism, and other non-kuru variables are involved; and it is for this reason that I urge you to get the specimens all properly catalogued as soon as possible. Please settle on the definitive spellings as I have indicated them. It is an immense problem locating misspelled names, and often an impossible one. We have so many different "series" of studies and specimen-collection underway that numerical-designation systems have not always been used. For the kuru cases, we have restricted ourselves to a single definitive name for each case; and if the spelling is not in error, our alphabetical file serves to locate cases for us promptly.

I shall shortly be sending to Melbourne, either to Roy or to Lois, new bleedings on several dozen kuru cases. These will be the final bleedings before the kuru study is temporarily discontinued; and it may be very difficult to get any more specimens from the field until we get the new, renewed field project organized—which may be a matter of months, perhaps a year. Thus, send your requests for specimens promptly, and stand by for portions of whatever I send down to Roy and

Lois. Please tell Roy, however, that some of every serum specimen must be saved for serological studies we hope to carry out on the specimens. Should you have any ideas of anything else to track down in the sera, please follow through every "hunch"; for the specimens will be refrigerated and fairly fresh, and we need every lead we can get.

Again, Cyril, my thanks. After mid-December you had best address me in the U.S. I shall send you my mailing address and dates in my next letter. For the moment, however, I hope to hear from you again while I am still here in Okapa.

Sincerely,
Carleton Gajdusek

DCG to ROBERTSON 13 November 1957
Dear Dr. Robertson,

I thank you for your letter of September 23rd. I left on an extensive patrol around the border of the Fore kuru region, through uncontrolled Kukukuku country, and eventually crossed the highlands and descended the Lamari, Subu, and Purari Rivers to the Papuan coast on September 26, and thus did not receive your letter until I arrived in Moresby on November 3rd. I am most sorry not to have been on hand to know of your pathological findings earlier. The report of the 16th August never was forwarded to me by any who may have seen it, and the report as I now see it would have influenced my thinking considerably. In the meanwhile, our American collaborators (at the National Institute of Neurological Diseases and Blindness of the National Institutes of Health) had cabled to me a report of more extensive neuronal degeneration—especially in the cerebellum, as you report, and also in basal ganglia, anterior horn cells, inferior olives, thalamus, and pontine nuclei. The neuronal degeneration in the cerebellum and extrapyramidal system, they say, is associated with extensive neuronophagia. Kinao, the first patient whose brain we obtained and sent to Melbourne, was a very typical case of kuru. However, our fixation may have been at fault, and we had originally been requested by the Hall Institute group to cut the brain into slices before fixing. Subsequently, both you and the U.S. groups requested fixation of the unsectioned brain; that is what we have now been doing. Two further brains were sent to Melbourne, and from your letter I see that at least two are in your hands. Did the third case reach you; can you identify the cases by name? Case I is Kinao, undoubtedly; which of the two subsequent brains you identify as Case II, I do not know.

We are most disturbed by the dreadful publicity kuru and our work is getting in the press and radio, and extremely annoyed that our "Confidential and Restricted" manuscripts in press with the medical journals have been shown to the press here in the Territory. Our first report is out, I believe, in the *New England Journal of Medicine*; our second report should soon be out in the *Medical Journal of Australia*. Unfortunately, I am now having a telegraphic battle with the editor of the Australian journal, trying to insist that he not rush to press before I have corrected galley proofs and before essential additions and corrections are made to the article in his hands. Without them, it will be a most inadequate and incomplete report, and of no credit to the journal, and hardly up to some of the lay press accounts. With the corrections I have already in the mail to him, the paper will be a substantial contribution to the kuru problem. However, he has informed me today by cable that he cannot wait for our corrections. Since I have not yet even seen proofs, I am understandably furious; and if the paper appears without the minor alterations and without my having seen galley proofs, I shall not easily forgive the journal its needless haste.

Vin Zigas and I are trying to wind up now our nine months of kuru study, in order to settle down and evaluate our immense stack of clinical and laboratory records and to plan—with the aid of our laboratory collaborators—a renewed field approach to the problem for 1958, during which I plan to return to the Territory for a further attack on kuru. I shall keep in close touch with you, Dr. Robertson, and shall await further word as to what you have found on neurohistological study. I shall also keep you informed of what the American workers report on the specimens they are studying. When we reassemble patients again from their distant and scattered hamlets, when the kuru field project is again opened—for as I have said, we are now "closing down"—I certainly hope that you can visit our Kuru Research Center and help us with this fascinating problem in which your pathological investigations are already of great assistance. I thank you for your interest and cooperation.

Sincerely yours,
D. Carleton Gajdusek, M.D.

DCG to SMADEL 14 November 1957
Dear Joe,

The enclosed two letters to be added to the Kuru files will tell you a bit more of what is going on. Actually, there are a dozen further studies of every sort I am trying to correlate and organize, and the entire job is fantastically immense. Coupled with the current attempt to reevaluate every living kuru case during the next two weeks before I leave and to collect the last batch of specimens, the correspondence and paper revision and correlation work has me working from dawn to dawn.

Vin apparently did not realize the absolute necessity of not letting press and newsmen see papers in print, and the politics-ridden Port Moresby administrative office also broke faith and released on every hand data I had given them in confidential and restricted manuscripts which were in press; I found myself, upon emerging from the Papuan jungles and swamps after almost two months of patrolling, surrounded by radio and press men and their "Laughing Death." To my horror, however, it appears that the local *Time* reporter—visiting the Post in my absence—got his hands on personal letter files and all manuscripts, finished and in preparation; and although I have not yet seen *Time*'s accounts, I warn you to expect anything and everything, for no one here has the slightest idea of the usual and ethical channels of scientific publication. I would not be surprised to see the restricted clinical film I have just prepared, directed, edited and scripted, and sent off to Canberra for final printing, advertised in sensational large print in the cinema ads in the *Sydney Herald*. These, then, Joe, are the odds under which I am now working. Sorry for it all, but what the hell can one do from the unexplored Kukukuku mountains in which I have been at work.

Everyone here is more interested in *Time*, *Life*, and Australian press accounts than in the careful investigation and elucidation of the disease; and as one emerges from these protected highland outposts to kuru-free regions, politics and national and individual pride take precedence over accurate analysis of data and cautious study of the problem.

In spite of it all, however, we may be able to get a good deal more done in the next two weeks. Lucy Hamilton is back with me; and we have sent off—with the two trace-metal–free brains and visceral tissues on two kuru cases—a stack of *pitpit* of the two alkaloid-containing varieties you requested. Already en route to you are stacks of specimens for toxicological study. The *erereba* trials I would take with a grain of salt, but in the next week I shall make new trials of it and send on reports.

[108]
Botanical checklists of all Fore diet and contact plants are being typed and will soon be off to you for distribution to toxicologists, botanists, and pharmacologists for suggestions. I will also send off such staples as *kaukau* (sweet potato), yams, etc. for trace-metal analyses, along with fire ashes from kuru houses for similar studies. Smoke from these fires may be important. Geologists tell me there is much arsenical iron pyrite hereabout; and whether this might get into food from the *mumu*ing of food (by steam-cooking in pits using hot stones) or in some other way, I do not know. Arsenic analyses are thus in order, although what little I know of it as poisoning has not led me to suspect it at all in kuru. Of course, some organic arsenical might act quite differently.

It seems rather certain that kuru progresses from early or moderate stages to fatal termination even in patients removed from the native village to our Kuru Center; and here almost all the "exotic" foods such as grubs, fungi, wild fruits, grasses, herbs, etc. are no longer components of the diet, and even smoke exposure is minimal. They do still get a good deal of local garden staples—but a good deal of rice and canned meat supplement. Cases removed completely from the kuru-area to Kainantu (and the case I dug up from history in Lufa) seem also to progress out of the kuru setting. Although this is not a conclusively proved matter as yet, it certainly must influence our thinking along possible toxic lines. A poison which produces a hardly perceptible initial lesion, and for which the lesion progresses even when futher contact with the agent is impossible, must be considered. These are few, as far as I know.

Has histology of other organs given any clue?

[109]
Enough for now. I have been at the typewriter for 24 hours with only a few breaks to organize the "rerounding-up" of our kuru patients and to visit those I have in the hospital. If it were not for the wonderful fun I have with these marvelous natives for whom I hope we may eventually bring some relief to their affliction, the bastards on the periphery of our problem would drive me from it. I am impatient and in a bad mood; excuse me.

Sincerely,
Carleton

P.S. Another post-mortem at 6 a.m. this morning—with all organs, sciatic nerve, brain, testes, adrenals, and all other viscera taken. Brain is in trace-metal–free formol-saline; but we have used the last such in the Territory, and it is hardly in two volumes of the fixative. It is for this reason that I want to know if the first trace-metal–free brain was well fixed, for we do *not* have 5–6 brain volumes of trace-metal–free formol-saline to use. We are, at the moment, out of all formaldehyde and fixative; but I hope some comes in to Kainantu from Lae by plane today, and we will have it here, with luck, tomorrow. If we have any deaths in the next 24 hours—and our patients have been dying almost daily now—we shall *not* be able to fix tissues. Thereafter, all will be OK again. At the moment, I have no Wright's or Leischman's stain; and unless a plane gets through from Moresby with it soon, I cannot do any further blood smears. These are just the usual problems; but when I come back, I hope it will be with all these solved.

DCG to CURTAIN   15 November 1957
Dear Cyril,

I just mailed out a letter to you when I received your letter of November 4.

I am not at all surprised by all these unusual serum-protein findings, for our preliminary work last year on Cape York aborigines and on New Britain coastal and highland natives and on Papuan natives revealed that all subjects had what would be, on our usual European standards, considered "abnormal" serum-protein patterns. Since it is still quite possible that you are on an important lead in kuru, you cannot escape getting involved in the problem of Melanesian and Southwest Pacific racial serum-protein patterns. However, for the moment I would strongly advise keeping away from coastal groups. First, you will get involved immediately in people who run close to a 100% spleen index in childhood (from malaria, presumably); and in many regions most are infected with filariasis, and the majority have the complications of both yaws and malnutrition thrown onto the problem in many groups. Considering that all groups run 1% leprosy or better (and some almost 10%), it is obvious that there are far too many immediately evident variables to sort out easily.

Malnutrition, kwashiorkor pictures in childhood—with various degrees of chronic liver involvement and strange dietary patterns—are involved in evaluating even malarial- and filarial-free highland populations, but at least near-universal chronic malaria and filariasis do not enter in as immediate problems. Here among the kuru-affected populations, even the malnutrition and kwashiorkor matters are ruled out; and such things as ankylostomiasis are also absent or minimal. Thus, Cyril, I could suggest no place in the Territory wherein the population was better suited for trying to work out the well-known and apparently rather universal pattern of so-called abnormal serum-protein patterns in Melanesian natives.

Without this matter being worked out, it will be absolutely impossible to evaluate findings in kuru patients—even if they are of great importance. Thus, Cyril, please continue to document to me all sera you are using and periodically refresh my memory, for it is quite possible that our opinion on an "active kuru" case may change in time—as also that on a selected "normal." Herewith I am documenting the sera in your list titled "Kuru samples apparently normal" in your November 4th letter. I also must make a correction on my letter now in the mail. *Namarufa* of your series (which I instructed you to correct to *Namaorufa*) was listed as "died of kuru" on my sheet. This is not the case. She should be listed as "atypical kuru (?)" and removed to a separate group, not included with the true kurus or those who have died of typical kuru.

Please, Cyril, revise your lists and grouping according to the information I supply, and let me recheck your revised lists eventually. Also, let me check all your "normals" which stem from me. Yes, it would be possible to do some lineage work on the subjects—but a really immense and time-consuming task. If your results warrant it, we shall do so. At the moment, please make sure that you have controlled your kuru results by use of "normal controls" of same sex, age, and race—or by specimens as close to this as possible. We shall shortly be shipping Roy Simmons further sera—some went out today—and with them I shall include Europeans bled here in Okapa and/or Kainantu, along with native control-subjects and further kuru specimens. You can get serum aliquots from Roy.

Enough for the moment.

Sincerely,
D. C. Gajdusek

DCG to WINTON   15 November 1957

Ronald M. Winton, Editor
Medical Journal of Australia

Dear Dr. Winton:

Enclosed you will find the fully corrected galley proofs which we have immediately corrected and rushed back to you. If your printers follow corrections on them carefully, we shall be fully satisfied with the article. The enclosed manuscript is a retyping of the corrected carbon-copy of the original manuscript which I rushed off to you a few days ago and on which I have now made those grammatical, form, and word-order corrections which I find you have made in editing the manuscript. Thus, this manuscript should correspond exactly to the galley proofs and can be used for rapid reproofreading of the printer's corrected galley proofs and page proofs.

We have just cabled you as follows:

> Corrected galley proofs mailed today Stop We must refuse permission for publication without these corrections Stop Please do not jeopardize scientific accuracy for sake of panicked rush to print Stop   Gajdusek Zigas

I am sorry that we have had to be so insistent. As it stands, your galleys reached Kainantu on November 14th yesterday, and thus, your note requesting corrections by mail by November 13 could not possibly be observed and cable of proof corrections, even if minor, would have been inconceivable. We have not held up the proofs having returned them to you in less than 24 hours of their receipt by us. We must insist on these corrections if the article to appear in your journal is to be a worthwhile scientific publication. As it is, the first preliminary report which has appeared in the *New England Journal of Medicine* has had incorporated into it the recently obtained dramatic neuropathological findings and the changes in attitude resulting from these. To our dismay our "restricted and confidential" manuscripts have had material from them released in Port Moresby to newsmen and we find the ridiculous and disturbing reports in the lay press contain mention of case numbers and pathological findings and hypotheses which would be more accurate and more up-to-date than the report would contain in your journal, if it appeared uncorrected. What would make matters worse, we should be forced to immediately revise and restate our results and conclusions in articles in other journals if the original article remains uncorrected. I was expecting time for a slow, leisurely correction of the galley proofs and thus had set the pathological findings aside for inclusion when the proofs arrived. I then noted that they were based on study of autopsy specimens obtained from a larger series of cases than those summarized in the text as it was first prepared, and therefore took the opportunity to revise the case series to the 154 level at which important changes in our view of the age-sex incidence pattern began to be evident. This is the background for the revisions. They do not demand any extensive breaking up of type. Instead, they demand only cautious correction of all existing type and deletion of only a few sentences in the entire proofs. In place of these deletions a few new texts, a bit amplified, are to be inserted.

Since your journal has been willing to publish so long and extensive an account of kuru with ample illustrations it is certainly fitting that it be a medically and scientifically sound and up-to-date report. To have it lagging behind, and inaccurately at that, the hideous accounts in the lay press and to have it contain material which we must immediately modify, retract, and correct in articles in other journals certainly reflects poorly on the article as published and the journal in which it appears. For these reasons, Dr. Winton, Dr. Zigas and I are adamant in insisting that time be taken to see and follow out proof corrections. If you and your editors follow the printer's corrections of the type, Dr. Zigas and I shall be willing to leave in your hands the final rereading of the galley and page proofs and we shall be satisfied with your checking of the map, illustrations and appropriate number of the captions for these. Please be certain that the photographs you received correspond to those which I selected and that captions are appropriate.

We thank you for this difficult matter of editing and for the added effort our corrections impose upon you. If the paper appears with the proof corrections observed we shall be most anxious and willing to publish some of the further reports on kuru which are in preparation with your journal, should they meet with your editorial approval.

Sincerely,
D. Carleton Gajdusek, M.D.

DCG to BURNET and ANDERSON   16 November 1957
Dear Sir Mac and Gray:

I write after a long silence to let you know what course our work has taken and to ask a few questions. After taking out two months for extensive patrolling—which included a long patrol through the Kukukukus and crossing the island to Papua and down the Lamari, Subu, and Purari to the Papuan coast—thence to a Kuru Conference in Moresby (which we organized with Dr. Scragg and Dr. Gunther), am finally back at Moke-Pintogori in one of the most rushed and hectic periods of the kuru study. I am trying to make a complete reevaluation of all 100 or so surviving kuru patients and to make extensive new therapeutic trials of drugs as well as to collect an immense load of urines and blood, CSF, and autopsy specimens for new studies—some of which I hope to get to shortly in the U.S. This, then, is a rushed month in which we are trying to recapitulate the first four months of kuru study and to extend our investigations further.

I have your letter, Sir Mac, of September 27. It has been waiting for me for well over one month. The first kuru report is out, I believe, in the *New England Journal of Medicine*; and the paper for the *Medical Journal of Australia* should shortly be out. However, we have had a great deal of trouble in our work with the latter journal in the last few days. This paper was written before pathological reports were available and when our total case summary was well under 100. As we studied the disease further and collected further cases, modifications of some importance in our view of the disease occurred and, also, the pathological reports became available. I informed the editor that I wanted to revise the paper somewhat to include these later findings; there was plenty of time, and I had already so revised the American paper, which has appeared with at least the most important revisions. I hoped to edit galley proofs leisurely and to make the necessary changes and additions to them. Now I find the editor is suddenly in an unpardonable rush to print. When I rushed airmail the revision on a corrected copy of the manuscript, he telegraphed back to my cable that he could not wait for them and was rushing to print this month. I cabled our demand to see and correct galley proofs first. They have now arrived, and on them he sets a return deadline—only a few days after their dispatch—of November 13. We received them only on November 14, and eighteen hours later on November 15 they were already back in Lae with full editorial corrections and revisions—including addition of pathological findings, revision of case numbers to coincide with those reported in pathological reports, etc.

However, since proofs arrived with his plans, stated in a note, to rush publication even before our corrections could be received in Sydney (which will be Monday or Tuesday, I suspect); we have cabled him our refusal to permit publication without receipt and editorial use of our corrected galley proofs. This is the first time in my life I have been denied the right to see and correct galley proofs. We are perfectly willing to lose priority in publication and have the paper even months from now. Furthermore, we would rather withdraw it from publication than have it appear without our revisions. Since the journal is going to give it immense space and publish with it 12 or 13 illustrations, it certainly is in the interests of ourselves and of the journal to have the article up-to-date, medically sound, and scientifically accurate. Should the editor not heed our pleas and publish without our corrections, we shall be unwilling to publish another word of the kuru story in Australian journals; and we shall be forced to modify, restate, and retract— in American and European journals—much of what he publishes.

I tell you all this, Sir Mac, in the hope that you can advise him to give us the benefit of a few days delay in this panicked rush-to-print, and to make all corrections and revision we have marked on the galley proofs. There is no hurry from our point of view. We can wait even months before the paper appears. I realize that the horrible publicity which the radio and press have given to kuru is the cause for his rush, but premature, unedited publication will only confuse the issue and reflect poorly on the journal. To our horror, manuscript material marked "confidential, restricted, not for release" (because it had been submitted for publication) has been released in Port Moresby to press and radio; and I read full quotations from our papers in press. You can imagine my anger when I see paraphrasing of my very statements—often printing without paraphrasing—in Sydney newspapers, when the statements were made only in confidential manuscripts submitted to journals for publication. In view of this hideous blunder by whoever dropped our manuscripts or their contents into press hands—a full breach of medical and scientific ethical procedure—we would like the Australian article to be at least as accurate and up-to-date as the *Sydney Herald* and *Time*. On this matter, then, Sir Mac, we would be most grateful to you for any help. We have made all corrections we seek to make, both on a revised copy of the original manuscript which should already be in the editor's hands and again on the galley proofs which he will have received before you get this letter. Enough of all this.

Kuru cases seen and studied by us now exceed 200. Our deaths from kuru among these 200 number over 50 already. The problem of mild, incompletely expressed cases of kuru is now receiving considerable attention from us, for we have a number of mild, uncertain cases which "recover" or remain stationary. Thus far no full-blown, advanced case has survived or appears likely to do so, and the same is true for fully expressed cases of moderate severity and duration—with only the few exceptions which we believe to have been hysterical mimicry of kuru. We have, furthermore, now tracked down another disease syndrome—distinct from kuru but extremely close to it clinically, and of quite a different duration and usual course. It is, however, undeniably a "close relative of kuru"; and since it occurs in a different but neighboring people, I am most attracted to the possibility that we have a "genetic variant" of kuru—much like the different patterns which many of the complex heredofamilial syndromes run from one family to another affected family, which probably accounts for the mass of eponymically designated syndromes in the books.

Cyril Curtain reports finding of what may be a new or abnormal serum protein in very significant amounts in our "control" sera, and he still gets—but inconsistently—very abnormal alpha and beta globulins in many kuru cases but not in controls. Thus, he is involved in the immense problem of the globulin and serum-protein patterns in Melanesian natives; and to sort out all the infectious, environmental (dietary), and genetic variables which may influence the picture will be a hard task. I think he may really have a lead in kuru; but I am most cautious, since until the full story of globulins in this and surrounding populations in "normal" individuals has been worked out, the rather frequent globulin "abnormalities" in kuru cannot be interpreted.

Our work on toxicology and ethnobotany and study of all dietary, contact, and bite exposure to insects, spiders, etc. is intensified—but no good leads thus far. We put a full boundary around kuru by well over 1000 miles of hard walking, interviewing for weeks on end, and mountain-climbing. I have tracked down—as well as I can in this brief time—all memory-recorded kuru deaths in the past and have place, age, sex, and family-history data on each.

We are most anxious to know of any results on the urine amino-acid studies which were done on further urine collections sent down to Gray for transmission, I believe, to the Women's Hospital. Furthermore, we would like any further word on neuropathological finds and on pathology of other tissues received. I have written my thanks to Dr. Graeme Robertson for his note. To our chagrin, we did not get to see a copy of his report until a *Time* reporter arrived carrying the full report in his notes from Moresby and showed it to Vin—to Vin's great surprise and astonishment. I was out on patrol and thankfully missed the press. However, they caught me in Moresby, unfortunately.

We are so rushed and pressed for time at the moment that I cannot write more.

If the Australian article comes out with our corrections, we shall be proud of it—for it has in it all we are prepared to say at the moment. Otherwise, we shall have to get the revised story off to other journals. Thank you for your interest and cooperation on the kuru problem. May we hear from you shortly?

Sincerely,
Carleton

DCG to SMADEL   16 November 1957
Dear Joe,

Enclosed is a letter to Sir Mac, on the second page of which is an item—marked in red ink—which you and others should know about. Yes, this is really a distinct new non-congenital and progressive CNS syndrome known to the people of the Kawaina-Nambaira regions by the name *tukeseina* or *tukesa*, and which runs a course moderately different from kuru—we even spotted the difference ourselves in the stance, tremor, and gait—and of which we now have one patient. I shall soon study him intensively and take motion-pictures of him. The course runs over several years, tremor is more constant and regular, gait and stance are impaired in a different way than in kuru. We have really not studied it well yet. The people affected are just beyond the kuru-border in an "uncontrolled" region where they still shoot arrows at us, and they have a full tradition of the disease; we shall look into it further.

An Adelaide University team will arrive to carry on clinical follow-up of our cases in mid-December. We are at the moment entirely on our own, however (as we always are in this study, except when it comes to talk in Moresby), and we are trying to do an immense job in the next few weeks. Please let me know what you think of the possibility of getting NIH help for a return expedition in under one year, if I get back shortly. I must return and shall return, but I would rather do so on an income with some support. Is there enough interest in kuru and the associated problem to make it likely that, if I get to work on it back in the lab, I can escape for a brief follow-up study before the end of 1958? I am making plans for that here, as I prepare to leave.

Enclosed are a few odd case-record copies to simply be put in the Kuru record file. However, shortly I shall be shipping immense piles of case-records, correspondence files, etc. Please, Joe, keep your eyes on them, for all of our results are incorporated in these files of case-records and correspondence; and it is on these that I hope to work immediately upon my return.

Sincerely,
Carleton

DCG to SMADEL   17 November 1957
Dear Joe,

Enclosed are extensive Fore food-lists which represent part of the fine work Lucy Hamilton has been doing with us on kuru. Lucy we stole from the Department of Health for kuru work, hung on to her tightly when they cabled for her return, and now, after a Kuru Conference in Moresby, we have her again for two weeks. However, she is about to be married and lost to both the P.H.D. and kuru—to the latter temporarily only, I hope. Since our time is short, we are rushing ahead with intensified studies and a feverish attempt to document, record, and summarize all of our findings before our project breaks asunder. Lucy has only a matter of ten days more with us; and she is working with me on a full write-up of her findings and correlations of them with our epidemiology, case-findings, and clinical studies. We have just had Kevin White, the Lae botanist, here assisting in field botanical work and collections; and he has helped in the revision and final preparation of Lucy's extensive lists, which I am submitting to you now for:

1. permanent file with the Kuru records (one copy);
2. dissemination through NIH to all who may be able to contribute (physiologist, pharmacologist, toxicologist, etc.) to the problem by survey of these records (one copy);
3. transmission (on loan) of one copy to the Smithsonian botanists and entomologists and mycologists (the third copy), for their comments on possible toxicology, alkaloid content, etc.; and, later, to any toxicologists and pharmacologists who may be able to contribute ideas; and later to other museum groups.

Lucy and I are now preparing a full epidemiological evaluation of the kuru problem as far as it has thus far been worked out. We are working it up section by section. Soon the section based on her Fore dietary survey will be written. Thereafter, the sections on food consumption and strange and exotic foods by the kuru and non-kuru Fore, and also a comparison of Fore with other surrounding linguistic groups as to food usage (including native salt preparations, body paints, drugs, etc.) will be prepared. We are also working out both a detailed epidemiology on the Moke community in respect to kuru over recent decades and the present, and the large-scale epidemiological pattern of kuru over the entire kuru-region. These and other individual sections will eventually be incorporated into a series of kuru-epidemiology papers by Vin, Lucy, and myself; it is in a mad rush to get the immense sheaf of data recorded in better order and (at least temporarily) analyzed that we are now rushing furiously ahead. Back in Washington, I shall work up the papers definitively and keep in touch with Vin and Lucy by mail. If we are successful, all the data for the final paper preparation should be ready for transmittal to the Kuru record file at NIH (which awaits me there, I trust) within the next few weeks. These lists of Lucy's are part thereof.

We are sending out via Qantas a shipment of plants which are summarized on the accompanying sheets. These are for immediate study. Kevin White is just completing a week of botanical work with us and has made these collections for your study and—particularly, toxicological—pharmacological analyses. These two plants have not been linked specifically with kuru, but of the entire Fore dietary list they appear as most suspicious at the moment.

We are proceeding on the assumption that women and children of both sexes—even older boys who no longer stay with the women in the gardens—differ from men in their exposure in that they (i.e., the women and children) may be prone and accustomed to eating some plant, insect, etc. uncooked in the gardens and bush as they go about their daily lives. Adult men may have already largely abandoned this propensity. Just what it might be, we have not been able to spot. The shipments en route are those of most toxicological and medical interest in Fore diet, and their possible relationship to kuru must be ascertained.

Although I now have further cases of women from low–kuru-incidence communities developing kuru in the kuru region, I have obtained these along with a host of new cases of kuru-region women developing kuru outside the usual kuru region, or in regions of low kuru incidence—but none have ever developed it really far away. One adult male patient got the illness working in a calaboose fully one day's walk outside the furthest boundary of the kuru region, although he comes from the center of the kuru region. In the calaboose it is most unlikely that he had any dietary exposure different from all the others. Thus, as with a number of other cases, we are forced to conclude that the lesion (if toxic) is inflicted early and without producing associated symptoms, and then proceeds over the course of many months to mounting neuronal destruction without subsequent exposure to the toxin causing the original insult. Cases such as the woman who progressed from earliest kuru to fatal termination in Lufa (miles away from any site where kuru was previously known, in a kuru-free population) and our current cases progressing in our Kuru Hospital (when taken fully out of their bush and village setting) and, it appears, even the cases we have had in Kainantu (fully out of the kuru-region). All these make it most necessary to consider some accumulated and then slowly released toxin (perhaps a metal or rare earth, etc.) or some material which continues over many months to act after initial exposure.

Finally, I can still see no sign of infection or post-infectious phenomenon; but with something as unique and unprecedented as kuru, a unique and unprecedented explanation is required. Thus, I continue to wonder whether some arthropod-borne vector might not be transmitting an agent causing an asymptomatic sensitizing infection (like von Economo's sleeping sickness, but clinically less apparent) and, finally, I wonder whether some neurotoxin akin to that of "tick paralysis" might not result from arthropod bites or ingestion (insects and spiders). Finally, is any arthropod ingestion likely to give rise to neuroallergic response? Cannibalism as a possible source seems well ruled out.

We are sending you also shipment of clotted bloods in sterile 30 ml venules from kuru cases, along with a few exactly similarly collected controls. These are the most critical specimens we can collect. I suggest designating them for trace-metal studies, including even rare earth if such things can be done, for we are really looking for the proverbial "needle in a haystack." If these arrive in such condition that serum specimens can be easily and well removed, wire me information and I shall then send similarly fresh kuru-blood specimens from a host of kuru cases as the final supply of specimens for the Kuru file. An immense serum file assembled in Melbourne should be underway to you already, and further shipments will also be off. These are designed for serology and any chemistries still worth doing. However, these specimens I now suggest sending directly as clots in venules, planning (as we now have) on a really tight air-express, refrigerated delivery schedule designed to provide fresher serum and clots for any more critical toxicity or chemical studies. If you wire me about whether the first trial shipment warrants a second, also wire about condition of fixation and receipt of the last shipment of two trace-metal–free brains. Another trace-metal–free formol-saline–fixed brain is here hardening and awaiting shipment.

[110]

Sincerely,
Carleton

DAVIS to GARLAND Paoli, Pennsylvania
20 November 1957

Dear Doctor Garland,

I read with interest in the November 14 issue the Special Article: "Degenerative Disease of the Central Nervous System in New Guinea . . . ."

I would like to ask the authors if there is any chemical evidence in the blood of a chronic magnesium deficiency? In their cooking or fires would there be any chance for chronic inhalation of zinc fumes or zinc ingestion? Is there any chance for chronic mercury-salt ingestion, like cinnabar-HgS? Have blood zinc and mercury levels been done on all the ill studied? Chronic zinc and mercury poisoning will give Parkinson's-like changes, and magnesium deficiency will also give rise to obscure similarities.

Sincerely yours,
Perk Lee Davis, M.D., F.A.C.P.

SMADEL to DCG 21 November 1957

Dear Carleton:

It is good to learn that you plan to return to the States in December. I hope nothing comes up which will interfere with those plans, since the National Institute of Neurological Diseases and Blindness staff is taking steps to appoint you as a Visiting Scientist for a one-year period beginning just after Christmas. The salary level will be equivalent to that for a GS-14 Medical Officer: i.e., $10,320 per year. I hope these arrangements are satisfactory to you. We are looking forward to having you here, and to the papers which you will put together from all of the data collected in New Guinea.

I am enclosing information on identification of some of the plants we received from you. We have received two shipments of plants, and chemical work on them is continuing. Currently Dr. Horning is checking on the occurrence of glycosides in these materials. Dr. Masland in NINDB has been having some difficulty finding anyone who is set up to perform the trace-metal studies for those elements mentioned in your letters. Copper is the only one being determined at NIH. However, he believes the laboratory at Fort Meade will probably be able to do the determinations.

Mrs. Beyersdorfer in NINDB has already informed you of the receipt of film and the production of prints. By now you should have received the 12 prints made from the latest batch of negatives which we selected for you. It has been our assumption that these 12 are for your use in writing a new paper on kuru. By now the *Medical Journal of Australia* must have received the prints sent from here. If this is not the case, perhaps you can use some of the 12 illustrations sent you by Mrs. Beyersdorfer—in which case we could then send you different prints.

Also enclosed is a copy of your paper in the *New England Journal of Medicine*. The reprints are not yet available, and you will probably be here before they are.

Although I appreciate conditions are difficult where you are, I am afraid you have been remiss in providing proper information to those groups performing anatomical studies for you. Some of the people in Australia were disturbed to learn that studies similar to those they were making were being performed by others simultaneously without their knowledge. They were particularly concerned because they have only limited technical help available. Furthermore, they were concerned that they had no acknowledgment from you of their reports on the anatomical studies. Please see that both groups are informed more thoroughly of what the other group is doing.

Please keep us informed of your time of arrival, in order that planning may proceed satisfactorily. NINDB will make space available for you for the writing of a series of papers on kuru. All the material sent by you is on file awaiting your return.

With kind personal regards.

Sincerely,
Joseph E. Smadel, M.D.

SIMMONS to DCG   22 November 1957
Dear Carleton,

Thank you for your long letter of November 12th with appropriate explanations. I go on leave December 9th to January 2nd.

Herewith the results of extras to Lot 9 and Lot 17. Tested 21 November 1957. I do appreciate that in a place like New Guinea considerable difficulties and upsets may occur. [111]

I notice you are in difficulties with the Editor of *MJA*. It would seem to me that if you rushed a manuscript to him earlier, you should be prepared to have the paper published on the publication date arranged for the journal. As the paper is yours, it should not be "inaccurate, incorrect, not up-to-date, and very misleading." I have always found the *MJA* fair to deal with. You must have rushed the paper!

There has been a huge amount of publicity given to kuru in the papers with even an up-to-date version in *Time*. I regret to see that while the Hall Institute has had much mention, the work at C.S.L. or at the Baker Institute did not warrant a single reference. As there will be more publicity, it would be fair to mention other institutes which have devoted time and money to assist, even in a small way.

A further thought about the *MJA* paper—I wonder how you would get on with the Editor of *AJPA* or other publishers under similar circumstances. I mention these things because your letter suggested that the *MJA* was being unfair to you.

Send further samples directly to Lois and not to me, please. If they arrive after I leave, she will hold my samples in glucose-citrate in the frig until I return. This may be my last letter which will catch you in New Guinea.

Please accept my best wishes, and may you have a safe trip home. Subsequent letters after December will be to the USA. Cheerio.

Regards,
R. T. Simmons

P.S. Have recently tried to help Curtain sort some names out.

DCG to SIMMONS   24 November 1957
Dear Roy:

Enclosed are copies of blood-specimen lists for the last few shipments to you. Obviously there are new kurus for the kuru study, and there are many other blood specimens designed for the serological epidemiological file (to Lois) and aliquots to Cyril Curtain for his work. Of particular interest to you will be a mounting series of blood from Kukukukus (all pure-blooded with documented pedigrees) and from Auianas—a Fore-adjacent, virtually kuru-free people of entirely different language. The major linguistic-cultural groups hereabouts I shall list for you:

Fore	entirely kuru-affected.
Kimi	only kuru-affected in one region; and Kimi bloods sent to you, in general, are from kuru-free populations.
Auiana	virtually kuru-free. Two cases in Fore-bordering villages decades ago; none since.
Kukukuku	kuru-free.
Iagaria	kuru-free on Lufa side, whereas a pocket which has lived and intermarried with Fore for generations has much kuru.
Keiagana	like the Kimi: part kuru-affected, part kuru-free.
Usurufa	linguistically identical to Auiana except for slight dialectical differences—intermarry with them, yet hold their tribal identity.
Kanite, Iate, Kamano	three groups with languages closely akin to Keiagana and to each other, but holding their tribal identities.

Virtually all specimens I sent you can be grouped into one of these groups, and I shall always try to provide adequate information to make assignment possible. A certain number turn out (after exhaustive questioning and identification of their parents) to be "mixed-bloods," with this clearly indicated on our lists. These specimens are still fine for serological file and for Cyril's work, but perhaps should not be bothered with in your work. They are not many; they are clearly indicated and they serve to suggest possible occasional marriage contacts in the distant past, showing the direction thereof.

Besides continuing to send bloods on kurus, I shall now only bleed those major linguistic groups for whom the total number in your hands is still statistically unsatisfactory—Keiagana, Auiana, and Kukukuku principally.

As you may have heard, our neuropathological findings continue to point strongly to a toxic factor. We are beginning to study the composition of the rocks used in *mumu*-cooking (steam-pit cooking over hot rocks) as we become discouraged over every other possible lead. However, more and more cases developing in Fore well outside the usual kuru-region are now in our hands, and we know that

1) the illness progresses relentlessly even if patient is totally removed in earliest stages from the kuru regions and the diet of these regions, and
2) the disease can develop in a Fore who resides outside of the kuru region for well over one year in a region inhabited by people who do not otherwise have kuru.

Obviously these two observations make the search for a toxin in the kuru region most difficult. It would either have to be something the Fore brought with them or found in the non-kuru region which those they lived with were not exposed to—or it would have to be a most unusual toxin, exposure to which months or years previously was all that was required to initiate a progressive neurological destruction.

Enough for now; and again, Roy, my admiration for your patience and tolerant coping with all the problems I present you with.

Sincerely,
D. C. Gajdusek

DCG to SMADEL   24 November 1957
Dear Joe,

Enclosed is a list of kuru-blood specimens off to you for trace-element studies in sterile venules. Controls can easily be collected similarly and sent along with other kuru bloods. These were delayed inadvertently and missed one plane connection, and thus we can hope for even prompter transmission on most future shipments.

Enclosed are summary lists of recently dispatched still negatives and cine negatives. All undeveloped.

Enclosed are some summary sheets for kuru and non-kuru blood specimens sent to Melbourne for work by Roy Simmons (blood-group genetics) and Cyril Curtain (abnormal hemoglobins and serum-protein study by Tiselius electrophoresis and ultracentrifuge, etc.). Curtain thinks he has found a new globulin in normal Fore, and real abnormalities of alpha- and beta-globulins in most—but not all—kurus. I am skeptical for reasons summarized in my last long letter to him, but it is the first real lead of any sort and it must be pursued.

We are collecting *mumu* stones (some being beryl-containing ores and some arsenical iron-pyrites, we are told) for very thorough study. When these go off to you—stones cannot be made a light shipment—they will go off with a section of our epidemiological papers in which *mumu* cooking and the use of these stones in steam-cooking is described. Although such *mumu* cooking is rather universal in the highlands, it is possible that local stones might vary greatly and that some stones could serve to contaminate the food. Why the peculiar age-distribution of cases? Well, perhaps foods directly adjacent to stones in the *mumu* are particularly involved or a certain type of food of all those heaped into the steaming pit, and these are particularly the fondness of women and children and some men—a wild possibility, but one which needs investigating. In any event, rare earth, beryllium, or arsenic in *mumu* stones, etc., etc., cannot solve the matter. The substance must be found in the body, and thus the trace-metal samples of autopsy viscera and the trace-metal–free brains are most critical items. Please get all we can from them. It looks as though further autopsy materials may be unobtainable. The natives have given up our medicine; they know damn well it does not work, and I am fighting (verbal battles in Fore), bribing, cajoling, begging, pleading, and bargaining for every opportunity to see a patient, and strenuously working tongue muscles for hours for every further day we get a patient to stay in the hospital, accept therapeutic trials, etc., etc. Vin is sick and tired of the "duress of personality" which is required to pressure every case into our care, and I do not like the effort. It means, however, that unless we start curing cases quickly, we cannot expect any clinical material—much less any autopsy specimens. I am willing to keep up the push, using every ruse short of actual duress by force and authority—that we cannot contemplate—but Vin is tiring even of the battle of personalities between us and the Fore which our attempt to do something for the kuru victims is demanding.

We have now accumulated many further cases which make the toxic possibilities ever more limited and ever more baffling: Thus,

1. Many cases develop outside the kuru region in Fore or kuru-region people who have moved into the kuru-free population to live or work.
2. The illness progresses relentlessly even if patients are totally removed in earliest stages from the kuru region and are on diet of the non-kuru regions.

This obviously demands that any toxic agent causing kuru must be something the

Fore bring with them or find in the non-kuru regions, yet which the local residents with whom they are living are not exposed to—or it means that a most unusual toxin exposure is involved, one in which exposure months or even years previously is sufficient to initiate a progressive slow neurological destruction.

Please see that a copy of the *NEJM* article gets airmailed to us—we will not see it otherwise. Excuse this barrage of correspondence. I hope you can find someone to read it.

Sincerely,
Carleton

[enclosure]   Additional Notes on Kuru, 24 November 1957

Watching the illness carefully again in the past two weeks and reviewing our epidemiological and clinical findings, the two items listed on the accompanying letter become ever more convincing. Namely:

1. Cases develop outside the usual kuru-area in visiting or immigrating Fore (from the kuru area) from weeks to years after they have left the kuru-region to reside with people who do not get kuru. We now have several such cases.
2. The relentless progression of the illness even when removed from the kuru-region and off the usual kuru-region diet, even when this occurs in very early stages of illness. Evidence that this is so is beginning to accumulate.

Even those newly discovered communities outside the boundaries we thought we had set wherein kuru has occurred in the past, we find, are impressive in that they are communities with extensive Fore contact (presumably also intermarriage), such as Fore moving in among them or their having moved into the Fore regions during the time of intertribal warfare.

The two items noted above, taken with this added evidence for possible genetic—or at least a distinct "cultural"—predilection, makes the toxic possibility look even more difficult to establish. However, if the neuronal destruction is not genetically predetermined completely, and if exogenous toxin is unlikely, is there any chance that we may be involved in a peculiar sort of autoallergic state or rheumatic-like self-perpetuating process or drug-reaction, exfoliative dermatitis-like reaction akin to Stevens-Johnson disease? What I specifically wish to ask is whether any of these hypersensitivity states might conceivably release a kindred cytological—if not neurocytological—lesion?

A number of wild possibilities come to mind—some can be dismissed already. Others cannot be dismissed by any means. Thus,

1. Cannibalism with human ectoderm or brain consumption might provide a year-long or life-long sensitization triggered off by consumption of raw or poorly cooked ectoderm or neuronal tissue.

    The Rub: Cannibalism has ceased, did not involve brain as far as we can find; and even if rare cases still occur, many of our youngest patients have rather certainly not consumed human tissue.

2. Consumption of animal brain—particularly if poorly cooked or uncooked—or, perhaps other neuronal tissue (ganglia, etc.) and/or ectodermal organs. Conceivably, prolonged exposure to raw or poorly cooked pig—especially ectoderm, ganglia, or brain—might lead to a hypersensitivity state which slowly progressed or was triggered by moderate new exposure.

    The Chance: Fore pig—the major meat eaten—is (as is other game meat) cooked in *mumu* by steam-cooking over hot stones, or steamed in bamboo tubes. In both cases it comes out notoriously undercooked. Whether the brain is eaten or not, we must still further investigate. It appears that the head is usually not opened in cooking; the brain therefore cannot be very well cooked. It is usually eaten later in the household of the owner of the pig and not by those feasting, who first consume the rest of the pig. Certain men and youths definitely deny that they eat or like the brain, and I have a hunch women and children get it mostly—occasionally men—and that it may or may not be further cooked again, a day later, in bamboo-tube steam-cooking. Thus, repeated long-term exposure to undercooked pig-brain, perhaps particularly by the kuru-affected age/sex group, is a distinct possibility. What are the chances of this?

3. Consumption of raw meat in childhood. As a *mumu* is being prepared, we often see small children given pieces of raw meat which they eat without cooking and without being stopped. Excessive raw-meat consumption in early childhood—perhaps particularly ectodermal (since skin is always eaten)—might be questioned.

Pig-eating and *mumu*-cooking are near-universal throughout the highlands, and kuru is not. However, we do not know of eating and cooking habits referable to the brain and ectodermal organs, and degree of consumption of uncooked meat. It might be a matter requiring *prolonged, repeated exposure,* as in brain reactions to brain vaccines; and the matter of individual reactivity might well enter in as a genetic component—as is the case with most immune reactions.

These are wild conjectures, I admit. But take them for what they are worth, and see if anyone can give us a ray of hope.

Finally, critical things like search for manganese (a most immediately essential matter) in brain and other tissues have not yet been done.

[112]

26 November 1957

P.S. Our much sought-after case has appeared!!!: a man who perfectly well walked to Goroka from the kuru region (Keiagana) for work and ended up five months ago on the coast as a laborer, who developed *there on the coast* kuru now of about three months duration—very typical, severe, and rapidly progressive. He certainly would not have started out for work with even the slightest suspicion he had kuru (and patients themselves invariably diagnose the illness before we do). Thus it is a case with onset far away and completely outside of the kuru region and full, very rapid progression in three months completely removed from regions wherein kuru has ever previously been observed. What, then, Joe, does this make of all our toxic search? You can see why I have been forced to lean genetically—for predisposition, at least.

BURNET to DCG   27 November 1957

Dear Carleton,

Since receiving your letter, I have had a note from the *Medical Journal of Australia* indicating that they had done their best to cope with the last-minute alterations and additions. It therefore seems inappropriate for me to do anything in the matter.

Yours sincerely,
F. M. Burnet

SIMMONS to DCG   2 December 1957

Dear Carleton,

Thank you for yours of November 24th giving details of linguistic-cultural groups and other data.

Enclosed find results Lots 18 and 19 for all samples to date.

There was a map of the New Britain paper Lois sent you for corrections. At that stage I stopped writing, when the paper was three-fourths finished. I have eight papers in press at the moment. I will proceed with the N.B. paper in 1958—do you wish it sent to USA for your final approval? You probably should see and add to it if necessary.

You have not given me a departure date from New Guinea—however, I go on leave on Friday 6th, and thus samples will be stored, hoping for the best. Please collect as usual.

Comments 1 and 2 on kuru in your letter could be consistent with a poison, although a year to develop the effect is a long time. A few years ago, there were a lot of thallium deaths in New South Wales—this was in rat poison and was administered in food by a loving spouse or good friend. One woman did about five. This was slow-acting and was invariably fatal, until someone woke up to what it was. I merely use this as a parallel. Could it be said that every kuru case has lived with the Fore people at some stage? I feel that food or method of cooking is possibly the best lead.

Regards—best wishes,
R. T. Simmons

DCG to SMADEL [early December 1957]
Dear Joe,

With the first jeep to go out in ages about to depart and leave me alone once again among the Fore and kuru, I am rushing off this note. Rains are blocking our tracks, and soon transport may be a dreadful problem. I lost a jeep over a cliff in a landslide in my last attempt to save time on the Purosa "road" to the south—to cut, by driving, a few days off the walking—and the rescuing of it cost me more trouble than the walking would have. However, we are now collecting the last of those specimens I shall get here; and in a few days, the first two of the four-man Adelaide team will arrive. I shall stay on to get them oriented in the problem and leave. They have two months for the work only, and I have no clue as to how well and elaborately equipped they will be. I hope, however, that they will have extensive laboratory facilities for stepping up some of our bedside clinical investigations.

Enclosed is a list of specimens going out with this letter and to be shipped from Kainantu as soon as the flooded strip permits a plane to land (by drying off a bit). The urines were really hell to collect; and full protocols on volumes and time of collection, and BAL dosage, will be sent in my next letter and will also appear on the patients' charts. Thus, *all* the necessary data will be in your hands.

I have saved three bottles (c. 90ml) of urine from each voiding—all collected in trace-metal–free water-rinsed kidney dishes with as little debris and wind-borne crud in them as is possible under these bush conditions (even our roofs have a constant "fall-out"). They certainly are the last such set there will be for a while; and I suggest settling manganese and other metal possibilities with them, if possible. Furthermore, if there is adequate volume of any specimens for any amino-acid or other studies, use them also for that, for they are refrigerated from very shortly after collection.

Bloods are sent on clots. Trace-metal studies, obviously, are still of principal interest. However, anything which anyone can think of which may help would be welcomed. Perhaps some may have sera adequate for further Tiselius and UC studies to confirm the extremely high beta-, alpha-, and gamma-globulin patterns which Cyril Curtain continues to report from Melbourne.

All bloods are collected with only ether-cleansing of the forearm. This may not always get through all the dirt and pig-grease which the needle must penetrate between us and the patient's vein.

The summary items which make the toxic possibilities more-than-ever intriguing and difficult to track down were listed in my last letter to you.

Brains on two patients are still hardening, and tissues are sent herewith for immediate study. The brains which have been trace-metal–free collected have been fixed with very limited volumes of formol-saline. We have very little of the precious stuff on hand. One is now sitting in a similar small volume of 5% formol-saline, and I shall let it fix for a long while. The other is in regular (not trace-metal–free) formol-saline (at 10%, at that) and will be hard as hell. If anyone at NINDB believes there is value in further pursuing the quest for trace-metal–free brains, they had best rush appropriate containers and formol-saline to us—for no stills (glass stills for such critical work) can be found in the Territory, and what we have we have had to get from Melbourne.

Thanks for all the steady flow of information. Anything else in pathology as yet?? Enclosed is also a Fore diet summary which Lucy and I worked up before she left us. She did all the weighing and food study; but from the limited data, nothing further could be calculated. This is to be part of a series of three papers on kuru epidemiology we are writing, and thus please keep it in the file.

Sincerely,
Carleton

MASLAND to C. G. BAKER   6 December 1957
Memorandum.   Subject: Trace-Metal Studies in Respect to Kuru.

At the suggestion of Dr. Cantoni, I had a telephone conversation with Dr. Bert Vallee. Dr. Cantoni had indicated that the latter was the most experienced student and investigator in this area. I described for Dr. Vallee the problem of kuru, pointing out that the distribution and character of the disability lent support to the thesis that this might be an intoxication or deficiency disease, but that we had no specific leads of what the nature of the deficiency might be. I then inquired as to whether he or members of his organization might consider it worthwhile to undertake analyses of the neural tissue which we now have available.

Dr. Vallee replied in the negative. He gave as an example his experiences with the single element *magnesium*. Animals deprived of magnesium for considerable periods of time will still show normal levels of this element in the tissues, when these tissues are analyzed in the usual fashion. In order to demonstrate the irregularities which occur under conditions of magnesium intoxication or deprivation, it is necessary to fractionate the neural tissue—that is, to separate the glial cells from the interstitial tissues and from the mitochondria. When this extremely tedious work is completed, then very significant changes in the distribution of this element can be demonstrated. Dr. Vallee cited this to indicate the extreme complexities of the problem of doing trace-metal analyses, pointing out that in this instance where we do not suspect any single element, it would be almost a life-work to carry out significant studies. Dr. Vallee further indicated that, although he has had repeated requests for such studies, in each instance where efforts of this sort have been made, a great deal of time has been spent collecting data which has subsequently been found valueless. It was his recommendation that these difficult and time-consuming studies should not be undertaken unless there is some specific clue as to the particular element requiring analysis.

In the light of the above recommendations, I am hesitant to make further efforts to obtain these analyses until Dr. Gajdusek has returned and we can confer with him about it. With Dr. Klatzo, I have examined the three specimens which we recently received in plastic bags. We have reported on the state of preservation of these specimens, and they will be retained intact by Dr. Klatzo in the proper containers until such time as a decision as to their disposition can be made by Dr. Gajdusek.

Richard Masland, M.D.
Assistant Director
National Institute of Neurological Diseases and Blindness

BURNET to REES   6 December 1957
Dear Dr. Rees,

May I join Dr. Anderson in thanking you very much indeed for the help your staff have been in regard to the kuru investigations. As far as the Institute is concerned, we have now closed down on any direct association with the work.

You may have noticed that a paper on the subject has appeared in a recent *Medical Journal of Australia*. I think the disease will probably turn out to be almost, or wholly, of genetic nature.

With kind regards,

Yours sincerely,
F. M. Burnet

DCG to SMADEL   7 December 1957
Dear Joe,

On rereading Bentley Glass's summary to "Copper Metabolism: A Symposium on Animal, Plant and Soil Relationships" (sponsored by the McCallum Pratt Institute of the Johns Hopkins University and edited by William D. McElroy and Bentley Glass, The Johns Hopkins Press, Baltimore, 1950), I am more than ever impressed that I may have been very naive in expecting trace-metal studies of a crude nature to reveal a trace-metal–induced illness such as kuru has many signs of being. Thus, the copper, iron, and molybdenum interrelationships are so acute and so complex that relative concentrations of each can determine excess or deficiency state of one—and absolute values for one alone could be most misleading.

Why do I keep suspecting metals? Well, Wilson's disease is still the nearest "ringer" for kuru; and it is a heredofamilial copper-protein complex deficiency, if I recall recent literature correctly. Secondly, I now am cognizant of the fact that ascorbic acid oxidase diamine is also copper-containing. We already know of heredofamilial biochemical defects occasioned by disturbances in tyrosine, phenylalanine systems (phenylpyruvic oligophrenia, to be specific). Finally, all the recent work on psychoses (schizophrenia) and other CNS ailments being associated with reactions which are specific for copper-proteins lets one think along these lines further. When one hunts for toxic basal gangliar or cerebellar-damaging agents, one is left with mercury, copper, manganese—as far as I can find out thus far. Animal deficiency diseases involving ataxias have been traced to trace-metal deficiencies (and here the copper–molybdenum complex interaction story has come to the fore). All these threads of thought keep bringing me back to the trace elements, and copper in particular. Others may act antagonistically or synergistically to copper—the molybdenum interaction is the only one I have thus far dug up.

Now, our study of Fore diet has revealed an immense consumption of ascorbic acid; the sweet potato staple is very high in ascorbic acid. This has led me to wonder—a rather naive and rash wonder, I suspect—whether excessive ascorbic acid, tyrosine, or some other normal metabolite intake might not put demands upon enzyme systems in excess of those which a normal, or somewhat low, supply of the trace metals required can supply. Glass's statement: "Moreover, ascorbic acid oxidase, tyrosinase, and all other copper enzymes become irreversibly inactivated during the course of the reactions they catalyze"—this brings this possibility even further to mind. Thus, I am now wondering whether copper enzymes may not have as a substrate some normally present product, either in excess in Fore diet or not handled by a marginal copper supply, and this metabolite in turn does the toxic damage. Allowing for the hundreds of other possibilities, this sort of thinking makes it apparent that simple hunt for excess of some trace metal is not enough. For this reason I wonder how we can manage to investigate such matters as serum ascorbic-acid levels in our patients. Serum phenylalanine and phenylpyruvic acid—can you get any of these worked out on blood specimens we send you? I am doing all in my power to keep them refrigerated from here to you, and succeeding to some extent. Could you check for the usual $FeCl_3$ test (I think) for phenylpyruvic acid in the urines you are receiving? Could your biochemists get into the serum for any lead? Or is the age of the specimens on arrival such that these analyses are hopelessly delayed and useless already. I only mention all this for it seems apparent that biochemistry does not work as simply as it would have to, were we to have a copper or molybdenum excess staring us in the face by microelement studies. These must be done; they may prove the matter to be far simpler than I now fear. But the more likely thing is one of these more complex effects. When I return, I certainly hope to be able to pursue such a line; but the thought came to me that there is some possibility the NIH or NINDB groups might get a lead from appropriate specimens sent now.

[113]

Since writing page 1 of this letter, your letter of November 21 has arrived. I shall answer all it brings up:

1. The position with NIH and NINDB is excellent and fully satisfactory, and I shall do my best to get there early. At the moment, the Australian Adelaide University team is to be here in a few days, is to stay for only a month, and is to get very little done unless I keep patients together and on hand for them—for it is ever more difficult to bring patients into our Center for study and to persuade them out of their bush hamlets when I find them there on case-hunting patrols. Once they are under way, I am off.

2. Enclosed are protocols on the last six cases BAL-treated, on whom urine specimens have gone off, as summarized in my last letter to you. These document time of urine collections and their volumes.

3. Zinc has been added to mercury and copper as a known basal-ganglia–ataxia producing chronic toxin; and I urge that zinc, mercury and manganese poisoning be ruled in or out definitively on the bloods, CSFs, and these urines—if at all possible.

4. Deficiency possibilities are going to be harder to track down, I fear. Perhaps

copper, molybdenum, manganese, and zinc—certainly Cu is "essential," and the others probably are. Levels might give a hint in this direction, but if it is a "relative" deficiency—i.e., relative to some other metal or to the supply of some metabolite—it will be much more difficult to track down.

5. The twelve enlarged prints are here. They are for the German article, which I have rewritten and will send off in revised form with the prints. It brings the kuru series up to two hundred cases and includes many new highlights of epidemiology. The German journal has accepted the article, and I hope will not mind the revisions. The journal is *Klinische Wochenschrift.*

6. The *NEJM* article is fully satisfactory; and I thank you, Joe, for all the help with it. [114]

7. Graeme Robertson's path reports on the three brains he had been sent had never reached me. Late in September or early October, while I was crossing the island on foot after leaving the Kukukuku peoples, the *Time* reporter arrived; and to Vin's horror, he showed Vin the path reports which we had never received. Although I have attempted to deal directly with all our collaborators and have done so (for nothing short of that can mean a thing, for we alone know what the specimens are we are sending out), prestige and publicity considerations have brought in numerous "intermediaries" at many stages. Thus, in the trace-metal reports, elaborate studies I had collected for, arranged for, and requested were lodged in Moresby and Melbourne files unknown to us—while those who did them could not have a clue in the world as to what the specimens were they had studied, or exactly how they had been collected. I found these reports by chance in going over files in Moresby. The same had happened to Graeme Robertson's original path reports. Yes, Joe, Australian feelings have been hurt by not having everything on kuru studied in their hands. The lead article in their journal, an editorial on kuru, proves that amply. Since they may be better set up for solving some of the trace-metal excesses or deficiency possibilities than anywhere else, I certainly hope that they may be able to turn up some leads as to pathogenesis. However, all collaborators we have had are fully credited in reports as they come out, and joint studies are all based on extensive correspondence. I feel that everyone concerned has been amply informed at every stage. Any objections you have received stem from dislike in not being "exclusively" involved in the kuru study to the exclusion of all collaborators, who, it seems, are looked upon not as co-workers seeking knowledge but as competitors.

8. Mrs. Beyersdorfer has sent a fine report of the Kuru Film file, and I only hope that further films keep arriving safely. I have what I hope are excellent follow-up pictures on many cases filmed in their early stages of illness. Furthermore, if the film now at Lae gets up here soon, I shall—just before leaving—make cine films of many cases that have been cine-documented well in the early phases of their illness and are now terminal. Please urge her to return to Lucy Hamilton the Kodachrome originals, after prints and copies are made for our file. I sent only a few of Lucy's best, so that we might have the advantage of these on hand. I myself will have a Kodachrome file of some thousand pictures to add to kuru documentation. I am sending a few of Jack Baker's Kodachromes which do not duplicate material I have been able to photograph. These originals should also be returned to Jack at the address I send with the slides. Thank you and thank Mrs. Beyersdorfer

for the immense job of film filing and processing. I think, if they are all kept together, that I shall be able to identify and describe every shot.

A merry Christmas and a happy New Year.

Sincerely,
Carleton

DCG to SIMMONS 8 December 1957
Dear Roy,
[115]
I shall be helping the Adelaide team from Australia—which has arrived to study kuru—get some patients in their hands, and shall give them what assistance I can for a few weeks before myself departing. Since their visit is brief, they will need to get a good start if they are to accomplish much—for it took us weeks and weeks to get under way at all. You have no doubt seen the paper in the Australian journal—*Medical Journal of Australia*—and the rather amusing editorial comments in their lead article. Except for a few misprints (none of which are very serious), I am satisfied with what they published—although since the article was written, our series has mounted to well over 200 kuru patients, over 100 of whom have died during our period of kuru observation. We have some autopsies behind us, and neuropathological reports on the entire series coming up. But the etiology remains a mystic matter unelucidated and, of course, the only matter of extreme interest and importance. Since we have failed as detectives thus far, we are hoping for a clue in any and every direction; and for this reason I urge you to suggest anything you wish—as far-fetched as it may seem—for kuru is a "far-fetched" disease from any and every point of view!

Toxic possibilities have been our major concentration and hope; but they have not materialized. We have not been able to get our hands on a Fore toxin used in pica, medicinally, in ritual, diet, or inadvertent exposure; and, to boot, we now even have cases of kuru under our very observation which have developed from earliest first minimal signs to severe disease fully outside of the kuru region, in Fore or other kuru-region natives resident "abroad." One (which we have not yet publicized) even developed as far away from "home" as the Papuan coast, where he was working. Thus, Roy, my first bet, which I have subsequently backed away from—namely, strong genetic predisposition at least—still looks like the best bet in spite of many possible objections to it. For that reason, your studies may eventually prove critical. I can assure you of accurate and cautious documentation of all the samples you have and are studying, and only beg you to be patient until I am at ease in Washington working on analysis of the records. I already have given you my address there. I may ask for kuru patients' specimens to be continually sent on to you, if you are willing, as long as a medical worker is here to collect them. However, such workers will be here but briefly, I fear, after my departure; and I shall not return in under six months—although within a year I hope to be back. [116]

Please send any criticisms, comments, or suggestions which the *Medical Journal of Australia* paper may bring to your mind or which have come to mind from our cooperative work and correspondence. I need them and shall welcome them.

Sincerely, and best wishes to you and your family for Christmas,

Carleton Gajdusek

DCG to Editor, *Klinische Wochenschrift*   11 December 1957

Editor-in-Chief
Klinische Wochenschrift

Dear Sir:

I am sorry for the long delay in returning to you the enclosed card authorizing publication of our Kuru paper. In so doing, I am taking the opportunity of submitting a somewhat revised and corrected manuscript from that which you have now in your hands. Kuru investigation has been proceeding so rapidly that it has been wise to make these changes and additions. Furthermore, we have now had our preliminary reports published in the United States and in Australia; and it is certainly in our interest and those of your journal and of your medical-scientific readers to revise the story up to date and to bring in additional material not yet covered by our American and Australian preliminary reports. For this reason we beg that you use this revised version of the paper, which brings our study up through our first 200 kuru studied cases.

[117]

I am sorry to give you these editorial problems. However, we have had rather unfortunate worldwide publicity in the lay press given to kuru and our work, and it is certainly essential that the medical literature be at least as up to date and accurate as the newspapers.

[118]

The only outstanding item in this paper is the second map, locating the kuru-region and the tribal-linguistic groups more accurately. This is an important feature of the paper, and I hope to get it into your hands shortly.

Our work here at Okapa is to be called to a halt temporarily, and in January or February I hope to be passing through Heidelberg en route back to the United States. I believe it is best to save any proof-correcting and editorial comment until such time as I appear in late January or February in your office.

I thank you for your kind editorial assistance and patience.

Sincerely,
D. Carleton Gajdusek, M.D.

[119]

DCG to SMADEL   19 December 1957

Dear Joe,

The Australian team is here with a £2000 grant for kuru study; but the first two, Dr. Donald Simpson (Professor of Neurosurgery from Adelaide) and Dr. Harry Lander (in medicine and toxicology at Adelaide), must leave after a brief ten days of looking over our "kuru menagerie." They are excited, impressed, and—having come to prove it a toxin—have swallowed hook, line, and sinker our genetic evidence and are trying to dissuade me from my current toxic stand. Professor Robson and old Dr. Cleland of Adelaide are due on December 26 or thereabouts;

and since their program will start in a shambles if I do not stay—perhaps I am over-exaggerating my importance as a cog in a wheel—I have again postponed my departure pending their arrival and moving in. They too will be limited in time, but they may have six weeks or so.

Simpson has given me a good deal of neurological training, corrected my numerous neurological errors, pointed out my neurological blunders and oversights, and straightened out all the neurological pedantry of the kuru situation with me. I am pleased to see that we were not as far out as we might well have been. In general, however, I find Vin and I are winning point by point in the "re-description of kuru," although Simpson has clearly shown that we have left misimpressions in our accounts thus far. Pyramidal damage can be detected; the clinical picture is more cerebellar than anything else. But, above all, it is a static and kinetic picture—with repeated change depending upon emotional mood, upon environment and physiological state. Thus, even knee-jerks can, during a single examination, be very extremely marked one time and disappear another (as with kinetic scoliosis); the nystagmoid eye-jerking can approach true nystagmus one day, and later that day have almost disappeared. Strabismus may occur in transient spasms or brief periods; and almost any reflex pattern (especially extensor plantars, which we have now occasionally found)—from patellar and ankle clonus, pectoralis after-tremor, to eyelid and orbicularis oris fibrillations—can come and go during one examination. Nothing in all this puts us a bit further ahead etiologically or pathogenetically. I still believe we have a strong heredofamilial, genetically determined predisposition to some as-yet undetected environmental variable. That the variable may be an infectious agent is beginning to seem more and more possible in my mind. Thus, I beg that your neuropathologists get all the protozoological, spirochetal, leptospiral, fungal, and viral histological studies and search done that they can. Toxoplasmosis was long in being discovered, and the Pneumocystis carinii stared pathologists in the face for generations before it was "seen."

Our case developing on the Papuan coast—and it is a certain case of rapidly advancing kuru—is soon to be fatal, though he was never yet back in the kuru-region since its onset; we are keeping the patient at Kainantu. He got ill while working on the Papuan coast, and was away from the farthest limit of the kuru-region for two months, at least, before the first subjectively evident ataxia appeared to him. Thus, Joe, toxic factors, if any, must be so subtle as to be hidden in the cloak of enzyme-defect or intermediary metabolite, etc., I feel. Otherwise, this case and much more of our epidemiology becomes uninterpretable. Finally, if we are "missing the boat," this might be by excluding infectious possibilities too glibly. The diffuseness of the pathology from one point of view, coupled with its specificity as a well-ordered "syndrome" hitting only certain systems of cells and not others—all this is still a possibility. Protozoas, viruses, trypanosomes, Borrelia, treponemes—these are all in my mind at the moment. How else can the patients "take the illness out of the region with them" when either in the earliest stages or before the disease has developed, and still progress or get ill in "kuru-free" regions, progressing to fatal deterioration? Enough.

[120]

Sincerely,
Carleton

MASLAND to SMADEL   24 December 1957
Memorandum. Subject: Comments on Dr. Gajdusek's letter dated November 24.

I have reviewed with great interest Dr. Gajdusek's most recent letter of November 24 regarding kuru. The data which he provides seems very effective in ruling out a genetic disorder, since this disease does not appear in closely intermarried communities outside of the Fore region. He also rules out a toxic factor in indicating that people emigrating from the Fore area develop the disease after they have left, and that the disease progresses in spite of the removal of the patient from the affected area. I wonder if we have adequately ruled out the possibility of infection. Dr. Gajdusek mentions that the meat which these people eat is usually grossly undercooked. I am thinking of a parasitic disease such as toxoplasmosis, but involving some other as-yet unrecognized agent. I know that efforts have been made to culture organisms from this material, but I wonder if adequate consideration has been given to a parasitic disorder. I must confess that the pathological findings which indicate a diffuse degenerative process without inflammation are much against this, but it could be that we are dealing with an unusual reaction to some parasitic invader. This is a wild chance, but evidently this disease is a wild one.

                                                          Richard L. Masland, M.D.

DCG to SMADEL   24 December 1957
Dear Joe,

Enclosed is a full manuscript, including associated correspondence of my revised article in press in *Klinische Wochenschrift*. This is a rather more definitive paper than the American and Australian preliminary notes; and it also corrects a few previous errors and misconceptions, as well as bringing case-figures up to the two hundred mark. Our "cases" for this year are now approaching 250!

I have seen the vanguard of the Australian team in and out. Lander and Simpson were refreshing, although brief, co-workers. In a few days, Robson and the elderly Prof. Cleland are to arrive. I shall turn over records and data to them, and then be off. However, the mass of records, data, and full correspondence files are now being packaged. Today I am sending the full case-records of our first fifty deaths, along with massive correspondence, laboratory data, and epidemiological files off to you. Since the entire study in the U.S. depends upon the successful receipt of this first stack of records (and of the next shipment, which will go off one week later along with my departure), I am most anxious to know of their safe arrival. If you can let me know, by cable by the 7th January, of the receipt of the large box of records, you can still reach me here. I shall try to depart on that day.

Two brains are off to you; and when you acknowledge receipt of the first stack of records by cable, it would cheer us a good deal to know of the arrival and condition of these further brains. One additional brain is hardening and will be shipped off before I leave. However, we have given one to Prof. Sunderland of Melbourne, who visited us; and another of our posts, including the complete spinal cord, went off to Adelaide with Simpson who was just here with us. The Australians will certainly be after all the others they can get, so those now en route and the one remaining may be the last. Vin and I, however, are still the only ones getting posts and collecting cases; and since so much of our hopes lie in the work going on at

NIH, we shall do our best to route at least another one or two brains along with spinal cords—for it is now fully evident that paralysis and muscle fasciculation are present as late terminal phenomena, and that cord damage as a late development is probably certain.

I am sending out further blood specimens; and before I leave, I shall send out fingernails and hair from controls and kuru patients—for As and trace-metal studies. We rounded up a stack of fingernails on patients and controls for Harry Lander of Adelaide for As studies, which he is doing; but since hair and fingernails are important in many metal toxicities, we shall send these off during the next few days. Bloods and urines, etc. will go off during the next week; and further specimens will go off after my departure. In addition, vast serum files in Melbourne—for study of an abnormal serum globulin found even in normal Fore natives, and for other studies—are being sent off to you by Lois Larkin for me. Thus, within about two weeks, the last of the stacks of records and specimens should be posted and I shall be bound for Washington.

Kuru is obviously more important than all expectations. That it is a new disease is certain. That it offers astonishment, intrigue, and imagination to the neurophysiologist (Prof. Eccles) and neurosurgeon (Simpson) beyond anything they have seen in their careers, we had expected; but this is interesting confirmation of the suspicions we had: that it is something which modern medicine cannot afford to pass by as a simple medical oddity—there is too much chance here for real advances in our understanding of human physiology and disease if we can "crack" kuru, genetic or otherwise.

Endocrine possibilities must be pursued. Pregnancy does seem to retard the disease, and it seems to progress more rapidly shortly after parturition. The adult sex-ratio and a suggestion of more rapid deterioration while on stilbestrol therapy all point this way. Thus, 11 oxy steroid and 17 ketosteroid analyses on the urines we are sending seem in order. Can they be done on the HCl-preserved 24-hour specimens? How about other hormone possibilities?

Protozoan parasite possibilities and errors of intermediary metabolism are all in the forefront of our attention. In the former case, only the pathologists and parasitologists can give us a lead. In the latter, we would suggest that some abnormal genetically determined pathway of steroid or estrogen or other sex-hormone metabolism may be involved.

Enough for now.

Sincerely,
Carleton

DCG to ROBERTSON   24 December 1957
Dear Dr. Robertson,

I am sorry to answer your kind letter of November 21st after such a long delay. Actually, I have been preparing to leave daily, and constantly have postponed my departure for new developments. The Adelaide team arrived in the persons of Drs. Donald Simpson and Harry Lander; and I set about to dig up cases for them, to open our records and stacks of data to them, and, especially, to learn from them as they surveyed two dozen of our 200-odd kuru cases. I have been set right on numerous errors in neurologic interpretation and examination techniques by

Donald Simpson, and he has been a most stimulating collaborator during the past two weeks. I shall now await the arrival of Prof. Robson and Prof. Cleland from Adelaide, and then shall be off. Simpson and Lander are leaving today.

The initial reports on kuru in the *New England Journal of Medicine* and the *Medical Journal of Australia* contain many minor errors and misstatements and a good deal of conjecture on which my opinions have slowly altered or shifted, which are all revised and amplified in a longer and more definitive kuru contribution off to the German literature. Such inexcusable errors as the appearance of "dysphasia" for "dysphagia" in the Australian paper are highly embarrassing, but nothing very major has been altered in our current thinking on kuru. I shall get reprints of the two publications off to you shortly, and the manuscripts of subsequent kuru publications will be sent to you also. Should you be able to criticize, correct, or comment on any of our work, I shall be most flattered and grateful to you.

Dr. Igor Klatzo (of the National Institute of Neurological Diseases and Blindness of the National Institutes of Health in Bethesda, Maryland) has seen a large series of kuru brains, but he has not yet sent us any definitive reports. Preliminary reports on some of them are briefly summarized in our Australian paper; and Dr. Simpson has a copy of the more detailed report on one case, which I have asked him to transmit to you. Furthermore, I shall be back in the U.S. working over the volumes of field-notes and field epidemiological studies and the stacks of kuru clinical records at the NIH within a month or two. From there I shall promptly send to you full pathological reports on all material we have studied, and I shall try to collect histological preparations on a large series of cases for your scrutiny as well. Donald Simpson and Harry Lander are taking with them a complete autopsy report and specimens, including full brain and cord, of one of our patients who died during their visit here. This will undoubtedly be the best material obtained in postmortem examination, for Simpson himself did the post with complete cord removal. Since a terminal anterior horn cell damage has gradually become more and more apparent to us in our clinical study of kuru, and since Donald has been able to give us further clues as to how to demonstrate this, cord pathology obviously becomes even more important.

I am most interested in seeing sections of the three cases you have studied for us. There is no chance of their arriving here before my departure, and thus I beg that you send them to me in America. I shall go over them there at the same time as I am working over the kuru records for more definitive kuru epidemiological and clinical reports. [121]

Thank you, Dr. Robertson, for your interest in kuru and your valuable assistance to us. I am sorry that the Moresby office unfortunately mislaid in files your pathology reports, bringing them to our attention only long after their receipt. I am also sorry that when I decided to shift much of the biochemical and other kuru studies to facilities offered to me in the U.S., you were late in being informed thereof. I had kept many of our Australian collaborators fully informed, but in your case I was unpardonably negligent. I am sorry.

May your holiday season be restful and enjoyable.

Sincerely,
D. Carleton Gajdusek, M.D.

P.S. In recent months the possibility of an infectious etiology in kuru has again come often under our consideration. We have absolutely no suggestion of an acute meningoencephalitis process, nor of any acute or chronic febrile disease. However, in view of the epidemiology which makes toxicity most unlikely—unless it be coupled with an inborn error in metabolism (enzymatic defect)—we are forced to repeatedly reconsider our findings. Infectious etiology, were it in the form of a treponeme, Borrelia, Leptospira, Rickettsia, or, especially, some obscure protozoan akin to Toxoplasma or Pneumocystis carinii (which long eluded the parasitologists, even on close scrutiny of the pathological material) must all be considered—although it is an unlikely suggestion, I realize. However, kuru is a highly "unlikely" disease. Thus, Dr. Robertson, I suggest with apologies that every thought be given to this wild possibility in examining the histological material.

DCG to SMADEL 31 December 1957
Dear Joe:

Enclosed are lists of specimens off today for all sorts of rather essential studies, for which I hope the NIH and NINDB groups interested in kuru can arrange. Enclosed is also a list covering some of the specimens sent in the last two shipments.

The first half of the case-records, correspondence files, and other kuru records—on which all my work and paper-writing, and future lab studies of kuru, must be based—are now off to you via Qantas. Please check on the arrival of this most important shipment, for without it we are really stalemated. In a few days, the second and final half or three-quarters of kuru records—maps, case-records, correspondence files, lab results and notebooks, and, finally, the epidemiological card-file covering all of kuru epidemiology—will be off to you.

I did two post-mortems between 6 p.m. and midnight yesterday—using the crudest of instruments, since I had sent all our instruments out to Kainantu for Vin to have them on hand for autopsy of a patient we have been keeping there "outside of the kuru region." I got full cord (performing a laminectomy from the low lumbar to high thoracic), as well as brain and a full sample of all viscera, as well as some of the peripheral nervous system, on both patients. These will eventually be off to you. The fresh, refrigerated tissue (muscle, liver, spleen, and kidney) are off to you today for trace-element and other chemical studies. Vin also got, yesterday, the brain and cord on the man who died of kuru in our Kainantu series.

Obviously, hospitalization on controlled kuru-region–free diet does not alter the progression of the disease. The man who had just died, in fact, developed his disease on the Papuan coast while there as a plantation laborer, and had progressed to its termination without ever being back inside of the kuru region. What do your "toxicologists" say to that? However, if kuru pathology refuses to fit in with genetic hypotheses, then the NINDB group had best get busy on parasitology and fungal, protozoal, etc. studies with real intensity; and if toxicity still is overwhelmingly the "first choice," what leads can they give us on an agent exposure to which must be asymptomatic and must be able to precede the first development of symptoms by many months, probably many years? The final possibility—that of some strange toxin found by the patients even when they are far outside the region in which kuru is known to have ever previously occurred—seems very, very remote, since those who develop kuru are all ethnically and originally from the kuru region we have fully demarcated now.

I trust you have received the revised manuscript in press in *Klinische*

Wochenschrift. All illustrations and the manuscript are now in their hands, except for the final figure: a detailed kuru-region linguistic-group map, which I am now trying to complete. I shall mail it to them shortly. Obviously, however, there is an immense further amount of writing to do on kuru, simply from studies of our records, once I get back. I shall stop by Kuala Lumpur briefly, very briefly, en route home and can collect mail there sometime late in January; write in care of Bob Traub. Obviously, Joe, if we are forced into the genetic corner with kuru, it remains one of the most challenging matters in world medicine, providing an opportunity for a breakthrough in the fields of progressive neurological disorders, enzyme-defect genetics, and in the field of human-population genetics—so important in these days of radiation consciousness. Thus, whatever the outcome, kuru looks like a good bet for future study. I shall be anxious to discuss all this with you and others at NIH shortly in person.

Sincerely,
Carleton

SMADEL to MASLAND 10 January 1958
Dear Dr. Masland:

This is concerned with Dr. D. Carleton Gajdusek, who is being considered for a position as Visiting Scientist in the National Institute of Neurological Diseases and Blindness for a period of time while he analyzes and prepares for publication the voluminous information he collected on the new disease *kuru* that he has been studying in New Guinea.

I have known Dr. Gajdusek since early 1952 when he came to work with me at Walter Reed Army Institute for Research. As indicated in the accompanying curriculum vitae and bibliography, Dr. Gajdusek has had excellent training and a productive experience. He is one of the unique individuals in medicine who combines the intelligence of a near-genius with the adventurous spirit of a privateer. In addition, he is an indefatigable worker who is dedicated in his efforts in medicine. Over and above this, he is an excellent laboratory man and a sufficiently good clinician so that he received his Boards in pediatrics before he came to work with me in 1952.

I have no hesitancy in recommending Dr. Gajdusek to the NINDB as a worthwhile candidate for position of Visiting Scientist in that Institute.

Sincerely yours,
Joseph E. Smadel, M.D.

DCG to SMADEL Kainantu, Eastern Highlands
12 January 1958
Dear Joe,

I am finally out of Okapa en route home. Prof. Robson and a general practitioner from Port Moresby are in for a remaining two weeks of their one month of study; but instead of bringing in the high-powered laboratory we had hoped for, they have arrived only to check family histories for one month, to calculate out some statistical probabilities of genetic etiology. Actually, we have the very same family-tree data both in our case-files and in the massive volumes and card-files of epidemiological data which I shall shortly be sending to you. However, I have most certainly not had time as yet for analysis of the data. It has been most pleasing, however, to find them getting the very identical family trees which we had recorded in the same communities; and, although I knew it, it required some such independent check before anyone would believe our story that the Fore natives were at least as good informants on case anamneses and on family background as are white patients in civilized communities.

Actually, Joe, the "political climate" is rather warm. Thus, Prof. Robson and others have obviously been appointed to urge me to squelch rumors that my work was not strongly and heavily supported by the Australian Territorial Government. They are right. It has been well and strongly supported; and Vin Zigas—who has shared every bit of it with me—is fully employed by them, and they and their administrative staff have provided funds and support from the very onset for all the drugs, building, etc., etc. that this ambitious project has entailed. However, in the early days I was a most unwanted intrusion into some rather nebulous "research plans" involving perhaps the establishment of a Territorial medical research director's post, and rather extreme political pushing *hin und her* in an attempt to maneuver certain candidates into the position. An American working in an obviously "choice" research project such as kuru almost immediately turned out to be a decided "fly in the ointment"; and for a matter of some six weeks, I was all but ordered out of the area—a suggestion I did not accept—and then subjected to a cross-fire from Sir Mac, ex-Director of Public Health Gunther (now Assistant Administrator) and sundry other "politicos." We only screamed for more collaborators and assistance, and got excellent support from Dr. Scragg and the Department of Public Health in several successive (three in all) grants for £100 each for research expenses and also additional funds for building, for drugs, and the obvious expense of running a precarious "jeep-lift" into isolated Moke over the rather perilous route. Thus, I can most honestly state that we have been most warmly supported from the start by the P.H.D.

All that clouds the picture—and still does, to some extent—is the warring factions in Australian medical politics. Thus, I see that there is now much talk and disturbed discussion that Adelaide University has crept in where Sir Mac and the Hall group thought they might. In fact, kuru still needs and could well use a high-powered long-term approach of basic research; and no group in Australia has yet been ready to give it that, nor does it look possible for a while. The Adelaide group is planning to rush out a few (probably many) reports to get the "Australians" into print on the matter (for some reason, Vin Zigas does not seem to count to them as an Australian); and—using largely the cases, case material, and data we have made available to them—a series of confirmatory Australian reports are due. However, I doubt much that the attitude expressed in the *Australian Medical Journal*'s editorial to our article is much changed on the continent.

Public Health Department is most concerned about my running off with even my files of personal correspondence with you—let alone with my personal field-notebooks, which no one else in the world could possibly interpret. Vin has an almost complete carbon-file of all case-records. The immense disorganized, uninterpretable (except by me from my scribbled volumes of field-notes) epidemiological file is of no use to anyone but me, and is now off en route to you in spite of repeated attempts to get me to duplicate the whole thing—a job of some two months' organization and another two months' copying at least—for deposit with them. Since it now appears that a group of would-be degree-seekers are already parasitizing the case-records, etc. for degrees which no one in the world but Vin Zigas should be entitled to receive, I am making no effort to leave full records behind; and Vin—now about to depart on a much-needed year of leave, with both a ruptured knee cartilage and chronic nervous and physical fatigue undermining his health—will leave in another month or so. After vacationing in Europe he is coming through the USA, and will stop in on us at the NIH; and I have promised to sit down with him and help him get a thesis out to Sydney University, where he can, it appears, get for himself the much cherished and fought-over M.D. qualification without which his career is very much hampered. In the meanwhile, while awaiting his arrival, I shall try to get the epidemiological papers and the definitive papers on case-analysis and correlation written.

This, then, Joe, should straighten out the constellation of rumors and cross-rumors which keep drifting in to us here. The Territorial government has supported our project—if reluctantly—very well. The fact of our study has enraged and chagrined and convulsed Australian medical circles; and various private interests from Canberra, Sydney, Melbourne, and Adelaide (and, I suspect, Brisbane too) have been drawn in, but thus far little has materialized.

The final four kuru brains will be shipped off to you by Vin after my departure. We obtained them each with complete spinal cord. In several cases, thyroid glands are also there; and pituitary within sella turcica is with every brain, and also a few cm of cervical cord removed from above. Do not let these get lost in unpacking the brains. We plan to send these final four autopsy specimens off to you very shortly. The visceral tissues on these cases have already been posted or are now being posted. Since we believe we have some real endocrine leads in the disease, we suggest strong concentration on the endocrine glands. Uterus and ovaries and testes are sent with the visceral specimens. Microbiological possibilities still appear, to us, unexcluded. With patients now developing kuru far outside the "kuru region" after exodus from it, and dying (as one of our recent patients did) of typical rapidly progressive disease without ever returning to the kuru region, latent earlier-acquired infection obviously looks more likely than toxin—in spite of the pathological urging to toxic thinking from the NIH. Harry Lander tells us that a few—not all—fingernail specimens had high arsenic levels; and I strongly urge that you look into rare earths, arsenic, etc. in the fingernails and hair I have sent you. More will be forthcoming.

[122]

Cyril Curtain is receiving my permission to report independently the abnormal serum-globulin component he found in normal Fore subjects on his Tiselius moving-boundary study of our specimens. The kuru protein data we shall try to work out once I get back to the U.S.

In the past few days, I have made a final tribal- and linguistic-areas map superimposed upon the kuru-region border on an immense scale, and the original thereof is off now to the editor of *Klinische Wochenschrift* for the cartographers of Springer Verlag to redraw into an appropriate figure for our German paper now in press. I think that all that we can possibly say about kuru has been said in that paper, and further publication must await the slow process of detailed analysis of our case-records, epidemiological reports, and field-notebooks. I am very pleased with the selection of the journal; for although it will not satisfy the Australian "politicos," it will remove the sting of having it appear in an American journal.

Please, Joe, have your secretary send me a copy of the January *AMA Archives of Internal Medicine* articles on my AICF test, if the papers actually are in that issue (January 1958) as I am told they should be. I have not been able to see the proofs; and revisions and corrections for me were handled by Lois and Ian [Mackay] at the Hall Institute, with nothing but my letters to go on. I am most anxious to see the papers, and could receive them in a few weeks at Kuala Lumpur where I plan to call very briefly on Bob Traub and the gang en route home. I shall not, of course, stay there any time (as I had previously planned), but interrupt there my forward progress as briefly as possible. I hope to receive first proofs of the German article there, as well as anything you wish to send me. I shall be most grateful if you will give me this opportunity of reading my own articles before I get to the wolf-pack in Europe. I really have forgotten what I have said. Furthermore, in paper II, Ian Mackay and I had an argument about keeping the discussion down to earth, instead of constructing ethereal castles of elaborate theorizing as my Hall-group collaborators were insisting. I do not know just what has come out. On last compromise, I seemed to have won out by firmly insisting on rather cautious discussion documenting our "stolen theories" to ancient German literature instead of directly to the "Burnet-Fenner hypotheses."

[123]
Finally, please tell the NINDB pathologists that the last autopsy specimen that I shall be sending them—brain and cord on Asisi, on whom they already have visceral tissues, peripheral nerves, etc. en route to them—had also his two eyes enucleated. They have been formol-saline–fixed; and Prof. Robson will take them to Adelaide, where they have, I am told, an excellent ophthalmological pathologist. Thus, we shall have retinal pathology—and I have sent it out to Adelaide to be done.

Peculiar white spots have been noted on normal retinae as well as on those of kuru patients. They are pinpoint sized, stationary, highly refractile white spots—some occurring even over the disc itself. They can be solitary or frequent. They can be in one eye or both, and in some patients are very numerous. Harry Lander first picked them up; and, once seen, we have been noting them ever since. They are not present in all kuru patients, and they are also found in some non-kuru normal patients.

Enough for now,

Sincerely,
Carleton

P.S. Please tell the NINDB that we have not lost their cine camera and will send it off to you. The Kukukuku patrol did "bugger" the light meter; but it is, I suspect, fully repairable. These will be sent off shortly.

DCG to SMADEL

Lae, New Guinea
24 January 1958

Dear Joe,

I am off today out of the Australian Territory for Hollandia, Dutch New Guinea, on my way home. I have just flown in from Menyamya, where I spent my last week here patrolling, collecting ethnobotanical specimens, collecting blood specimens, and studying the kuru-free Kukukuku populations just east of the kuru-region.

Since the Fore–Kukukuku border is the sharpest kuru/non-kuru discontinuity we know of, I felt this advisable. The Kuks on the border still are at warfare, some cannibalism, and without any government control as yet. In spite of this, we have been long among them, collected blood from them, and studied them; and two of my boys, young lads who have assisted in field expeditions, are from such Kukukuku communities. However, the chance to see the central Kukukuku, where government and mission control is now complete, was the reason for going into Menyamya. They are the "same" Kukukuku, all right, although there are about Menyamya some five distinct language-groups, all of which differ from the "wild" Kuk languages I know of. I have documented these additional linguistic groups; secured all the definitive cartographic, linguistic, and ecological data available; sent most of it off by mail to you today for the Kuru Map file; and made rather extensive medical-vocabulary lists among four of the language groups. In addition, I collected 131 blood specimens (which are already off, refrigerated, to Melbourne, for blood-group genetics and abnormal hemoglobin studies), the serum from which Lois will ship off to you shortly, if it all arrives in good order.

Obviously, this is a most excellent series to contrast with the kuru-region studies—it being from a totally kuru-free population, yet bordering on the kuru region and thus potentially one of the large populations "threatened" by kuru. I have all the available population data, and estimates also.

One of my pet projects for the future is an intensive anthropological, ethnobotanical, ecological, toxic-exposure, etc. (including marriage-pattern, etc.) contrast between the kuru-affected Fore and the kuru-free Kukukuku, for here the line of discontinuity is sharpest. In addition, I am prejudiced in favor of the Kuks, liking them extremely—although most others react to their sly, sometimes haughty sometimes unctuous, ways adversely. I am already well under way with such a study, but its completion must await the return field-study.

Obviously, I put more hope and stock in the possible results of the further laboratory study of kuru, for I believe that microbiological histology for a possible spirochete, protozoan, fungus, etc. is of utmost importance now—along with a fine-tooth comb through the histopathology of all systems, including the CNS, with some preference given to the endocrine system. When that finally fails (as it probably will—yet it must be done), the genetic analysis of our family-tree data and peripheral-region epidemiology will be required. Here again, nothing possibly conclusive can be expected—only such-and-such probability that the illness is genetically determined, with only such-and-such reliability of the estimate.

In the last resort, unless we are extremely lucky in the infectious-disease approach, I believe we shall be forced to push particularly the study of blood, serum, urine, and body tissues for trace elements, toxins, and, especially, intermediate metabolites and enzyme systems. This obviously requires a different approach to specimen-collection than we have usually been using. I hope to be able to push along that line, however, as far as available specimens will permit, and lay plans for the appropriate field-methods in the return field-attack.

Vin Zigas, as I told you, will visit us in Bethesda after his European vacation, arriving there in all probability in September. Vin needs pushing; but unless he writes a thesis on kuru and tries to get the cherished British-style M.D., his career here or anywhere in the Australian/British area is not very promising for the future. With others planning the same by full "parasitism" of our records and studies, it is only fair that Vin be the candidate to get the thesis out; and I have suggested a complete analysis of the vast clinical data, with an evaluation of the therapeutic trials in a presentation on the clinical picture of kuru and therapeutic regimens tried. Since I shall be doing this preparation anyway for our definitive kuru papers, he can manage it rather easily when he arrives; and since the huge thesis required by the University of Sydney would require a standard but complete abstracting of all clinical charts, etc. (in the appendix, at least), there is plenty for him to work on and do. He hopes to spend a few months in the U.S. working on the thesis. I am enthusiastic about this, for it offers a possibility of plotting out beforehand a return field-approach and, in addition, planning with Vin for some preliminary arrangements in Okapa, where things will have died out into complete inactivity, I believe. I am anxious to help Vin with the thesis. He is on vacation pay and thus can scrape through, for he has a year of leave; but I would like to do all we can to make it possible for him to live a bit less frugally (Australian salaries are not as much as the U.S.) with some study-grant for a few months or something of the sort, if we can dig one up.

[124]

Lucy Hamilton has received her original Kodachromes, and wants to thank you. I am working here with her, arranging for an intensive last push on the Fore ethnobotany. We have Prof. St. John (Department of Botany, University of Hawaii), world authority on pandanus. I have just brought out a huge collection of the five species of wild pandanus which the Kukukuku eat, and am arranging for new specimens of the Fore pandanus to come down here to Lae for him. They are all or most "undescribed species," and he will take time out from his vast Dutch New Guinea and Australian survey of the genus to describe first our kuru/non-kuru–area specimens, in order to give us some definitive names to rest our work on. He and Kevin White, a botanist here at the Lae Forestry Department, are helping with a list by world experts everywhere (from Belgium, Greece, Italy, and Germany to England, the U.S. and Australia) on various genera and New Guinea plant groups; and we are getting the duplicate collections of Fore ethnobotanical specimens off to the appropriate experts, requesting high priority of new descriptions, since these require them. A large shipment of pressed plants, all labeled with their appropriate destinations within the U.S., will be sent off to you shortly. If someone in the NIH can sort them out and redirect them to appropriate experts in the particular genera or group concerned, all this will be indicated on the shipment and we shall be under way. I send them through the NIH as a "clearinghouse" in order that we get the data together once again and get them through "plant inspection" by the already-arranged channels. Prof. St. John tells me that at worst the Department of Agriculture may demand that they be fumigated, which will not hurt their use for taxonomic study.

[125]

Thank you again for all the encouragement and help, Joe. See you soon.

Carleton

DCG to SMADEL   Hollandia, Netherlands New Guinea
                     28 January 1958

Dear Joe,

I am having a fascinating time here in Dutch New Guinea, and already I have spotted a half-dozen new problems—some good!! But, Joe, don't get worried, for I am leaving for Borneo in a very few days, and from there I head for Kuala Lumpur for only a brief stop there en route home. Kuru remains every moment my major interest. You can see from the enclosures that even here, rushing around the "wildest half" of this "world's wildest (and second largest) island," I am still working and writing and always thinking of kuru. Enclosed is a copy of a letter Voors had written to Vin Zigas just before I met him. I have discussed with him the suggestions at length. He is sharp. However, he did not know of our two recent cases—in men, both developing fully outside of the kuru-region and progressing to death without ever going back home or returning to a kuru-region diet. He did not know of our epidemiological and other data. However, his comment did point out that I had left out a sufficient discussion of the "deficiency" possibility in dealing with pathogenetic possibilities in the *Klinische Wochenschrift* paper now in press, and I have just added this enclosed section to augment the already immense paper. It is important mostly for bringing the veterinary literature into better focus, for it is the veterinary people (see *Proceedings* of the Seventh International Grassland Congress, 1956, Wellington, New Zealand) who have really opened and solved many problems in this field.

Furthermore, in studying veterinary literature, I have come upon the astounding, still somewhat mysterious but most suggestive work of Lee and others on *Phalaris tuberosa* toxicity in cattle—which produces a staggers-syndrome in sheep, which is as close to kuru as a sheep can get, I suppose. Since it also is in the Australasian part of the world, it cannot be ignored. I beg you to get someone to dig up the world literature on this for me, to see if anyone in botany and agriculture can tell us if any of the ethnobotanical specimens in the extensive lists I have sent you are related to *Phalaris tuberosa* even remotely; our people do eat grasses—cooked, however, but not always. The *Phalaris* study is naturally very unlikely, in view of our epidemiological data—but so is everything, including kuru. Furthermore, its neuropathology must be looked into if reports are available, for it is too inviting a suggestion. I have stumbled upon it here by accident, reading veterinary literature on trace-metals and pasture poisonings.

[126] Soil specimens for as intensive soil analyses as you can arrange for at Department of Agriculture—including, of course, molybdenum, copper, sulfate, etc. levels—will soon be sent to you along with fire-ash samples from kuru houses and control men's houses for similar studies. Metal analyses are especially needed. Hair and nails will be forthcoming, along with the three further brains of children on whom we also have complete cords which I left hardening in Kainantu. Vin will ship them to you. Those brains I shipped just before leaving Kainantu should already be with you. [127]

Sorry, Joe, for again having so many requests. I hope that you will pardon my repeated letters of this type. As you know full well, I cannot accept kuru as a minor problem of a stone-age people. It is more than that, and if solved will certainly give to medicine important new leads. Furthermore, it offers the best chance of solution, I still feel, of any of the degenerative disorders of the central nervous system.

Carleton

Appendixes

Appendix 1. Sorcery Among the South Fore, with Special Reference to Kuru

The following comments refer to a portion of the results of an anthropological survey carried out in three Native areas or districts in the South Fore region of Kainantu Subdistrict, Eastern Highlands. The three areas concerned were Wanto'Abarasa, Atigina, and Kimi.

Fore Native Areas

From the point of view of Fore social arrangements, the use of the term "native area" requires some mention. These areas, such as Wanto'Abarasa, Atigina and Kimi have no reference to Administrative arrangements, but are a matter of local Native organization. The Fore themselves consider their Territory to be divided into named areas, each of which recognizes a somewhat shadowy allegiance to a headman, known in the local language as *Kamintawa* (which means simply "leader," the same term being used for the leader of a Fore local grouping).

In former times, the main function of area *kamintawa* was as fighting leaders, and it was through their skill in this role that they achieved their rank. The position of *kamintawa* was, therefore, a shifting one depending on an individual's maintenance of his ability in fighting and war-leadership. With the coming of European peace, only a few years ago, positions of leadership appear to have become fixed as they were at the time of cessation of fighting, since there was now no means of displacement of elderly *kamintawa* by younger and more virile war-leaders. The Fore do not seem yet to have had time to adapt themselves to new circumstances in this respect or to evolve new means by which indigenous leadership (as distinct from leadership through Government appointment) could achieve recognition. For this reason, where former fighting *kamintawa* have survived they are still pointed out as the leaders of their various areas, though they are in many cases aged and do not play an active part in community life. In a number of cases, however, prestige remains to them as skilled advisers and arbitrators in community affairs, and to enlist the interest and support of such men can be of considerable assistance to a visiting research worker.

In the case of two of the areas mentioned above, the identity of area *kamintawa* is clear. The Atigina area recognizes Etagi, of the Wanitavi local grouping, as its *kamintawa*. Manevu, of the Iasui local grouping, is area *kamintawa* of Kimi. Both these men are now aged and lead somewhat retired lives, but both still have considerable influence on their people's reaction to visitors. I found them extremely cooperative and helpful.

In the case of the Wanto'Abarasa area, the position is not clear, two men now being indicated as area *kamintawa*. They are Suguia, of the Kamata local grouping, and Nasumbi, of the Anumpa local grouping. Seguia himself said that Wanto'Abarasa has two *kamintawa*. I did not meet Nasumbi, and was not in Anumpa, so I do not know the reaction of the locality on this point.

Just as area *kamintawa* had their main function in the field of warfare, so the areas themselves had little other function than that of fighting alliances or units. In ordinary day-by-day political, economic and other social matters, concentration was, and is, on local groupings within the areas. Even as fighting units, the areas do not appear to have been well integrated, and local groupings within an area occasionally fought against one another.

However, from the observer's point of view, distinctions between the Native areas have some importance in that there are variations in a few aspects of customary views and behaviour, as between one Fore area and another. It should be emphasised that these variations are in a few aspects only, and affect matters of detail rather than those of broad practice. For example, though it was until recently the custom for all Fore groups to practice ritual cannibalism, involving the eating of their own dead kindred, the somewhat elaborate kinship and other rules governing this practice differed slightly from area to area. Thus in the Atigina and Kimi areas I was told that only females and young uninitiated males had the right and duty of eating dead relatives, and that the same rule applied in Wanto'Abarasa, in the latter case with special emphasis on the duties of aged females in practicing this cannibalism of kinship. In certain other Fore areas, however, it appears that the duty of eating various specified portions of the bodies of dead relatives extended to all males and females in the kinship groups concerned.

In the same way, there are differences from area to area in the details of magical practices, including those coming under the heading of sorcery. Belief in such types of sorcery as Tukavu and Kuru extends throughout the South Fore area, but beliefs concerning the necessary techniques vary slightly from one area to another. This, however, does not affect the general South Fore attitude to sorcery, and involves no more than differences of opinion as to essential ingredients (leaves, vines, stones, etc.) required for sorcery rites.

The number of people included in any such Fore Native area is not great, ranging from about 180 in a small area to approximately 1600 in a large one. 1954 census figures for the three areas with which I dealt were as follows:

Wanto'Abarasa	1569
Atigina	1498
Kimi	992

Other South Fore Native areas are as follows: Iagusa, Amusa, Kemo, Ketebi, Mania, Magorti, Ifufurapa, Purosa, Awarosa and Iresa. With the exception of the one local grouping of Karu, in the Native area of Ifufurapa, I did not visit areas other than the three referred to above. (It might be noted that, for census purposes, Wanto'Abarasa is included in the North Fore region, but the division between North and South Fore is one of Administrative conveniences and does not indicate any distinction in social conditions—such small distinctions as do exist being between the Native areas themselves. Wanto'Abarasa, therefore, can very well be socially considered in conjunction with its southern neighbour Kimi and Kimi's southern neighbour Atibina).

Fore Local Groupings

Within each Fore Native area are a number of local groupings, such groupings forming the effective units for political, economic, and other social purposes. These are the units which are operative in day-to-day affairs, the interests and activities of members of a local grouping being closely united in a manner in which those of members of a whole Fore Native area are not. From the point of view of social integration, a Fore local grouping can be considered as comparable to what are called "village" in some other parts of the Territory. However, as Fore local groupings are each composed of a number of territorially dispersed hamlets, the term "village" could not be correctly used for them.

To take the Atigina Native area as an example, this includes seven local groupings as follows: Kamira, Wanitavi, Amora, Wanimkanto, Kume, Okasa, and Kasoru. These local groupings vary in size from the smallest, Okasa, with a population of 140, up to the largest, Kamira, with a population of 316. Some of the local groupings are made up of members of one clan only; others include members of two or more clans. Okasa, for instance, includes only members of the clan Okasa, while Kamira is composed of members of the clans Kamira and Kutena'visa. In each case the name given to a local grouping is the name of a clan, not of a place. In all Fore local groupings, even those containing comparatively small numbers of people, members are dispersed in several scattered hamlets, to which are given the names of the areas of ground on which they are situated. For example, the hamlet arrangement of Kamira, together with the number of people in each hamlet is as follows:

	Hamlet	Clan	Number
Native Area: Atigina	Apa:Nati	Kutena'vina	35
	Kiarupamenti	Kutena'vina	32
Local Grouping:	Ara.Vainapinto	Kutena'vina	7
Kamira	Wagiri	Kutena'vina	54
	Wai'oti	Kamira	41
	Arinoguti	Kamira	77
	Kundenti	Kamira	70
			316

Distances separating the hamlets in a Fore local grouping range from a few hundred yards up to about a mile. In a number of cases, hamlets are built on sites which appear to have been deliberately chosen for their inaccessibility. In this respect there is some difference between the local groupings of Wanto'Abarasa and Kimi on the one hand, and those of Atigina on the other. Wanto'Abarasa is in the neighbourhood of the Government Patrol Post at Okapa, and the Kimi area is nearby, while Atigina is more remote. This seems to have had some effect, in that the hamlets of Wanto'Abarasa and Kimi are often located close to the main road, and are almost always connected by reasonably good tracks, while those of Atigina are generally remote from the road and are often not connected by cleared tracks at all. Also, there are still hamlets in Atigina which possess various defensive devices against the entrance of sorcery—for example, high palisades with complicated entrances and various "tabu barricades." These are to be seen particularly in the local groupings of Kume and Okasa, where the people openly said that though defences were no longer usually necessary against physical attack, they are still needed as protection against sorcerers.

However, though the influence of Okapa Patrol Post appears to have been effective in bringing a number of nearby hamlets "into the open," and reducing inter-hamlet suspicion by facilitating freedom of movement between local groups, the belief in sorcery appears to have remained as strong in Wanto'Abarasa and Kimi as it is in Atigina. Particularly is this so in connection with the type of sorcery known as Kuru.

Residence rights in a hamlet and clan membership are determined by patrilineal descent. Marriage is patrilocal, with the usual few exceptions when a husband goes to live in a wife's group.

There is a good deal of marriage between the different local groups, but this does not appear to have had a great deal of effect in reducing intergroup suspicion. For example, in the Atigina local grouping of Kume a noticeably large number of wives have come from the Kimi local grouping of Keakasa, yet Keakasa sorcery is particularly feared by the Kume people—this so much so that Kume men have deliberately obliterated the track connecting them with Keakasa, and openly state the reason for having done so. Visits still take place between the two groups, but are made in daytime by the longer route along the "Government road," and Kume considers itself no longer open to night-time operations by Keakasa sorcerers.

Just as each Fore Native area has its recognised leader (*kamintawa*), so each local grouping within the areas has a *kamintawa*; and it is of considerable assistance to a visitor working in the area to win the cooperation of these men. Generally, it is not difficult to identify *kamintawa* of local groupings in Fore, as care appears to have been taken to select the actual indigenous leaders for appointment as Luluais and Tultuls. If the *kamintawa* of a Fore local grouping has not himself been appointed to a position of Government leadership, it will usually be found that his son or some other man under his sponsorship has been so appointed. In this way, Fore differs from some other regions in which distinction must be made between Government-appointed Native leaders and the genuine indigenous leaders, and in which it is sometimes difficult for a visiting European to discover the latter.

Fore Sorcery

Though belief in sorcery plays a considerable part in the lives of the Fore, this is not necessarily to say that it has more influence than in the majority of the Territory's localities. In fact, the number of different methods of practicing sorcery is smaller in the Fore than in most other groups known to me. For example, while a comparatively sophisticated people such as the Rabaul Tolai can, even now, name more than twenty different types of sorcery, careful checking among thirteen Fore local groupings failed to reveal more than six types. And each of the local groupings, without any hesitation, named the same six types.

The Fore have names for ten different types of disease or ailment, which they distinguish from one another. Under some of these names they group more than one kind of disease. For example, both yaws and leprosy are grouped together under the name *karu'iena*, and both are attributed to the sorcery method which bears that name. Of these ten types of ailment which are recognised by the Fore, four are thought not to be due to sorcery. Three of these cover groups of minor ailments which are said not to cause death. These are as follows:

Nanontana: a general name for headaches and other minor aches.

Nantagavavina: headaches, with vomiting.

Kagai: coughing, which can at times be severe, and has in a few cases been known to precede the death of aged persons. It is not, however, taken very seriously; and the general attitude is that aged persons can be expected to die anyway.

The fourth of these sicknesses which are not explained by the manipulations of Fore sorcerers is *korankinu*, or dysentery.

Fore views on *korankinu* are interesting and, I think, may perhaps be an indication that it takes some time for sorcery-explanations of new diseases to be evolved. As far as I could gather, dysentery had become known to the Fore only within the last ten years (perhaps less than ten years). From their experience they realise that it is a very serious illness which has killed large numbers of people, in this way being as much to be feared as the various illnesses which they attribute to sorcery. However, they are still at the stage of simply saying that it is something that "came from Kamano" (the group further north, on the road to Kainantu). They do not believe that any Fore man knows anything about it. Some Fore say that it might perhaps be due to a form of Kamano sorcery, but they are not sure about this. At present, serious though the Fore know *korankinu* to be, it remains to their minds unexplained.

My view on this point is that, if belief in sorcery in general persists in sufficient strength in Fore, a sorcery-explanation for *korankinu* will eventually be evolved, and when cases occur they will be attributed to the activities of Fore sorcerers themselves. This may be some indication that *kuru*, the type of illness and type of sorcery at present most important in Fore thought, has existed in the region for some time—at least for a sufficient period for detailed and widely held views as to its cause to have been worked out.

The six types of sorcery believed to be practiced in the Fore region are as follows:

Karena: In this form of sorcery, a special sorcerer's stone, called *nami*, is used (the stone is known as *nami* in the Atigina Native area. In Kimi it is known simply as a *karena* stone). No tabus are observed by the sorcerer, and no spell is uttered. The stone itself is the sole essential ingredient. *Nami* stones are said to be full of potency, and can be dangerous to the users themselves—an idea which is in accord with the general Pacific view that stones and other objects can contain continuous charges of "power" (or what is generally called in anthropological literature *mana*). A *nami* stone should not be held in the bare hand, but must be kept wrapped in leaves and handled with tongs made of bamboo. In its use in *karena* sorcery, a small piece is broken off, pounded to a powder and mixed with the food or tobacco of an intended victim. This leads to severe internal pain after the food has been eaten or the tobacco smoked. The Fore claim that this is due to "the victim's liver going wrong." Some deaths are said to have occurred as a result of *karena* sorcery, but many cures are known, both in cases where sufferers have been sent to hospital at Okapa or Kainantu and in those in which treatment has been given by local Fore practitioners. In the Kimi local grouping of Iasui, for example, the Tultul Kiebu and another man named Aiwantaia gave the names of nineteen people they could remember as having died from *karena*, as compared with eleven remembered cures by local practitioners. The method of cure given was that the following ingredients should be collected (the names are in Fore): *karena'igena* (a vine); the bark and leaves of *takai* (a tree); the bark of *kau'ia* (a tree); and the leaves of *kenata* (a tree). The first three ingredients are treated and mixed together by the practitioner and then eaten by the sufferer. After this, the *kenata* leaves are eaten alone. The treatment causes vomiting and, as has been noted, in some cases cures are effected. In my opinion, the name *karena* is probably used for more than one type of illness, as I have mentioned in referring to Fore sorcery in general. On the whole, fear of *karena* sorcery does not at present appear to play a very big part in Fore life. The nineteen remembered deaths attributed to it in Iasui was an unusually large number in comparison with the other local groupings with which I was dealing. In some local groupings, the people claimed that they could not remember anybody dying from *karena*, the whole concentration of thought in such places being on *tukavu* and *kuru*, for which lists of remembered deaths were invariably long.

Ka'i: In this, portion of the same *nami* stone is used; but in practicing *ka'i*, the stone is not placed in food or tobacco. The stone, wrapped in leaves, is held in bamboo tongs and pointed at the intended victim, while the sorcerer calls his name and calls on the *ka'i* sickness to enter into him. While in *karena* an actual poison might possibly be used, this could not be so in *ka'i*, which essentially resembles the Australian "pointing of the bone." Clearly *ka'i* is an attempt to explain or rationalise sickness, and perhaps one type of sickness only, since descriptions of symptoms were in this case always more definite and more invariable than in the case of any other Fore sorcery except *kuru*. Death is said to occur very rapidly after *ka'i* first becomes apparent. Blood is said to run from the nose, mouth, and ears; and various parts of the body become swollen, particularly the penis, testicles, and legs. The same description was given in all Atigina local groupings, and there was little variation in either Wanto'Abarasa or Kimi. Generally, the people said that if a man started to suffer from *ka'i* in the morning, he would be dead by sunset. This rapidity of death was to some extent borne out by the case of Inunkaia, a man of Iasui. One Thursday he had visited me at the rest house at Iasui, and later returned to his hamlet. When I asked somebody to take a message to him the following Monday, I was told that he had died from *ka'i* the previous day. He had become ill only on the Saturday morning, and was dead within approximately twenty-four hours. Descriptions of his symptoms conformed to the general descriptions I had previously obtained: his penis was swollen, his tongue was swollen and protruded from his mouth, his head was twisted to one side, and he could not speak. The people of Iasui said that no attempt to treat him had been made by a local practitioner and that, in fact, no cure was known in the neighbourhood of Iasui. However, in other local groupings I was told that there were known cures for *ka'i*. These varied from place to place. For example, an aged man named Nantati, in the Atigina local grouping of Amora, claimed that he had been taught an effective cure by a Kume man named Wane. Nantati's cure was to obtain a vine called *asana* and mix pieces of this with a berry known as *u'a* and a grass (*enani*). These were put together in a small slip of bamboo and pushed up the sufferer's nose and down his throat. In the Wanitavi local grouping, I was told that the leaf, not the berry, of the *u'a* tree was used, and that this was eaten mixed with the bark of the *nagaru* tree. The Wanitavi people claimed that this cure was invariably effective if the mixture was eaten before the blood had commenced to run from the nose and mouth of the sufferer. They could not remember anybody having died from *ka'i* in Wanitavi, and claimed that their Tultul, Kuna, and another man, Waruata, had quite recently been cured of *ka'i* by the local practitioner. Numbers of deaths attributed to *ka'i* were generally small, the largest number remembered being eight in the Atigina local grouping of Amora, while some groupings such as Wanitavi claimed not to remember any deaths at all from this form of sorcery.

Karu'iena: This is the form of sorcery which is primarily believed to cause yaws and leprosy, though I believe that various types of ulcers and sores are also attributed to it. There was some vagueness concerning the method of practicing *karu'iena*, but the most general version was that food leavings of a victim should be obtained and wrapped in a special leaf (the name of which I was not able to obtain). This is placed on stones near a fire, the victim's name called, a spell uttered and the *karu'iena* sickness ordered to go into his body. A cure given in Wanitavi is to obtain the bark of a particular tree (whose name the people claimed not to know), mix this with the leaves of the *warebu* tree, and rub the mixture on the sores. I was told in most Fore localities that *karu'iena* was at one time an important form of sorcery, and one much feared. The noticeable success of medical treatment of yaws, however, seems to have altered this situation completely, and *karu'iena* has now lost most of its prestige and is little thought about.

Kesena: The ailment known as *kesena* involves aching and swollen joints, but is not thought to lead to death. To afflict anyone with *kesena*, one places a small amount of ash on a piece of *ano* (grass), with a twig placed across it. A spell is uttered over this and it is placed on a track just before the intended victim is expected to pass along it. When he steps on or over the arrangement the *kesena* sickness goes into his body. Many people claim to know cures for *kesena*, and these vary from place to place. All, however, involve cutting and/or anointing affected parts of the body. Fairly typical is the cure given me by Nantati of Amora, an elderly man whom I found to be on many points a particularly cooperative and helpful informant. Nantati's cure is to obtain some bark of *poia* (tree), leaves of *kena* (a green food-plant), the bark of *tagai* (a vine) and the bark of *ivo* (a tree). These are mixed with the blood of a pig, cuts are made in the patient's joints and the mixture rubbed into the cuts. *Kesena* sorcery does not appear to play a large part in Fore thought, though the illness described as *kesena* is recognised as being extremely uncomfortable and unpleasant.

Tukavu: When one comes to the form of sorcery known as *tukavu*, one is dealing with one of the two most important forms of Fore sorcery. However, whether or not belief in the practice of *tukavu* can always be connected with sorcery beliefs or may at times be associated with actual physical violence, remains a moot point, for this is the well-known practice widely known in Pidgin in many parts of New Guinea as "*shutim nel.*" The technique is that a victim is made unconscious by physical assault (this is in the case of Fore; in some other parts of New Guinea a victim is thought to be made unconscious by magical manipulation). Long pieces of thorn are then thrust into his elbows, knees, ankles, back, neck and other parts of the body. The thorns work into the body, and the victim is revived, made to forget what has occurred, and sent on his way home. Shortly afterwards, he begins to feel pains in the parts of his body where the thorns have been placed, being particularly affected by the thorn thrust into his neck, since this works its way into his throat and prevents him from speaking or eating. According to the Fore, victims of *tukavu* have rarely been known to recover and, with one exception, I did not meet anybody in the locality who claimed to know a cure. The exception was an aged man, Barata, in the Kimi local grouping of Keakasa. Barata claimed to have cured a man named Taketo belonging to the same local grouping, and Taketo supported him in this. The method of cure was to obtain the bark of two trees called *ikanti* and *aipewa* and mix this with two types of grass, *matasa* and *ioni*. To this was added pig's fat and the mixture was given to Taketo to eat. As well, an extremely large cut was made in his abdomen and some of the mixture was rubbed into this. He was able to show me the scar left by the cut, but when asked about marks where the thorns had been supposed to be thrust into his body, he explained that *tukavu* men are skilled in removing all such marks. He was not willing to name the enemy whom he believed to have practiced *tukavu* on him. I think it is reasonably clear that beliefs associated with *tukavu* (as with most types of *shutim nel* activity) are generally attempts at subsequent explanation of an illness which has developed, but that occasionally efforts are actually made to carry out the physical processes of *tukavu* practices. There have, in fact, been some complaints at Okapa Patrol Post from people who claimed that they had been physically assaulted for purposes of *tukavu*, but had escaped before the procedure could be carried through. Also, in fact in at least one other region of the Territory (the neighbourhood of Telefomin) it is known that thorns and other substances have been inserted in the victims' bodies, and that at times *shutim nel* has really been carried out. Lists of remembered deaths from *tukavu* are longer than those for any other ailment except *kuru*. Such lists of course, probably have little relation to actual happenings, but serve to indicate the place various types of ailments and different sorcery beliefs at present have in the minds of the people.

Kuru: The illness called *kuru*, and the sorcery beliefs associated with it share with *tukavu* the main place in Fore thought concerning sickness and death. The means of practicing *kuru* are best illustrated by describing a specific instance. One of the first *kuru* sufferers seen by me was Namugoi'ia, wife of Au'ia, Luluai of the Wanikanto local grouping in the Atigina area. She was trembling spasmodically, and was not able to walk without assistance, her leg movements being jerky and uncontrolled. Her face seemed to have a rather fixed or "rigid" appearance, and in speaking her words were slurred and indistinct. In discussion with her husband Au'ia, he told me that she had been suffering from *kuru* for about two months, and that her condition was steadily becoming worse, although she was still being helped down to her garden each day. Au'ia's story was that when he and his family had moved into a new house his wife had left a fragment of her skirt behind. This had been obtained by two men, Iam and Kavarakea of the Kamata local grouping in the Kimi area, both of these men being known to Au'ia as bitter enemies of his. Iam and Kavarakea had taken the piece of skirt back to Kamata, where they wrapped it in a parcel composed of the leaves of the *karuku* and *kigi* trees, together with some *kigi* bark. The parcel was bound with *kemkeranta* vine. A fire was made and a sorcerer's stone was heated. When the stone was sufficiently hot, the parcel (called in Fore *asaina*) was placed on top of it. Iam then proceeded to beat the parcel violently with a stick, at the same time calling Namugoi'ia's name, and saying: "I break the bones of your legs; I break the bones of your feet; I break the bones of your arms; I break the bones of your hands; and finally I make you die." After this, Kavarakea threw the parcel into a nearby swamp, where it sank below the mud and water. It was at this moment, according to Au'ia, that Namugoi'ia had commenced to tremble, at first only slightly, but as time went on with increasing violence. As soon as the trembling started, he realised that his wife must have left some personal possession in her former house and that his two enemies from Kamata must have obtained it for use in *kuru*. His wife subsequently confirmed this by claiming to remember having left a fragment of her skirt in the old house. In this particular instance, the two men, Iam and Kavarakea, were described as being involved in a partnership for the purpose of practicing *kuru* together, but in most other cases it was thought to be a matter of one sorcerer carrying out the necessary rites alone. Iam and Kavarakea of course denied having practiced *kuru* at all (I met only one man who admitted having practiced it), but Au'ia remained convinced that they were responsible for his wife's illness, and that they were also probably responsible for the death of his eldest daughter, Kima, who had died as a result of *kuru* several months before. In most *kuru* cases, it seemed that as soon as the illness became apparent, relatives of the sufferer immediately and almost automatically settled on some known enemy as being responsible. There is, however, a means of divination when uncertainty is felt regarding the identity of the sorcerer. This is to collect a large number of rats, give each the name of a particular local grouping, and then cook them for a fixed period. They are removed from the fire and opened. The rat whose internal organs are least cooked indicates the local grouping in which the sorcerer is to be found. The process is then repeated, on this occasion giving each rat the name of a probable sorcerer in the local grouping concerned, the least-cooked rat once again indicating the agency responsible for *kuru*. It has become well known to the Fore that all methods of sorcery divination are frowned upon at the Administration Patrol Post at Okapa, because they have at times led to physical attack on localities or people believed to be responsible for sorcery. Very few people, therefore, would admit to me that they had carried out divination rites in connection with cases of *kuru*. Invariably, however, they would give me the name of the person or persons they considered responsible, merely saying, rather vaguely, that they knew "it must be so-and-so because he had always been an enemy."

The people are convinced that there is only one possible means of curing *kuru*. This is to discover the correct parcel of leaves (*asaina*), open it and throw the contents into running water. If one discovers and opens the wrong parcel, it is said that one has cured some person other than the one intended, perhaps someone quite unknown to one. This is used as an explanation or rationalisation of those instances in which recoveries have occurred without the relatives of the sufferer themselves finding a *kuru* parcel. In actual fact, extremely few recoveries are remembered. In the local grouping of Wanitavi, for example, I obtained a list of eighty-five names of remembered deaths ascribed to *kuru*, but only one remembered recovery. In Amora, there were seventy-two remembered deaths, and seven remembered recoveries. In Okasa, I was given the names of seventy-one people considered to have died from *kuru*, but nobody could remember any recoveries at all. And the situation was approximately the same in other local groupings. As with *tukavu*, these figures do no more than serve to indicate the position *kuru* has achieved in Fore thought, particularly since it is now probable that the people are attributing all sorts of deaths which occurred some years ago to *kuru*. In only one of the few cases of recoveries was an actual claim made that the *kuru* parcel had been recovered from its swamp, opened, and the contents thrown into running water.

This was in the case of a man Neni, of Wanitavi. I was told that Neni had suffered from *kuru*, and that it was generally known that Iaragi, of Kasoru, had been responsible for his illness. Iaragi, however, had relented and had himself recovered the *kuru* parcel and removed his own sorcery. Later, Iaragi confirmed this, openly admitting that he had practiced *kuru* and had subsequently cured Neni. (This was, no doubt, a fairly obvious means of enhancing his prestige and standing as a sorcerer.) In all other cases of recoveries, it was said that some unknown persons must have removed and dealt with the relevant parcels.

The present position is that the Fore consider *kuru* to be beyond the scope of European treatments. In this, the *kuru* situation is different from that in connection with all other types of Fore sorcery beliefs, since people tell of even some types of *tukavu* having at times responded to treatment either at Okapa or Kainantu. On the whole, the attitude towards *kuru* is one of resignation and despair. The only sound remedy for this attitude would appear to be that a significant number of *kuru* sufferers should be cured. I have already mentioned the effect of medical treatment of yaws in reducing the importance of *karu'iena* sorcery in Fore. A similar reduction of prestige would no doubt occur in the case of *kuru* beliefs if the disease could be successfully treated. Until that can be done, it seems improbable that any amount of explanation or propaganda will achieve much in removing what is the main cause of suspicion and insecurity in an otherwise unusually harmonious group.

Charles Julius, Government Anthropologist
(report submitted to the Public Health
Department of the Territory of Papua and
New Guinea, February 1957)

Appendix 2. Kuru
[An Administrative Report to the NIH]

Kuru is a localized progressive degenerative disease of the central nervous system, occurring at any age in certain groups of natives, predominantly female, of the eastern highlands of Australian New Guinea. It resembles the heredofamilial neurological disorders, such as cerebral ataxia, paralysis agitans, and hepatolenticular degeneration. However, reflex patterns are usually normal, and there are no sensory changes or systemic involvements. It is rapidly fatal, rarely lasting over one year.

No gross pathological lesions have been found, even in the brain, and chemical tests have yielded no information as yet.

All treatment thus far has been ineffective in altering the course of the disease. Nonaddicting analgesics, sulfonamides, antibiotics, adrenal cortical steroids, antihistamines, anticonvulsants, nutritional factors, and BAL have been tried with no success.

Transactions with Dr. Gajdusek:
1. Correspondence apparently began before 1 May 57.
2. 1 May 57—Brain-handling instructions sent to Dr. Gajdusek.
3. 7 May 57—Summaries on Wilson's disease sent to Dr. G. by Dr. Bailey.
4. 6 June 57—Approximately 35 urine samples received by NIH, sent by Dr. G. on 28 May 57.
   Six rolls of 8mm color film to be developed and copied were received around this date, sent by Dr. G. on or before 28 May 57.
5. 9 June 57—Small pathological specimens from Dr. G.'s first autopsy of a Kuru patient, male child, had been received and examined. Brain and major portions of viscera had been sent to Melbourne.
6. 18 June 57—Received from Dr. G.:
   a. Manuscript on Kuru
   b. History of a woman Kuru patient autopsied
   c. Complete autopsy material, including brain, from the woman above. This is the first brain sent by Dr. G. to NIH.
7. 25 June 57—Letter from Dr. Smadel to Dr. G. acknowledges receipt of items mentioned in #6 and the films. Studies on chronic cyanide poisoning sent to Dr. G.
8. 7 July 57—The following equipment was shipped to Dr. G. from NIH:
   a. 1 16mm motion-picture camera (Bolex H-16) with case
   b. 1 heavy-duty tripod
   c. 1 exposure-meter
   d. 8 100-ft. rolls of 16mm color film
   e. 4 100-ft. rolls of 16mm black-and-white film
   f. 36 one-doz. pkgs. of Keidel-type vacutainers
9. 8 July 57—Received from Dr. G. (sent 29 June):
   a. Autopsy material from two female Kuru patients
      1. two brains
      2. one complete post
   b. Histories of these patients and others
   c. Two rolls of 8mm film
   d. Three strips of 35mm black-and-white stills
10. 15 July 57—Sent to Dr. G.: First six reels of developed film.
11. 23 July 57—Received from Dr. G. (sent 10 July):
    a. 1 brain
    b. 2 cervical cords and 1 pituitary
    c. 2 sets of autopsy material other than CNS (formalin-fixed)
    d. 2 sets of autopsy material other than CNS (Zenker's-fixed)

      e. 1 manuscript: "Kuru—Clinical Study of a New Paralysis–Agitans-like Syndrome"
      f. One history
12. 26 July 57—Sent to Dr. G.:
      a. Black-and-white prints of 35mm film sent to NIH on 8 July
      b. Bond paper
      c. Letter from Dr. Baker summarizing materials received from Dr. G. thus far
13. 29 July 57—Received from Dr. G.: Telegram requesting help on freight problem.
14. 29 July 57—Sent to Dr. G.: Telegram in reply to one of previous day.
15. 31 July 57—NIH. Showing of first six reels of film sent by Dr. G. Drs. Masland, Sarason, Gladwin, Shy, Klatzo, Dastur and Ness attended.
16. 2 August 57—Received from Dr. G.:
      a. 2 brains
      b. One set of autopsy material other than CNS—formalin–fixed
      c. One set of autopsy material other than CNS—Zenker's–fixed
      d. Liver, kidney, spleen—Formalin fixed—case of cirrhosis?
      e. Liver, kidney, spleen—Zenker's fixed—case of cirrhosis?

Summary of specimens, material, etc. sent or received to date (5 Aug 57)
1. Received by NIH:
      a. 6 brains
      b. 2 cervical cords
      c. 8 sets of tissue obtained at autopsy (some sets duplicated when sent in another fixative. These were not counted as additional sets.)
      d. 36 urine samples
      e. 11 rolls of 8mm film
      f. 3 strips of 35mm black-and-white stills
      g. 2 manuscripts
2. Sent by NIH:
      a. 1 16mm motion-picture camera and case
      b. 1 tripod
      c. 1 exposure-meter
      d. 8 100-ft. rolls of 16mm color film
      e. 4 100-ft. rolls of 16mm black-and-white film
      f. 35 one-doz. pkgs. of Keidel-type vacutainers
      g. polyethylene bags
      h. developed film from the first 6 reels sent by Dr. G.
      i. black-and-white prints of 35mm film sent by Dr. G.

H. A. Imus
(for J. A. Smadel)

Appendix 3. Report on the Kuru Disease

At the invitation of Dr. Scragg, Director of Public Health for the Territory of Papua and New Guinea, we visited the peoples that are afflicted with the disease that is known by the natives as Kuru.

We arrived at Kainantu on Wednesday, October 30th [1957] at 10:30 a.m. and were met by Dr. Zigas, the District Medical Officer. After visiting Kainantu Hospital and examining two Kuru patients there, we were driven by Dr. Zigas to the Patrol Station of Moke in the midst of the Fore linguistic group, arriving at about 3 p.m. On the way there we examined one advanced case of Kuru in a native village. At Moke Hospital there were nine cases of Kuru assembled for examination. The whole of Thursday was devoted to a thorough clinical examination of those nine cases, which were at various stages of the disease. We left Moke at 7 a.m. on Friday, November 1st, and have had since then the advantage of a prolonged discussion with Dr. Gajdusek and Dr. Scragg at Moresby, and we have also seen there the film prepared by the National Film Unit. Furthermore, we have closely studied the papers written by Drs. Gajdusek and Zigas and the very careful and extensive case reports that they have prepared.

We wish to report that we have confirmed that Kuru is an unique neurological condition well worthy of continued and intensive investigation. Undoubtedly it is a degenerative condition of the extrapyramidal system involving primarily the basal ganglia but also probably the cerebellum as well.

The intensity of attack on the Fore people can be judged by the estimate that Kuru accounts for about half of the total deaths. If the nature of the disease could be discovered and its incidence controlled, it would be a magnificent achievement of medicine both in its scientific aspect and in helping these courageous and underprivileged people in the severity of their affliction. We feel that Australian Medical Science should be given the opportunity to co-operate effectively with the American workers that are already engaged on this investigation, and so to help peoples for whom we are responsible under the United Nations Charter.

We have been greatly impressed by the work that has already been done. There has been a systematic survey of the whole area of the Fore linguistic group and of the adjacent groups, some of which also have Kuru. The survey has extended beyond the furthest boundaries of the afflicted area. The clinical course and the uniqueness of the disease have been well established by investigations of several hundred patients. It has been shown that in the great majority of patients there is an inexorable progress of the disease to a fatal termination in 6–9 months. There is much information dealing with the peculiar age and sex incidence, and the possible familial incidence of the disease. Furthermore a wide variety of therapeutic agents has been tested, and an enormous amount of material has been secured, both from living and dead patients and despatched for examination, but as yet very few reports on this material have been available.

We wish to report that the investigations conducted by Drs. Zigas and Gajdusek are wholly admirable and worthy of the greatest praise. As yet the nature of the disease is unknown, but we would agree with the conclusion of Dr. Zigas and Dr. Gajdusek that it is very unlikely to be occurring as the aftermath of a viral infection. A number of possible causes of the disease can be suggested, but these must be regarded merely as providing grounds for further investigation. Such investigations have already been initiated by the very thorough dietetic survey conducted by Miss Hamilton and by the large amount of material, such as blood, urine and brains that have already been despatched by Drs. Zigas and Gajdusek for trace-metal estimations.

We further wish to report that it is necessary to continue the investigation intensively so that the magnificent work that has so far been done may be used to full advantage. It is urgently necessary to define the course of the disease by following up the known Kuru cases that are still alive and continuing with special therapeutic measures such as BAL and Versein medication. For example there are a few cases in which the advance of the disease has been checked, and it is very important to follow these cases closely and to watch for further exceptions to the inexorable progress. It is also essential to survey the area in detail in order to discover if there are subclinical cases and to detect new cases as early as possible, so that the clinical course of these cases may also be fully defined. Finally the range of the neurological signs of normal subjects needs to be investigated.

For these purposes it is essential to send an expert clinical team to be based on Moke so that a systematic survey can be made on the Fore peoples for a period of at least 8 weeks. It will further be necessary to provide adequate assistance for them. Dr. Robson, Professor of Medicine at Adelaide University, is prepared to lead this team, and to be available at the beginning of December 1957. The estimated cost of this project is £3000.

> H. N. Robson, Professor of Medicine,
> University of Adelaide
> S. Sunderland, Professor of Anatomy,
> University of Melbourne
> J. C. Eccles, Professor of Physiology,
> Australian National University, Canberra

Appendix 4. A Chronology of the Kuru Area

1900–1930	European influence on the coast, with no knowledge of the existence of highland populations. The sweet potato may have been first introduced during this period.
1900–1910	Cannibalism appears in the North Fore and Keiagana, entering from Kamano and BenaBena areas.
1917–1918	German soldier-adventurer Detzner walks among the "Kukukuku" to the east of the kuru area. Cannibalism introduced in the South Fore from the north.
c. 1920	Kuru first appears in Uwami, Keiagana, and Awande, North Fore.
1929	Rowlands (a missionary) enters Kainantu area. Lutheran Mission established at Raipinka (near Kainantu) and at Onarunka. Introduction of some trade-items, and possibly maize and casuarina trees.
1930	Gold-prospectors Leahy and Dwyer enter the highlands to the north of the kuru area, and pass by Mt. Michael and down the Purari River. Taylor and Leahy patrol into the highlands.
1930s	Kuru spreading rapidly throughout the South Fore.
1932	First planes seen by Fore: Government and New Guinea Gold Fields planes from BenaBena. First patrol posts in highlands. Measles and mumps epidemics among the Fore.
1934	Ashton brothers, gold prospectors, enter northern fringe of kuru area from Kainantu.
1936	First European into the kuru region proper: Ted Ubank, gold prospector.
late 1930s	Father Tufanel walks from Goilala to the Eastern Highlands. First steel axe circulates in South Fore, demonstrated by man from Takai village. "Tall black man" leads patrol (gold prospectors?) from the south to Kamira (South Fore), carrying cargo boxes, hurricane lamps, axes, etc.
early 1940s	World War II. Australian observation posts in the highlands. Japanese infiltration from Markham Valley. Ted Ubank conducts whites fleeing Japanese to Papua along the Lamari Valley. Many planes flying over the highlands. Japanese plane crashes near Awarosa (South Fore), and two survivors are helped to leave in the direction of Kainantu. Sheep and goats probably introduced to highlands around this time.
1943	Dysentery epidemic throughout the highlands.
c. 1943	American plane crashes above Awande and Miarasa, killing the passengers.
late 1940s	First ANGAU (Australian New Guinea Administrative Unit) postwar patrols; first patrols into the kuru region. Cargo cults in North Fore.
1947	First government patrol to North Fore. Patrol Officer Skinner demonstrates firearms; people spontaneously "line" up.
1945–1950	Certain southern South Fore communities move south into uninhabited malarial forest. Trade-salt introduced. Steel axes begin to replace stone axes.
1949	Tarabo Lutheran Mission established (native evangelists). Okasa pine forest surveyed. Usurufa and parts of North Fore "derestricted." Whooping cough epidemic. Second government patrol (to North Yagusa and Okasa Fore) met enthusiastically.
1950	First resident European missionary at Tarabo.
1950–1955	Return of Fore from southernmost extent of their settlement (Abonai) because of malaria deaths. With deaths from kuru, men become progressively involved in women's tasks in gardens, food preparation, and

	infant and child care. European vegetables introduced into gardens: potatoes, peanuts, tomatoes, cucumbers, cabbage, onions, lettuce, pumpkin, peas, haricot beans, and new varieties of sweet potato. Domestic fowl introduced.
1950–1957	Gradual disappearance of many cultural practices: warfare, cannibalism, infanticide, suckling of piglets by women, institutionalized premarital sex with cross-cousins, nudity of male children and old men, use of *wati-mabi* (penis display) except in jest in *singsings*. Polygamy and child-marriage decline, due to official discouragement. Goats introduced by Seventh Day Adventists into some North Fore villages.
1952	Australians Ronald and Catherine Berndt do first anthropological fieldwork in kuru area: two three-month periods among Yate, Usurufa, North Fore. Cargo cults active.
c. 1953	Kunimara Patrol Post established at Tarabo, with detachment of Royal Papua and New Guinea Constabulary.
1954	Okapa Patrol Post established at Pintogori, with European Patrol Officer. Introduction of penicillin.
1955	North Fore laborers in Kainantu, Goroka, and BenaBena. Road from Kainantu to Tarabo and Okapa finished. Mage (North Fore) village school established. Road to Atigina completed. Seventh Day Adventist Mission established at Okasa. Coffee and other cash crops planted in North Fore.
1956	First North Fore person to see the coast, taken by the Administration. First plane lands at Tarabo airstrip. First vehicles to Okapa and Tarabo. Gonorrhea epidemic at Moke village (North Fore). Kerosene and matches become available. Shovels and hoes begin to replace digging-sticks for gardening. Anthropologist Charles Julius working in South Fore.
1957	Kuru Research Center set up in Okapa; medical study of kuru begins. Road to Purosa completed. North Fore epidemics of whooping cough, influenza, and measles. First South Fore person to see the coast. Two kuru patients taken to Port Moresby. First Fore to leave main island taken to Rabaul. First vehicle reaches Purosa (South Fore). Yaws eradicated with penicillin in North Fore. Village residential pattern shifts to family households (from communal men's houses surrounded by separate women's houses).
mid 1950s–1957	South Fore and Gimi (Kimi) refugees from inter- and intra-village fighting seek shelter with Pawaiian people.
1957–1960	World Mission Station established at Purosa; Seventh Day Adventist Mission at Mugaiamuti. Establishment of medical aid posts throughout the kuru region.
1958	South Fore, southern Keiagana, and Gimi areas declared "derestricted." Cloth increasingly used in dress. Yaws almost gone in south; scabies and impetigo beginning to decrease.
1958–1959	Lutheran Mission reveals men's secret bamboo flutes to women. Government school established at Okapa. Mission schools at Tarabo and Purosa. Lutheran native evangelists move into South Fore. Treatment of leprosy begins in 1958 (virtually under control by 1964).
1959	New Tribes Mission established in Negibi (Gimi). First Fore laborers sent throughout New Guinea. Money economy begins in the kuru region. Trade-stores open in Okapa. Summer Institute of Linguistics linguists enter North Fore and Kanite areas. Kuru incidence begins to decline in younger age-groups.
1960	Seventh Day Adventist Mission School established at Keiakasa (South Fore). Tinned meat and fish enter diet. Young men speak Pidgin, play kickball and basketball, and accumulate wealth. Cash crops spread into South Fore. Last traditional men's house destroyed.

1960s	Increased marriage between linguistic groups and with people outside of kuru area (Anga, Awa, Chimbu, and Kamano). Emergence of practitioners using *"driman"* (dream) and "smoke" curing sorcery for various diseases. Emergence of new group of wealthy young men with European goods. Traditional initiations largely cease. Belief in sorcery as cause of yaws and leprosy disappears. Malnutrition in infants and toddlers declines.
1961	First Fore laborers sent to outer islands of New Guinea.
1961–1962	Saave (a kuru healer) working at Uvai (Gimi). South Fore anti-sorcery *kivungs* held.
1962	Influenza epidemic. Coffee-growing training school at Okapa. New government hospital completed at Okapa. Lutheran Mission hospital for kuru patients established at Awande. Okapa linked to Lufa-Goroka road.
1963	Pig-anthrax epidemic. Triple-antigen immunization of children.
1964	First elections for Administrative Council. Highland Christian Mission established at Yagusa (South Fore).
1964–1965	Decline in kuru incidence evident in all age groups.
1968	Disappearance of kuru among children under 9 years of age.
1974	Disappearance of kuru in 10–19 age-group.

Appendix 5. Letters from Lois Larkin

[The letters below were discovered among DCG's personal correspondence as this book was going to press. Although it has not been possible to place them in order among the other letters, we have appended them here, deleting only a few paragraphs containing news of shared friends from Melbourne.

Lois Larkin was microbiological technician to Sir Frank Macfarlane Burnet ("the Boss") in 1957. She was continuing the auto-immune complement fixation (AICF) work begun by DCG while he was at the Hall Institute in 1955 and 1956, and much of the correspondence below relates to AICF laboratory work and publications. It also provides an informal inside view of activities at the Hall Institute before and during the period in which DCG's kuru research in New Guinea was a source of conflict between him and Burnet.

As can be inferred from these letters, Lois Larkin joined DCG at Okapa in June, 1957, where she spent a month assisting with the kuru investigations. While there she met Patrol Officer Jack Baker, whom she later married.]

LARKIN to DCG   12 March 1957
[first of four letters mailed together c. 16 March 1957]
Dear Carl,

So much has been happening that I hardly know where to begin.

First, Mrs. James' sera, both pre- and post-operative specimens, give little or no reaction with antigens made from her own organs. I have enclosed all the results in detail in case you see something I have missed. The results with the sediments are very interesting, but quite beyond me. Ian [Mackay] and the Boss are intrigued with the results, and as usual FMB[urnet] has a host of theories on the subject. One is that as the body produces antibodies to its own organs, they are absorbed by some mechanism in the spleen, and, since in this way the antibody would not come into contact with the nuclei of the cells, this could possibly explain the results obtained using the low speed sediments, as these sediments would contain nuclei, broken nuclei and a few mito- and microchondria. He has also swung around to *your* theory of antibody production after damage to tissues, burns, fractures, etc., and mentioned this fact at the discussion on Saturday. Frank Fenner was down from Canberra and he gave a talk on recombination using *Vaccina* virus. We had a double feature program and after a tea-break Ian gave a talk on Macroglobulinaemia. I had to explain the complement fixation tests and simplify the results, and, honestly, I have never ever been so nervous before; but all went well.

Yesterday one of the lupoid hepatitis cases died (L3). She was only a low titer positive, but I got all the usual antigens and tomorrow I will be back in the antigen production field again. If this keeps up, I shall soon be able to set up a butcher's shop.

Mrs. James has recovered sufficiently to allow her to go home. She has not been very cooperative, and I have been able to get only three serum specimens since the operation; so far all have reacted with normal antigens to the same high titers as the post-operative ones.

Today I inoculated two bunnies with living antigen. I intend to bleed at 7 day intervals, giving a booster injection at consecutive 14 day intervals, and also to do the same with kidney antigen. Is this all right? I am also in the midst of making up the chick embryo, rat, and mouse antigens.

There is a surprise for you, it knocked me for a six. FMB and I were in the postmortem room inoculating mice on Friday when out of the blue he asked me if I would be interested in a trip to New Guinea. After recovering from the shock I told him I most certainly would be. It appears that they (who "they" is I am not quite sure, but I think it means Dr. Gunther, FMB and SGA[nderson]) are interested in a child study project. (A little out of their field, don't you think?) From what I can make out, technicians are hard to come by and the Boss thinks that I could get a lot out of a 2–3 month trip. I am not banking on these plans coming to anything, as he changes his mind so often these days, but it would be wonderful and I am living on hopes at the moment.

Tonight (it is now 1:30 a.m.) I have just seen the Union Theatre's last play for the season, "Look Back in Anger" by John Osborne. It is a play dealing with the modern approach to marriage and life in general—very enlightening, intense, and dramatic, and as usual well done. I think you would have enjoyed it.

Before I forget: FMB has finally, much to his satisfaction, received an invitation to visit Stockholm. . . .

No more news, so until the next installment of results is ready,

Love,
Lois

LARKIN to DCG   13 March 1957
[second of four letters mailed together c. 16 March 1957]
Dear Carl,

Just ready to mail letter one when I received another letter from you. I am rather surprised that it did not burst into flames on the way down; you sound, to put it mildly, furious. I don't blame you. I have tried to discreetly find out what is behind their silence, but the security in high places around here defies even my spy system.

As far as I can ascertain, Carl, Ian [Mackay] is the only one that knows anything about the child study plan, and I only found out because of the Boss's reference to it when he asked me about New Guinea. As for the peculiar illnesses amongst the natives, that is behind the curtain also. John Dineen did hear something about it discussed at the foot-and-mouth lunch, but this was well after you had *gone*. This type of thing goes on around the Institute all the time, and I think it is one reason why the Institute is stagnant at the moment. As you have always said, the reason for having group research is the great value of discussion of plans, work, and results. I know this incident has marred your stay here and I would have given anything to prevent that from happening. The only consolation you have is that you are not the only one in the dark about their plans.

I most certainly will not let anyone get their hands on the results of experiments for [AICF] Paper III, but I do wish you would stop worrying about it. You must have heaps of work to do and if you keep worrying about it you will end up having a coronary or something. . . .

Now it's my turn to get furious; in your letter you did not even mention whether you had received the mail and books.

Am still making up antigens.

Lois

LARKIN to DCG   15 March 1957
[third of four letters mailed together c. 16 March 1957]

Honestly, Carl, if this keeps up I shall never ever get this letter posted. Must apologize for getting furious in letter two, as your letter from Cairns has just arrived.

With regards to the paragraph on atypical pneumonia and reference 81 [in AICF paper II], they are not mentioned in any of the copies of the paper around the Institute. You will have to wait until the galley proofs arrive. *JAMA* has acknowledged receipt of the paper. Now that I know what to do with the disseminated lupis erythematosis [DLE] sera, I shall send them off with the last shipment, which goes next Wednesday. Ian [Mackay] is still getting details on the new high titer sera patients, so I will pass on the information when I get it.

Nothing was altered in the paper to *Nature* [AICF paper I]. Judith has received the map.

By the way, if there is any more talk of gratitude I shall blow a fuse.

Ross called in last night and we went to see the Mozart film at the Savoy. It was very good, but only about *The Magic Flute* period and the death of Mozart; I would have thought from the title that it would have been his life story. One of the shorts was a Russian climbing expedition in the Pamirs. Three hitherto unconquered peaks were climbed and called respectively Peak Pravda, Molotov, and Stalin.

Today as yesterday I am making up antigens.

Love.
Lois

LARKIN to DCG   16 March 1957
[fourth of four letters mailed together c. 16 March 1957]
Dear Carleton,

Thank heaven tomorrow is Saturday and there is no mail. This is the first time in my life I have written more than six pages to anyone, and I hope you appreciate the honor bestowed on you. . . .

Your information about Goroka and Wau was very interesting and I had no idea that it was so civilized. There is nothing much I can say about the Aussies' attitude to the natives except that it would make me mad. I think the trouble is that, deep down, lots of people want to act like kings in their own domain, and in New Guinea there are the opportunities for this type to succeed. Possibly the first settlers adopted the old English colonial ideas and as yet very few have the gumption to buck against established ideas as they are scared they will be odd-man out. Let's hope that in a few years time our attitude will change.

Well, today marks the start of my 24th year on this earth and I am in a philosophical frame of mind, so I had better close down before I bore you to death. Ross has presented me with a copy of *Ulysses*, and I bought myself a copy of Kafka's *Wedding Preparations*. I have finished *Growing up in New Guinea* and enjoyed it immensely. . . .

Love,
Lois . . .

LARKIN to DCG   3 April 1957
Dear Carleton,

You are most certainly not the white-haired boy down here at the moment! I thought that in the year spent down here you would have realized that the Boss likes to think out and arrange things himself. You know what a stuffed shirt he is, and I suspect that the reason he did not answer your letters was because he was none too pleased with your suggestions that he do this and that. Although he considers that kuru is the most exciting thing to happen in New Guinea for years, he and Anderson don't appear to be doing much about it. There was a rumor drifting around that Anderson is leaving for Moresby next week.

There seems to be a hell of a mix-up with the arrangements about the kuru investigations. Last Thursday week FMB was on the phone to Moresby and arranged to have all tests carried out there. You obviously were not informed of this arrangement, and when your first package arrived the Boss nearly had a fit. He considered you were cutting across Moresby in sending the CSF's down here, and that in doing the test he would be indirectly doing the same thing. After he read your last letter and discovered you were working with a Department of Health doctor he cooled down considerably. The CSFs were taken out of my hands, and all I can tell you is that, of those in the first shipment that were fit to test, all, except one with a weak Wasserman reaction, were negative. Anderson was not interested in trying any isolations, and French was not given a chance, as FMB considers this field to be Anderson's.

The cells and serum on the specimen from Ereio I sent to Simmons, and the results are:

1. | ABO | MNS | Rh | C^w | P | Le^a | Sy^a | K |
    |-----|-----|----|----|----|----|----|----|
    | B | Nss | Rh_1,Rh_1 | — | — | — | — | — |

2. Direct Coombs test negative.
3. No atypical antibodies in serum.
4. A_1 titer 1/20 A_2 titer 1/10

The last shipment, which arrived on Friday, I took out to Simmons yesterday. He is doing the blood and serum tests, and has arranged for Cyril Curtain and Shaw from the Baker Institute to do the abnormal haemoglobins. I also took out the map of the Gulf country for his inspection. As he has mentioned the fact that the B gene has come down from New Guinea, he wants a small portion of the N.G. coastline and Daru Island included. When I have fixed this up I will get it photographed and send you a copy. While I was there he told me he was trying to make a decision as to whether or not he could manage to test blood from the normal populations you mentioned in your letter. He is very busy and is not sure he will have the time that this type of testing takes up. Anyway you should hear from him soon.

In my last letter I mentioned that one of the lupoid hepatitis patients had died (HL4, not L3 as stated). I carried out the usual experiments and have enclosed the results. (Firstly I had better tell you that Mrs. James had little or no antibody to her own organs). This one has antibody to her own organs to the same low titer as her serum gives with normal antigens. Ian [Mackay] and FMB are a bit dubious about the low titers and are inclined to think the results are nonspecific. I don't, so I am going to test about 200 normal human serum with normal L and K and HL4 and K. If it is okay with you, I shall alter the numbers for normal sera in table I of Paper I, and if they are all negative, as I am sure they will be, it will help bring down the percentages of positive normal serum. . . .

Nothing much has been happening around the Institute. [Joshua] Lederberg has postponed his arrival, and instead of coming in July will not arrive until September-October. The change over to immunological work is going to be very slow, and so far nothing has been done about the macroglobulinaemia work. I have had another high titer positive serum from a suspected myocardial infarction, but apart from this and a few low titer positives, nothing much is going on in the AICF field. The trouble is that it is so hard to get follow-up bleeds on most of the patients.

In view of the Boss's hint that he might send me to New Guinea (last letter) I have been going to lectures on New Guinea and the Pacific. The lectures are given by a bloke called Davidson who has a coffee plantation in the Waghi valley on the Jimmy River side. I was hoping that he would give a few details on the anthropology of New Guinea, but so far his lectures have been mainly tall stories and reminiscences. Can you suggest any book I can read which gives a broad outline of native customs and life in New Guinea?

How much longer will you be at Moke? It is about time that social letter you promised arrived.

Love,
Lois

P.S. Nearly forgot: Just for fun I did an AICF test on Ereio serum and got titer of 16+ with both liver and kidney. I am going to do the last set of serum tomorrow so I will let you know what happens. I *did not* tell FMB about this result. As neither he nor Anderson are interested in the serum you sent down, I am going to send them back to the States.

LARKIN to DCG   26 April 1957
Dear Carl,

Have just received your last letter and parcel. It seems I am still missing you with the mail. I mailed a letter to Moresby last week, but as you probably will not get it for a while, I had better tell you the most important news again. . . .

The letter to *Nature* has been published and is currently appearing in the 30th March issue. The Institute's copy has not arrived, and the airmail copy from the Medical School is out on loan, so I have not seen it yet. Will send you a copy when it arrives.

The *Pneumocystis carinii* paper is out in the April publication of *Pediatrics*. My letter in Moresby contains one from the editor of *Pediatrics* which says that "he regrets that your additions to the paper arrived too late to be included, as the journal was already in the press."

Ian [Mackay] has mailed you a copy of [AICF] Paper II about two weeks ago. On Wednesday he had a letter from the editor of *JAMA*. The editor considers that the papers are too long and have too many tables, and although they are extremely interesting, are of a more specialized nature than the usual papers in *JAMA*. He has shown them to the editor of *Archives of Internal Medicine* and suggests that you and Ian submit them to that journal. Is this all right with you? I think that it is probably the best thing to do, as it will save all the time and work of condensing the tables and avoid further delay in their publication; *Archives of Internal*

Medicine is a pretty good journal. If you have any objections send me a telegram as soon as you can. If I don't hear from you by Thursday, May 2nd, I will take it that it is okay with you.

I was in touch with Simmons before Easter and he says that so far there has been nothing out of the ordinary in the results on the blood cells and serum. There was no abnormal haemoglobin and the only discrepancy Curtain found was that some of the serum had an elevated β. He wanted some normal New Guinea sera to check as controls, so I have taken a sample out of what I think and hope are the Orikaiva serum. If you have the serum book with you, will you check and see if the T series are Orikaiva? . . .

The results of the James experiment are exciting. I have about .5ml of sediment from the liver antigen, so will titrate the post-operative serum. Just for you, my sweet, I will check the liver, muscle, and spleen results. I have simmered down a bit and have decided to pardon your skepticism. Something else I have decided is that I am not going to send you the results for [AICF] paper III, as in my opinion you have more than enough to do without worrying about paper III. Unless you have any violent objections, I will keep the results here and send them to you later on.

Anderson is getting ready for a trip to New Guinea. He was to have left on Wednesday's plane, but the Qantas people are still running around in circles as a result of the pilots' strike, and he has delayed his departure for another week. As usual no one is quite sure what he is going to do when he arrives but I suppose you will run into him sooner or later. There has been no talk of kuru lately except that FMB had a letter from Ian [Burnet] who thinks he has a few cases of kuru in his area. Have you been there yet?

I am glad FMB has come around. I thought something like this had happened as his attitude has changed completely. The politics and petty jealousies behind the New Guinea work are very complicated and I cannot make head or tail of the whole mess. One thing I do know is that the chances of my turning up in New Guinea in the near future are practically non-existent. I have not heard any more about the trip and have just about given up all hope of it even coming off. I am tempted to take some holidays and come up while you are there. You have no idea the mental hassle that goes on every time I pass an airline booking office. I would love to see you and Dr. Zigas at the *singsing blong kiss* and flirt with the lecherous old men of the village. Thanks for the advice on the books. I have read Mead's books and am going to tackle Simpson's this weekend. The chap who makes the lectures I have been attending is in the Waghi valley at the moment. His coffee trees needed some special fertilizer or something. . . .

I have enclosed the results on 143 normal sera titrated against L + K, and also the results from the only positive kuru sera.

Have you heard that the USA and USSR are going to resume their friendly relations missions and that there is some talk of a group of Public Health people going to Russia in October? . . .

That's all I can think of at the moment. I am writing this in the kitchen and Mum is cooking a big panful of mushrooms; as Murray would say "aren't you jealous?"

With love,
Lois . . .

LARKIN to DCG   8 May 1957
Dear Carl,

What it is to be famous! Ian Mackay asked me today if I could put in a request for a picture of you. Colored, of course! It seems that Sir Ian thinks it would be a good idea if they had a slide of the discoverer of the AICF test to show at lectures, etc. Make it a good one. Get someone to take a shot of you surrounded by the wildest-looking natives you can find.

I was not going to worry you with the results of experiments for AICF paper III, but one of the animal experiments is a trifle odd. The results are enclosed. I would like some advice in a hurry, as I am in the middle of a repeat experiment using kidney antigen in place of liver. Rabbits were inoculated fortnightly with liver antigen and bled weekly for a period of seven weeks. As you can see there is a high nonspecific titer in the preliminary bleeds, and Rabbit II gives very little rise in antibody production and some peculiar results. . . .

I am not in a very chatty mood tonight, for I have just discovered that I mislabelled a tube of stock virus on Monday which was used in about six different experiments. It was supposed to be pathogenic WS– but was really non-pathogenic aWS–. I have not been able to confess yet, as the Boss was away today.

John Dineen has just finished an interesting experiment in which he inoculated a rat with a homogenate of its own liver and was unable to detect any production of antibody. . . .

Cyril Curtain (the bloke doing the haemoglobin studies) was here the other day, and he was wondering, if you think it is worthwhile, if you could send down six blood specimens from normal people so he could check their serum for elevated β. . . .

Have you seen Anderson? He left last Thursday and as far as I know is at present in Moresby.

Lois

LARKIN to DCG   8 August 1957

[written at top:]   Love to Vin and Gloria

Dear Carl,

It is absolutely awful being back in Melbourne, and I find myself thinking about New Guinea all the time. . . .

Nothing much was said about the extra two weeks' holiday, and so far the Boss has shown no interest in my holiday at all. He is extremely busy and it looks as though he will have to become a real director and cut down on his lab work, as he has a mass of paper work to do and is planning a trip to the USA, U.K., and USSR sometime next year. Lederberg arrives at the end of this month; he is going to work with us for a few weeks. The paper on somatic mutation has been published. Ian [Mackay] was quite agreeable to the [AICF] paper II title. He has written to the *Archives* suggesting that they make some decision about the AICF papers immediately.

There was a mass of work waiting for me on return. We received 8 macroglobulin sera from the States but unfortunately none of them were positive. Nothing very interesting has turned up so far, although we had more positive hepatitis and disseminated lupus erythematosus serum. Have received all the specimens of kuru safely and sent them out to their respective destinations. . . .

Would give anything to be with you on the seventeenth; have a wonderful party. Write soon.

Love,
Lois

P.S. What is the Pidgin for "In a while, crocodile"? The reply is "Bihain mi lukluk, pukpuk."

LARKIN to DCG   30 August 1957

Dear Carl,

You are going to blow a fuse when you read the next part of the letter, so be prepared, but it is just as much your fault as it is mine.

In the last set of blood specimens, which arrived Monday, you sent down some eleven specimens on missing numbers in the blood group table of the schoolboys. Apparently, as I found out next day from Simmons, you especially wanted cells from these to go to him. *Why the devil didn't you let me know?* I could not make out why you bled the odd numbers, and was unaware that Simmons' table was incomplete, so I am afraid you will have to send down more samples, as I took off the serum and *raused* [threw out] the clots. Sorry too much.

Regarding the maps—I have sent you rough maps only.

1. Kuru Map: This was done from memory and probably is full of mistakes. If you correct it I shall duplicate it and get glossy prints made. How many do you want?

2. Have sent up copies of the only New Britain–Papua maps I can find at the University. Pick one and fill in what you want marked. If you have a better map, send it down. Will get one day prints made, okay?

I thought it might save time if you sent down covering letters for the kuru map for both the German and American Journals so that I could send the maps on from here. . . .

Ross has managed to get in to Ormond College and is very pleased with life at the moment. He has started to work for the exams and should be all right this year. Murray is waiting for his exam results so he can write and tell you how good he is. I gave him a set of bow and arrows, the *kunda* drum, and a *pass-pass* [armband] and he thinks he is wonderful. You should see him trying to shoot an arrow—he can hardly budge the string on the bow.

I have to show the pictures next Monday to all at work, and will post them up to you after that. They are really good; I think you will be pleased with them.

Did you read Simmons' last letter properly? From what he has told me it seems he paid for all the equipment he sent up for the Kukukuku patrol *out of his own money*. It might be an idea if you reimburse him.

Things are very quiet around the Institute at the moment. FMB has been holidaying in the hills for the last week, Gus [Nossal] is in Sydney, Pat [Lind] in Adelaide, and several other people have gone to the snow for skiing.

Hope everything goes all right on the Kukukuku patrol.

Love,
Lois . . .

LARKIN to DCG   16 October 1957

Well, my dear, in spite of all instructions and corrections I still managed to spell Mt. Michael the wrong way on the kuru map. Unfortunately the error passed unnoticed and the map was sent off to the States and Germany. I am sorry, Carl, but I have been so busy lately I am not sure whether I am coming or going. I am Lederberg's assistant for the duration of his visit and his time is so broken that I find myself responsible for the experimental work, and I am sick of recombination experiments! We have been recombining flu strains and using serum and inhibitors to select for genetic markers, etc., etc. etc., so far without much success. He is an extremely pleasant person who keeps everyone on their toes with his questions. Physically he is very rotund and he surreptitiously devours sweets all day. Actually, he reminds me of Martin Luther.

We have received all the blood specimens sent so far. The Yar sera which were left in the icebox were very haemolysed, but I managed to get enough cells for Simmons to test. There was a bit of a mixup about the Yar specimens. The first group of Yar that arrived I thought were from kuru patients and Simmons has grouped them with kuru results in his Lot 13. Will you sort out which are kuru and which Yar, and let him know so he can get his records straight? You know what a fuss-pot he is. I have enclosed a list of sera sent to P.H.D. Melbourne for anti-strep titers. Cyril Curtain took samples of the F series (F1–108) to use as controls for his electrophoresis patterns, and says he will let you have the results in a couple of months.

Kuru has really been in the limelight as you can see by the press cuttings. It looks as though your predictions about the Administration have come true. But for the Russians and "Sputnik," their satellite, kuru would probably have been front-page news. FMB tells me representatives of *Life* and *Time* magazines have been to see him, and that they are sending reporters to Okapa. I hope their articles are a little better than those enclosed.

Down here we are going around with stiff necks from satellite watching. I have seen it twice now; it is really amazing to think that the tiny moving star is a man-made sphere. The papers are full of it and of talk about the Space Age; the toy shops are crowded with space-ships, suits, etc.

Dr. Wood and Ian Mackay are in Sydney attending a BMA conference. I am giving a paper on (guess what!) autoantibodies. . . . [Ian] wants to use some of your results in the definitive paper on "Differing reactivity of antihuman tissue antibody with autologous as compared with homologous antigen," and will be writing to ask you if it is okay. (I will send you the final draft when it is ready). . . .

The Union Theatre is operating again. It has been completely renovated and is now similar to the Little Theater in South Yarra. They are doing banned and what are called pale-pink (Communist-flavor) productions. Their next play is to be "Cat on a Hot Tin Roof" by Tennessee Williams.

How did the Kukukuku patrol work out? I have been awaiting the promised letters, but so far to no avail. I am seriously considering taking the advice offered in a recent hit-parade tune, and I quote: "I'm gonna sit right down and write myself a letter." I guess you have been very busy.

Write and tell me *all* the news.

With love,
Lois . . .

LARKIN to DCG 29 November 1957

Dear Carleton,

Have been meaning to write for days but kept putting it off, and now I have only 30 minutes to go before the mail is collected, so this is going to be short.

The galley proofs of the AICF papers [II and III] arrived last Monday, and were duly read, corrected and returned. The papers will be appearing as page 1 articles in the January 1958 issue of the *Archives* [*of Internal Medicine*]. . . .

Will you do the following things as soon as possible:

1. New Britain map and page for paper—Simmons is getting a bit sick of my stalling him and he would like to get that paper out of the way. He suggests leaving the Orikaiva until you do the papers on the N.G. blood groups.

2. Write out a list of the names of kuru patients in Lot 13 and send list to Simmons.

Saw Cyril Curtain on Thursday and I suggest you reread his last letter, as you seem to have missed the point in his asking for sera from coastal natives.

Nothing very exciting has been happening. Will write in more detail next week. Love to All.

Lois

Notes

Unbracketed notes are direct excisions from the letters; those in square brackets are editorial summaries of deleted material.

1. [The enclosed memorandum from W. W. Watkins, Secretary for Law, Crown Law Office, Port Moresby, dated 20 February 1957, states that the liability of the Administration for Dr. Anderson's safety "would be under the provisions of the Worker's Compensation Ordinance or Acts." It then proceeds to define "worker" and inquire into the terms of Dr. Anderson's association with the Administration should he undertake to do research in the Territory. It concludes: "I should think it highly probable that when the information requested above is available, it will be found that Dr. Anderson or his dependents would not be entitled to claim compensation under the existing legislation. In this event, I would suggest that either an insurance policy be taken out or a special contract be entered into by the Administration to cover the objections raised by Sir Macfarlane Burnet. However, if you will let me have the additional information I shall consider the matter further."]
2. :pneumoencephalography, dye studies, electroencephalography, clinical chemistry which is reliable, endocrine evaluation, etc.—which must all be part of a proper clinical workup of such a case. Specimens either collected here or sent would hardly be adequate for many tests. Moresby could do them no better.
3. Should you be able to think up even a nominal salary or grant for me as long as these studies are underway, or to reimburse my maintenance expenditures while I am undertaking this, please let me know. My address is: c/o Director of Health, Department of Public Health, Port Moresby, Territory of Papua and New Guinea. My current location for a cablegram is the same, or more directly to: Okapa Patrol Post, Kainantu S.D., c/o A.D.O., E.H.D., Territory of New Guinea.
4. Greetings to everyone at the NIH who may remember me, and to the WRAMC crowd. I hope I will be seeing you all later this year. The pneumocystis carinii paper should now be out in *Pediatrics*, much as you saw it last year. The parasite has now been found in Melbourne and in Sydney in infants' lungs—how about the USA? If and when this is finished, I have a brief field-survey of possible child studies in Dutch New Guinea on tap. Shall I send sera to Kuala Lumpur?
5. We could use further laboratory supplies, for I am ready to do all the field laboratory work possible. We really need a new hemocytometer—this one is badly scratched—with exact data on ruling area, etc. for absolute calculations. I could use some heparin or oxalate tubes for drawing blood for sedimentation rates, or exact data as to quantity of these anticoagulants to be used for standard Westergrens. We shall try some with our crude supplies on hand. Some new Pandy's reagent would be invaluable; I have little confidence in what we have, though we are using it both cool and heated. Being essentially a laboratory clinical investigator, I am willing to set up and labor in the field over any laboratory tests which can be done here, when reagents and equipment are on hand. New Fehling's solution (ours is mighty old) and a protocol for the drop, 5 or 10 quantitative CSF sugar test would be advisable. I have forgotten, I am sad to say, whether 1 cc of mixed reagents plus 1,2,3,4,5, etc. drops CSF successively. In Boston I would simply check my notebook.
6. We are awaiting supplies of medicines for treatment, especially such things as cortisone (for a brief trial and a few selected cases), anticonvulsants, tranquilizers, and long-acting barbiturates or other drugs you might suggest. As a shot in the dark, testosterone might be tried, since female predominance is so marked; but I must admit it is a most questionable approach.
7. Among supplies we should very much like to receive are a good neurology text and the Ford pediatric neurology book previously requested. A good laboratory procedure manual would not hurt.

8. Thus, Joe, please let me know whether you can initiate or carry through even a temporary project grant, on the promise of a paper which is already in draft from the work. Do you not agree that such a situation is rare and not to be missed as a possible new clue in nervous system diseases, about which we know so little as yet? From here, without reliable mail or communications, I can do naught. To leave here would be fatal to the project I have fallen into on my way to Dutch New Guinea and Malaya. I shall end up after this in Kuala Lumpur, ready to write and work further.
9. Send full instructions for fixation, for removal, slicing, and packaging, etc.; and send the necessary Customs clearances. Address any mail, telegrams, supplies, or instructions to me as follows: c/o Department of Public Health, Port Moresby, Territory of Papua and New Guinea. Add: at Okapa Patrol Post, via Kainantu.
10. [Included 80% of the April 20 letter to F. M. Burnet.]
11. Again, Joe, thank you for your help and interest. Let me know if anything comes of your note to the NFIP. Greetings to everyone in Washington, especially to Mrs. Smadel, Betsy, Nancy, and the crowd at WRAIR.

    After leaving here, I am still going to Dutch New Guinea, but by that time will have exhausted any bleeding venules (I am carrying with me about 150 sent months ago from WRAIR). Should there be any real interest in specimens from Dutch New Guinea groups, send venules to me care of Public Health in Hollandia, promptly. Otherwise, I shall restrict my work to field-surveys for possible future work on child growth and development.
12. [handwritten at top by J.T.G.: "Dr. Scragg to see and return for my reply."]
13. [marginal note by R.F.R.S.: "[illegible] not so. Lufa and Mendi, not Okapa."]
14. [marginal note by R.F.R.S.: "NO"]
15. [marginal note by R.F.R.S.: "?Jan. 7"]
16. BURNET to DCG, 24 April 1957. [Enclosure referred to has been lost.]

    I discussed this letter with Mackay and Dr. Wood before sending off the answer enclosed. I feel that under the circumstances there was hardly anything else we could do. I hope it is agreeable to you.

    We hope to have a chance to see the film of kuru patients soon. A letter from Dr. Scragg suggests we ask Kodak for a preview before it returns to New Guinea.

    Sera from Kala azar cases arrived from India today in fair condition. Lois will get on to them next week.
17. [22 April 1957 note from Smadel, enclosing recommendations (prepared by Dr. J. Godwin Greenfield, then visiting investigator at NINDB) for the removal and fixation of the brain under field conditions. Also similar instructions (forwarded May 1 by Smadel) prepared by Dr. Pearce Bailey, NINDB director. In mid-May, Smadel forwarded a summary of literature on Wilson's disease (hepatolenticular degeneration), prepared for Bailey by Dr. W. King Engel.]
18. Zenker's-fixed tissue—in Zenker's with 5% glacial acetic acid added for 36 hours—washed with one-hourly changes for 36 hours, placed in 50% ethanol for 24 hours, and then transferred to 70% ethanol, in which they were shipped (no iodine added yet to remove the excess Zenker's). Bottle No. 1 contains: ovary, pancreas, meninges, and heart. Bottle No. 2 contains: heart, aortic wall and valves, kidney, adrenal, and ovary. Bottle No. 3 contains: kidney, lung, spleen, and liver.
19. A few letters have reached me to let me know that the AICF-test paper is now out in *Nature* and that my pneumocystis paper is out in *Pediatrics*. The series of definitive papers on AICF reactions will probably appear in the *Archives of Internal Medicine*, an unfortunate result of coupling the experimental paper with a follow-up clinical coordination which would not fit well into experimental journals. However, I think our work is good and will stand up.
20. 1) On page 5, I took out two brackets and put a comma between Kokominjan and Yir-Yoront—does this express it correctly? I presume the names are synonymous!

    2) On page 6 we say *koko* meaning "speech"; does this mean "language spoken," "type of speech," or just "speech"?

    3) On page 8 we say *wik* meaning "speech." The same questions apply.

    4) On page 9 we say *wik-waiya* meaning "speech bad," and this seems fair enough! Is it necessary to alter or be more descriptive than simply using "speech" on page 6 and 8, or is "speech" in each instance sufficient once correct?

21. BURNET to DCG, 27 May 1957:
    No doubt you will have had a letter from Dr. Rivers. This is just to let you know that I have the cheque for $1000.00 from the National Foundation and that we will hold it until we have your instructions on how to act.
    I sent you a letter only a day or two ago, so that I do not think there is anything further for me to add.
22. Should you be able to get an airmail shipment of venules off to me—especially those adequate for trace-metal studies on their contents, i.e., chemically cleaned interiors—I shall do all in my power to get blood, tissue, etc. off to you for file. At the moment, liver and renal function tests, serology, etc.—rather crude, but the best we can get—are being sent off to Moresby and Melbourne.
23. We are going to try to dig up our good 8mm cine film of kuru, which we loaned to visitors and never got back; and I want to send it to you for duplication. I am told by those who have seen it that it is a really precious document. We were visited by Dr. William Smythe, from Rabaul, who has been shooting more 8mm film for us. I am thus taking the liberty of sending you four 25-foot rolls of 8mm cine Kodachrome, undeveloped—three taken here in Moke and one en route here by Dr. Smythe. Three have a good deal of kuru in them, which should be easily recognized. Danger: one *longlong* (village idiot) doing a foolish dance is also on the last of the four films, and is not to be mistaken for kuru. If you can get at least one (better two) copies of these made, and the originals eventually returned by airmail (to Dr. Zigas, Medical Officer, Kainantu, E.H., Territory of New Guinea), we shall have had the complete use of this cine photograph which is irreplaceable if it comes out okay. In addition, I should like to have your permission to carry out the same scheme with future cine film by visitors. Furthermore, if any good 16mm equipment can be dug up (as I wildly suggested in my last letter) I should still be most interested, for there is not much better a thing to film in cine in the modern medical world than kuru.
24. Trace-metal studies in the chemistry lab of the Department of Agriculture in Port Moresby gave slightly higher serum coppers—but not dramatically so, and not even higher than some controls.
25. A second kuru patient has died, and we managed to get a postmortem examination. Again the brain and viscera were grossly normal. Spleen rather small and shrunken. We have taken extensive organ specimens into 10% formaldehyde and also into the Zenker's–acetic acid which you have furnished to us. These we shall keep here for a while—washing, etc.—and eventually send specimens and brain off. Would you be willing to again receive portions of various tissues for further histopathology, as we have already without warning sent to you such specimens from our first autopsy?
26. As far as trace-metal possibilities: copper is one which Wilson's disease brings to mind, and mercury is listed as another which causes basal gangliar–type tremors in some forms of chronic poisoning. Can your associates suggest any other leads, for it is a bit unbelievable to find one to two percent—in some population nuclei up to five or ten percent—of a population sick at the current time with this astonishng, rapidly progressive, fatal disease. Cortisone therapy did *nothing*. Well-being was enhanced, perhaps, but tremors, ataxia, and incoordination were unchanged. BAL therapy is now instituted using "Injection of B.A.L.; B.P. oily injection of Dimercaprol" prepared by Boots Pure Drug Co. Ltd., Nottingham, England.
27. The complete autopsy materials, formalin-fixed brain (whole) with formol-fixed and Zenker's-fixed tissues, will be off to you shortly. Any histochemistry for heavy-metal deposits, etc. would be much appreciated. Furthermore, control specimens of the formol-saline used in fixing tissues are sent for blank analyses.
28. Tables summarize only first two days of BAL treatment. Patients are continued on the same dose for another three days or so. Dose is high, since patients are mostly thin and frail—and, you will note, most are small children!
29. I have received, in addition to the fine abstracted summaries of dose schedules which Gray brought, a complete review of current methods of treating Wilson's disease with such other methods (in addition to BAL) as chelating agents: penicillamine, calcium versenate, oral potassium sulfide, etc., prepared by Pearce Bailey of the U.S. Department of Health. Thus, we used probably a better-controlled BAL regimen, and previous workers have been able to use a bit more of it.

30. We sent a set of chemically clean trace-metal–free containers of bottled CSFs to Moresby for trace metals. I trust you will get aliquots for mercury, lead, manganese, etc. studies.
31. I ran out of paper on an aerogramme, so I continue my last letter on a new sheet. In a note from Tom Rivers, I learn that he has sent a check for $1000.00 U.S. made out to me from the National Foundation for Infantile Paralysis (which is an extension grant they have given me) to you. Will you please have your secretary post it to me registered mail c/o Dr. Vin Zigas, Kainantu by airmail? I can manage to handle it here. This is not to be confused with the $1000.00 additional grant to the Hall Institute to cover some of the expenses of my last year of work with you. I never learned whether Hughesi had arranged fully for it or not.
32. Dr. Price has done masterly work in getting together with the agricultural chemists—blood-copper determinations done for us. Nothing remarkable has turned up as yet.
33. Please greet any of my friends you may meet in Geneva. I am still planning to stop in Dutch New Guinea, Borneo, and Malaya and to then cross the Middle East to Europe; and I still hope to be there by autumn and back in the U.S. before winter.

Please ask the new librarian to send on reprints of the *Pediatrics* and *Nature* papers to me when they arrive, and inform her that a paper on Cape York aborigines has been submitted to the *American Journal of Physical Anthropology*. If the Institute wants copies thereof (Simmons, Graydon, and I have written it; and I am listed as of Hall Institute, where I did the work) and to include it in its bibliography, I shall be happy. Simmons can make any necessary arrangements. The New Britain–New Guinea survey paper will shortly be off also.

I have not yet fully digested paper II of the AICF series, but brief study leaves me quite satisfied. I hope we meet with no further editorial difficulties. Please ask Ian [Mackay] to send on to me any further editorial communications promptly. I should like to be able to look over page or galley proofs of the first and second papers, but there is no need to hold things up for this. If such proofs can be sent to me, any real errors I might discover could still be handled by cable.

A third paper, based on the reams of yet-unassimilated data we accumulated, is being written now; and I shall submit it to you once an adequate draft is ready.
34. We have our hands full with 24 cases on Dilantin, Tridione, and cortisone—various groups at the moment—and we are waiting to see whether intensive BAL treatment of ten has done any good or not.

Please keep me informed of the receipt of any of the pathological specimens, etc. Did the first set of fixed tissues (the brain of which went to Melbourne) ever reach you, and did you get the BAL urines?
35. WOOD to DCG, 14 June 1957:

Many thanks for your letter. Sir Mac left for England yesterday before your letter arrived, so I am answering it on his behalf. We have the various things under way and I shall answer it in detail next week. Many of the questions will be answered directly to you by Anderson.

Enclosed in this letter is the check for $1000.00 which we are sending as requested by you in your letter of 31 May.

All the very best of good wishes to you, and again very many thanks for all your news.

P.S. Mr. Hughes has received the $1000.00 grant towards your stay here, thank you very much.
36. Are all those I reported as Lot 1 and Lot 2 kuru cases? This also applies to Lot 3, which report today will be Lot 4.
37. I think, Joe, that this cine material, if it is okay, may be one of the most valuable documents of the entire work. Thus, I burden you with it. Enclosed with this letter is a black-and-white 35mm film No. 1. With the cine films come black-and-white 35mm films No. 2 and 3; these are already developed and form an important part of kuru documentation. I have several hundred Kodachromes on file already; and I shall ask my brother to bring you a set he has, which I sent home to have developed. Please keep them in sequence and in your hands or his. The black-and-whites I would very much appreciate having printed. If some convenient size—say, postcard-size or any standard small enlargement—could be made of each frame on the three films, and filed serially with you in a

Kuru file, I should be most pleased. There is no need to send copies to me; just have a Kuru file established, if you believe it justified, as I certainly do. In addition, certain pictures are obviously of importance and usable already in publications that I hope to start working on. Therefore, I should like to request two copies of enlargements suitable for publication to be made of the special frames designated on one of the enclosed sheets. One enlargement is to be kept in your file, and the other sent to me for selection for the papers; a set of papers is in preparation, and manuscripts will go off to you when they are ready. Something like 8×10, or near it, is perhaps best for the enlargements. One enclosed list documents all frames on the three black-and-white films. Another lists only those desired in duplicate large-sized blowups, one for us here and one for you.

38. I have not yet seen the *Pediatrics* article, but the *Nature* article has come up and I am most satisfied. The correspondence response is very satisfying, but I only wish I could stir up the U.S. editors on the two definitive papers which should be out and are not yet even accepted: *AMA Archives of Internal Medicine*—I almost feel like taking it out of their hands and sending it to *Clinical Investigation* or *Journal of Experimental Medicine*, for I know the work is good.

39. My address remains the Kuru Research Project, Okapa Patrol Post via Kainantu, Eastern Highlands, Territory of New Guinea; and airmail reaches me rather promptly.

40. P.S. Will you please send us a list of those patients on whom you have done hemoglobin electrophoresis. We can thereby determine which additional specimens to direct your way for such study. Thank you.

41. I found nothing to disagree with and was completely satisfied with the paper, and did not read your letter cautiously enough to realize that you awaited further word from me.

Your questions, I admit, required answering, but in surveying them and the paper, nothing startling came to mind; in thinking that all best be left as it is, I failed to state this to you.

To get to the questions you raised about the paper:

1. The comma between Kokominjan and Yir-Yoront in parentheses is fine, as you have arranged it.

2. *Wik* means "speech" (according to two non-linguist anthropologists who have worked there), meaning "language-spoken." However, no real linguistics has ever been done—nor can ever be done in the future, I fear—in this region; and just exactly what shades of meaning it has and its full grammatical form I cannot give nor could we find out, I fear. The English word *speech* is itself probably broader in meaning than the aboriginal word, and as such should lead to no confusion.

3. *Koko* is synonymous with *wik*, I was told; and thus whatever we use for *wik* we should use for *koko*. I believe "speech" is sufficient.

4. *Wik-waiya* meaning "speech bad" is okay too, I believe. Even in English, as in all Indo-European tongues, the word *speech* can be used for an individual's vocalization or for the tongue of a people; and I have yet to find the primitive people who lack this equivalent. For our Cape York aborigines, I do not know for sure; but I should be mighty surprised if what worked for "speech bad" would not apply to "speech" which is "language spoken." Here in our six different linguistic groups, the same would apply in New Guinea!

42. Please excuse our cable requests for a typewriter. We are using one which does not belong to the P.H.D., and which we have borrowed and must return. Furthermore, manila folders, filing cards (several thousand), and a cabinet or boxes to contain them are essential if we are to preserve the extensive epidemiological records from chaos. The latest cable from Moresby tells us that these supplies are coming, and we are therefore most grateful. As for the typewriter, we shall try the District Commissioner at Lae, as suggested.

43. I hope you will not mind my boldness in summarizing herewith the supplies and drugs which we see no chance of getting through usual channels of order and supply: 1. BAL (oily injection of dimercaprol), ten to twenty boxes of twelve 2.0 ml ampules each; 2. testosterone propionate, either as methyltestosterone or as the propionate; 3. any of the modern tranquilizers, none of which we have been able to get our hands on for a trial in severe kuru; 4. Dilantin-sodium, enough for a trial of one to two weeks' full-dosage medication on some ten patients.

44. [Summaries of six letters concerning the publication of the first kuru paper:

1. A covering letter (17 June 1957) for the manuscript to the Editor of *Science* from DCG, with a listing of his credentials. It includes the following description of the work:

   The scope of our work is ever broadening and the implications ever widening. Thus, it is now evident that the definitive preparation of our data for publication will be greatly delayed—for at least many months, perhaps a year—and that a large series of highly technical papers will be presented in the medical literature. In view of this, my collaborator and I are submitting to your journal this first report of this most astounding new illness in the hope that it may be published at the earliest possible date and gain thereby the attention of geneticists, biologists, ecologists, chemists, toxicologists, nutritionists, and anthropologists—in addition to the medical audience which will read our later publications—for the problem of kuru is a wide one which should interest many allied basic scientists. Furthermore, it demands immediate attention from those diverse disciplines who are qualified to contribute to its investigation.

2. A letter (1 July 1957) to Dr. Smadel from Dr. Graham DuShane, editor of *Science*, stating

   It is possible that *Science* might be a suitable place for the announcement of a new disease of this kind, but I should appreciate your advice on this point. I should also like to know whether the work seems to you to be sound, and whether the National Institutes of Health is giving financial support to this work.

3. and 4. Letters exchanged between DuShane and Smadel's assistant Dr. Carl G. Baker, pursuant to a telephone conversation in which it was decided that the paper would be more appropriately published in a medical journal; the letters merely recorded Smadel's intention of submitting it to the *NEJM*.

5. Smadel's cover letter to the *NEJM*, in which he stated:

   I have talked with Dr. DuShane, who agrees the article is more suitable for a medical journal. . . . Dr. Gajdusek is a brilliant clinical research man who, with support from the National Foundation for Infantile Paralysis, has spent the past two years with Sir Macfarlane Burnet in Australia. On his way back to the States, he planned to obtain a few blood samples in New Guinea for work on viruses upon his return. When he learned of kuru there, he became so interested in the disease he decided to stay and attempt to learn more of the disease and help those suffering with the condition.

6. A letter (7 July 1957) from Dr. Baker to DCG informing him that the paper had been withdrawn from *Science* and submitted to *NEJM*.

45. The enclosed letter from Qantas, with the reply I have just cabled them, will let you know what has happened to the photographic supplies we urgently require and most anxiously await. Since I suspect that they were not dispatched with a $250-odd shipping charge waiting for me on this end, I suspect Qantas is, as usual, in error. Especially since the venules arrived okay with the U.S. Government address-and-information label obscured by various other "Collect" labels which could easily be misinterpreted. However, to make sure Qantas does not take a month or more straightening out the mess while they sit on the camera, etc., could you ask the initial carrier in the U.S. (probably PAA) to cable Qantas-Lae some clearing order, just in case things get really snarled up at Lae—as they are apt to, unless, of course, they have already heard from Qantas by the time you get this.

46. Many new developments are on hand, and I shall shortly try to prepare a new review for you. However, we beg you to push for early publication of our first report in *NEJM*.

47. SMADEL to DCG [telegram, no date]   Carleton: The pathologists have examined the first brain and found marked degenerative changes in the cerebellum (Purkinje cells and one of the adjacent granular layers); also in the basal nucleus. They are most interested to see whether the other brains, which are being processed, will show the same changes! Joe.

48. SIMMONS to DCG, 5 August 1957:

   I had yours of June 30 and July 4. You would have had a letter from me dated 19-7-57 and many blood containers of two different types.

   I will do what tests I can for you; and naturally I am interested in the Kukukuku group, although I expect their gene frequencies to be mainly no different from others in New Guinea.

   I mailed off the Cape York paper to Washburn [editor of *AJPA*], and he advised me by return mail that it had already been sent off. I have three papers in *AJPA*; they will appear in September and December and the Cape York may not appear until March but December is possible.

I have sent all blood samples to Curtain, and have sorted out the name confusion with him in the earlier lots.

Enclosed are the results of various lots to date. In the latter lots I have added tests for Wr^a (Wright) and Di^a (Diego) groups.

As with the earlier lots, I want a statement from you as to which are kuru and which are not kuru. I would also like information on the group said by Dr. Anderson to be children. My records must be kept in order for later sorting.

49. [Smadel to Garland, editor of *NEJM*, regarding the addition of photographs to the first kuru paper. Dated 6 August 1957, the letter included the following:]

    Dr. Gajdusek is anxious to add to the manuscript several photographs of patients with the classical attitudes assumed in the newly recognized kuru. Prints of pictures which he took were sent to him several weeks ago, and I assume that he will choose the photographs which he thinks best suited and return them to me within the next week or ten days. Should this not come to pass I shall select several photographs from the group suggested by Dr. Gajdusek which are among the materials that we have here in Bethesda. I trust that you will be patient in your dealings with our colleague who is impeded in correspondence because of his remote location in the New Guinea jungles.

50. The first six reels of 8mm film are here, developed, and I am awaiting a projector and battery with transformer which is being brought in for our viewing of the films. Thank you. Keep the originals and second print on file at NIH. The camera and film you have sent is still at Lae, but I suspect from your last telegram that Qantas will now dispatch it to us promptly. They have been trying to collect—and succeeding in collecting, too—postage and freight rates from me, the Department of Health, and everyone they can; and they completely ignore my explanations that it is all handled by U.S. Government bill-of-lading. I hope you will get their main office to set a firecracker under the Lae office.

51. Enclosed also are a few venules of clotted blood which should be adequate for whole-blood trace-metal studies, with particular reference to manganese, copper, etc. Also, there are three CSF specimens from kuru patients for similar studies.

52. Also thank everyone involved in securing the 16mm camera and equipment for us. Today or tomorrow I may be able to get a look at the 8mm films you have returned to us. There are six reels here. In addition, there should be some later additional reels now in your hands. I hope they have arrived.

53. If your photo labs can handle any of the printing I have requested, we shall be most grateful. I am getting a map off to the *NEJM* for the paper they have—in case they decide they might like some cartographic localization of this entity—and I leave it to you to select from the black-and-white negatives now in your hands any photos of kuru strabismus, flexed posture, or facies to submit to them. Copies sent to me will immediately be captioned and captions returned to the journal. If, however, it is too late to illustrate this first report, well and good. We shall use illustrations for the second paper off to the Australian journal [*MJA*]. I do not want to hold up the *NEJM* article at any cost! Detailed epidemiological cartography is our current problem, and I hope to have that handled and discussed in our next report.

54. SMADEL to DCG, 13 August 1957 [telegram]   Prints of strips one and two airmailed you 26 July. No other still film received. You select and forward prints to Australian journal. Identify for me prints for American article. I will forward these.

55. We have received also $1000.00 worth of camera supplies and film as a result of mention to Joe Smadel that we would like to have followed our cases by 16mm cine. A Bolex H-16 camera, 800 feet of Kodachrome and 100 of black-and-white film, lenses, tripod, filters, and exposure-meter are here; and we will only have the excuse of our own technical inadequacy if we don't turn out some pretty good films, for the material is plentiful and excellent. We hope to view tonight or tomorrow some 10 rolls of 8mm kuru cine films we have taken.

56. Enclosed: copies of recent letters to Dr. Price (P.H.D., Moresby) and Dr. Michael M. Wilson (Public Health Laboratories, Melbourne)

57. Controls thus far supplied are: 1. Haneo, F/adult with congenital tremor, a differential

diagnosis problem when studying kuru; 2. Tarato, *Luluai* from Okasa, from where many kurus come; 3. Tove, M/17 from Okasa. Others will come with the further shipments of kuru bloods. This should be ample for a start. We have no other kurus in the ward at the moment. Specimens are sent for trace metals on the following: [list of 14 kuru patients from whom specimens were being shipped].

58. In the box containing the first two reels of 16mm cine film is a developed reel of Kodachrome 8mm cine film. This was our first kuru film, which I took on a borrowed 8mm Kodak "box Brownie–type" cine camera. It is far better than the subsequent six reels of 8mm cine film our guest Smythe took and which you have developed and sent copies to us. It shows kuru well, and I urge you to study it if you have the time. An accompanying sheet summarizes the contents of the 8mm cine films you have received. The current developed film, the first, is given the number 0. Then, I have renumbered in proper sequence the first six 8mm films you had developed and copied; and I beg that you have whoever is looking after them renumber them as indicated on the accompanying sheet, which is a brief "contents" table to be used later in editing an 8mm kuru film. We obviously have enough (in spite of the two initial very poor "travelogue" reels that Smythe apparently included) to make a really good presentation of kuru on 8mm already. Since the date of this new first reel (number 0) was early April, we have in the series pictures of the progression of the disease in individual patients. Many pictured in the first reel have already died, or are now terminal cases with less than a month of life expected.

Important note: In addition to telegraphing us information about whether our first two reels of 16mm are a success (and if not, what to do), please advise us as to whether we should shift from standard silent 16 frames-per-second to 24 frames-per-second, which is required, I believe, if a soundtrack is to be attached. Since it might be well to plan on a later soundtrack (we can get tape recordings of the foolish laughs, dysarthria, echolalia, etc.— along with plenty of background sounds, to boot), I have wondered whether we might not use 24 rather than 16, and plan definitively on sound-projection later. If so, I would like concurrence from you in NIH and we shall promptly switch all our shooting from 16 to 24 frames-per-second.

59. Also in the box with the 8mm film 0 and the 16mm films I and II are five developed rolls of 35mm Leica still film in black-and-white. These contain some of our most valuable kuru documentation, and illustrations of publications will come largely from these. They are unreplaceable films, for they show stages in patients pictured earlier and subsequently, and they show many patients in their native village settings as we discovered them on arduous patrols. Please, Joe, can we beg to have these five rolls printed carefully into prints approaching postcard size, for file copies from which to select publication frames. If such a set of prints were sent to us here, as we are hoping will be the case with the three previous rolls of black-and-white still film I sent to you, we can immediately select appropriate frames for the two publications in press. If you will take the responsibility for so doing for the *NEJM* paper and send us copies of what you select, we can check what you have selected for accuracy and promptly slip in captions. Summary sheets summarize the contents of these five rolls of black-and-white still film (35mm), which are films numbered 4–8, now in your hands. Films numbered 1–3 have been shipped much earlier.

60. [Contents of a letter (21 August 1957) from DCG to Simmons:

1) summary sheets of the series of "schoolchildren" bloods (numbering 101) previously forwarded to Simmons. The summary includes mention of those children who are not "pure-bloods" of the linguistic group to which they have been assigned and those who are not local to the kuru area. Various other information regarding these and several previous series of bloods is provided to help Simmons get his records straightened out and to give him more background on the populations bled.

2) the news that the first two reports on kuru were both in press and that a more definitive paper was already off to *Klinische Wochenschrift*.

3) a suggestion that certain of the sera from the schoolchildren series be provided to Curtain for his serum protein and abnormal hemoglobin studies.

4) arrangements for the handling of planned future blood collections.

5) the news that the two definitive AICF papers were to appear in *AMA Archives of Internal Medicine* in December and January.

6) plans for the map to be drawn up by Lois Larkin for the New Britain blood-group genetic survey paper.]
61. [The enclosure, entitled "Cycas Nuts and Amyotrophic Lateral Sclerosis," is an excerpt from Whiting's July 1955 pre-doctoral fellowship application to the NIH. It discusses the use of treated cycas nuts for food on Guam, and the possibility that the toxic glucosides that had been found in two species of cycad might be a causal factor in the high incidence of ALS on Guam.]
62. Since the first large shipment of blood specimens for trace metals, the following additional have now been sent to you: [5 kuru patients and 3 controls listed]. CSF on the above-listed kuru patients for serology and chemistries has also been sent, along with venules of blood for the kuru-serum file. Please make certain that Roy Simmons gets the clots on all of these for his genetic work and abnormal hemoglobin studies.

Sorry, Gray, to be so rushed, but mail is standing by. I shall soon send off the manuscript of the paper which the *Medical Journal of Australia* has accepted without change for the preliminary report of kuru. We are trying to get 8 photos to them in time for the article, for they have agreed to publish these also. There is another preliminary report off to press, and shortly I shall mail you a copy of the long, rather more detailed report we plan to publish in the European literature.
63. Enclosed are protocols of the last three films; copies were also sent with the films, and in previous letters the earlier films were abstracted.
64. If the National Institute of Mental Health and Blindness [sic] can see to agreeing with us, from the material in the eight first rolls now en route to you and the next four coming, please tell them to rush to us further film—particularly Kodachrome, since light here is usually too bright for Tri-X with our available filters—and we shall do our best to improve and correct anything that they suggest we do. We have a great deal more to show and to study on film!

An extensive patrol into Kukukuku territory is to start in early-to-mid-September (probably on Sept. 10–15); and if any film can get to us by then we shall be most happy, for there may be much worthy of medical documentation on that trip.

If further Tri-X film is sent, would you ask the NIMHB [sic] group to also send us the appropriate gelatin filters, as suggested on Tri-X film instruction sheets. These are: Kodak ND 3, Kodak Wratten Filter X2, Kodak Wratten Filter G, Kodak Wratten FilterA.
65. [list of name corrections and clarifications of sample labeling]
66. Blood specimens we sent back to you and the strep culture tubes did not arrive until after we had left this region.
67. and two [Government census] books and two *Luluais* since different groups went to Mani (Oriei) and to Misapi to be "first censused" in 1956.
68. *Asunauveri* is the ground name for the region just being opened and settled by the group. They call their place *Uarevana* as well (but further information leads me to question this now).
69. They are Te'hei-speaking; and *Te'hei*—which in the language means "man"—is what they call their language, in spite of the Fore calling them *Tavia*. They list a number of settlements south of here, related to them and which also speak unchanged Te'hei. These are, so far as I now can make out: So'o, Wa'u'wa, Hoi'uru'a'a, and Yarevana.
70. They do use the name *Tehei* (which also means "man") for their language. However, as a group they identify themselves by the term Yar (or Iar); and they are thus the Yar people.
71. [contents of the book:] Village Constable Urahau (wife Senau).
Village: Iari [Yar] people.
Village: Sojotu.
District: Gulf.
Native District Purari River and Subu River.
Issued 15/11/54 by A.D.O. A. D. Allen.
V.C. Urahau: reported at Beara this day that shortly after his return to his village (possibly in November 1954) the Bore'e people came to his village and killed a woman named Ta'ati'a and a man named Hari.
Names of killers not known.
C. Healy A.D.O.

29/4/55.
Eposi of Sojotu given Councillor's medal this day.
C. Healy.
29/4/55 Paid to 30/6/55.
C. Healy.

72. The Yani River he calls *Wi'iri*, and prefers to call this current village site *Wi'iri* to other names.
73. Census attempt done at Wi'iri (*Wi'ir* is the name of the Yani River in Yar language), this settlement of Yar people (Te'hei people) who have just left their previous permanent settlement of Waibibi (Sojotu) across the Lamari from the Lamari-Yani junction to this sago-rich site on the eastern shore (with two bridges and some extension onto the western shore) of the Yani River, about one day's walk above the Lamari-Yani junction. It is uncertain that these people will settle here (where they are now collecting plenty of sago) permanently, although they have now started gardens but not permanent houses. They mention the possibility of returning to Waibibi (Sojotu)—but claim they left there because of five recent deaths (in the past four months) attributed to *shutim nil,* and intended deserting the old village permanently at the time they left because of these illnesses.

The current site is ground known to the Fore as *Asunauveri,* and to these Yar (Te'hei-speaking or Yar-speaking) people as *Mu'i* according to some and *Wi'ir* according to others. They usually refer to themselves as Yar—the Fore call this Yarevana—and although some insist on using the word in their language for "man" (i.e., *te'hei*) as the name proper of their language, their village constable, Urahau, refutes this and calls them only *Yar* people speaking the *Yar* language. In the book issued to them from the Papuan government patrol post of Beara in 1954 when Urahau visited there, they are listed as *Iari* people. They list as speaking the same languages the following groups situated along the river south of here, below the Yani-Lamari junction and already down on the navigable portion of the river, the following villages: So'o, Waabowa, Hoiuru'a'a (Ho'urua), Maiara (Beara). They list a place called Soiyanadudau as a previous location of their settlement, south of here and now deserted.

74. No Europeans or Government Patrols are known to have visited these people recently. Most of the older men have worked on the Papuan Coast, usually at Port Moresby, Kerema, or Kikori. Their book was issued from the Gulf District, Beara Patrol Post.

So'o, the first settlement south, is said to be a bush-walk of three–four days south, followed by a few days on rafts down the river. In this season of floodwaters, they claim they would not make the trip.

75. At Wi'ir (Asunauveri) they have built two vine suspension-bridges over the raging Yani: two–three thin poles wide for footing and probably at least 200 feet long. These are *not* negotiable to cargo; and Fore cargo-boys fear them, the stream beneath them being very forbidding. The Yar people carried over our cargo, cut down to individual loads, the bridge tolerating only one person at a time.
76. They refer to the language spoken at Beara as *Tau* and claim all those living south of Beara speak Tau—as do many of the Yar people here. These Yar people—the men who have worked on the coast, at least—speak some Motu, but none speak good Pidgin. A few have a smattering of Pidgin and a few words of English from their work on the coast. To get work they always go south via Beara, Beara being the first place where they make contact with the Europeans.
77. Our trek down the easterly route and back the westerly route proves that the route is long and difficult and little traveled, but likewise proves that Kimi and Fore know it and make occasional use of it. Intermarriage with the Kimi and Fore is already an accomplished fact, and genetic interchange is already present. Whether it is present south in So'o we do not know, but these Yar people marry with those from So'o; and the wife of one man who died only a few months ago has already remarried into So'o.
78. Waibibi, across the Lamari from its junction with the Yani, lies just south and below the Moraei hamlets. There is a track or a rough route to Moraei, but the two Moraei Kukukukus we have with us on this trip (Waiajeke and Agurio, boys of 14 and 15 years) claim to have no knowledge of this Yar settlement. Neither do the Agakamatasa people, north of Moraei and west of the Lamari, although these Agakamatasa Fore do have

contact—including marriage contact—with the Moraei Kukukuku. Any trip to these people from the north should be based on the erecting of bush camps for one week (7 days) with food and supplies accordingly. Sago, in limited amounts, is the only food one can expect from the Yar people. Without a guide and track-cutters, the route—either one—is impossible.

79. The route is direct; supplies can wait for us at Uvai if we have not yet arrived there. Since Sinoko wants to make this trip, he might come with the supplies; and we shall send at least one of the *dokta bois* we have with us back to you. [Lists of the required trade items, medical and personal supplies follow, along with the reiterated details of the planned itinerary.]

80. Our schedule is still stated as above, and we rather urgently need the supplies at Uvai Rest House Thursday, although we may not leave there until late Friday. Should they be later in getting to us, they need but follow us on the road from Uvai to Mani. I will have so much data to collect that we cannot travel very fast.

81. These we are sending herewith go to Roy Simmons in Commonwealth Serum Laboratories, Melbourne. However, if they went to Hall Institute it would be okay, since Lois will get them to Roy. Thus, send them directly to Roy unless Vin has difficulty getting such a shipment out. Then simply use the old Hall address.

82. Address of Roy Simmons for Vin: Roy Simmons, Blood Grouping Laboratory, Commonwealth Serum Laboratories, Melbourne, Victoria. Contents sterile non-infectious blood specimens of no commercial value but of great importance to medical research. Refrigerate, but do not freeze, at all times.

83. [instructions to Zigas re shipping of exposed film to Smadel at the NIH]

84. I also have your letter of the 21st August, together with five kuru and three control bloods for metal estimations. The male aged 30/Tiarana was named Aganaga, and the male aged 30/Tamogavisa was named Keugeu.

85. If proofs are ready and mailed before mid-October, I should get them in time for editing and prompt return. Mail is only 10–14 days, usually, in reaching here (sometimes even faster, if I am around and not in the bush—and I should be on hand then). Finally, an additional review of kuru—far more extensive than the previous one, and more "up to date"—is off to you along with the next air shipment via Qantas. It is a rather lengthy manuscript which all who are following kuru should read (but which is, obviously, already a full month and one-half out of date), and it replaces all earlier summaries. With it I am sending a stack of carbon-copies of case-records of deceased patients, some of whose names will be recognized from the pathological material in your hands.

All items summarized as sent to me or received from me in Dr. Carl Baker's letter of July 26th are correct.

86. The first two 35mm still films have been printed up, and prints have arrived here. From them I shall select photos for the *MJA*, taking care that none used in the *NEJM* are duplicated in this second article. Thus I have requested number designations of those you sent to *NEJM*. I am having a map, a line map which is rather simple and documents the kuru region, rushed from Melbourne to the *NEJM* for the kuru paper. If this gets there in time, it also will add to the paper. If not, well and good. In any event, I am writing to the journal begging them to send you and to me also, here in New Guinea, copies of the proofs. I shall, if possible, get them back in time.

87. Enclosed is a carbon-copy of a few letters written back to our bases during our last patrol, of most minor importance but perhaps of interest to anyone following kuru blow by blow.

88. [DCG to Smadel, 19 September 1957: captions, for the *Medical Journal of Australia*, for photos referred to by frame-number.]

89. The prints being supplied are the only ones we have on hand. They are not all appropriately enlarged, and none are retouched to remove the blemishes inherent in the use of small negatives. We are writing now to the National Institutes of Health, Bethesda, Maryland, USA, where our negatives are filed and where our photographic work is being done, requesting that appropriate glossy-print enlargements with adequate retouching for publication be made of all 12 of these photographs and forwarded directly to your office. Upon receipt of these finished, retouched glossy prints, you may proceed with publication

of the article. The accompanying sheet provides captions for each of the photographs. You may number them and arrange them as you see fit.

Four separate photographs all labelled "C" have been supplied to you. All four are of the same group of 8 kuru patients. Obviously, only one of these is to be used. We are leaving the selection of which of the four is to be used to the photographic department of the NIH, where such selection will be made upon the basis of which of the four enlarges and retouches to a better finished print for publication. When you receive the retouched enlarged prints of these 12 photographs from the USA, will you please return to us the prints we are sending you herewith?

90. [DCG to Ingalls, Associate Editor of the *NEJM*: captions for the illustrations supplied for the *NEJM* paper.]

91. In the pack of pathological materials is one trace-metal–free brain, packed in plastic, which has never even come in contact with the cotton protecting it and has been only in trace-metal–free formol–saline in plastic containers since removed. This should be used for pathology and for all trace-metal studies. It is from Tasiko, a girl of about 6 who failed to respond to BAL therapy. Crude liver extract and BAL were the only treatment she received. Dozens of case-records are in the paper pack enclosed with the shipment.

92. This, together with the 16mm stuff we have shot (and hope to continue with) will give, I hope, a rather complete picture. Please let us know if the 16mm film we have taken is useable and adequate. I have asked for more film in my last letters to you. Any that arrives will be used with caution and care. It can never be wasted documenting late and terminal stages of the disease in patients studied on film in the earlier stages a few months ago.

93. Please excuse again the poor carbon-copy, but I cannot easily retype the four full pages. You or Gray may find it worth deciphering.

94. Enclosed is a list of the "contents" of the last (i.e., the 12th) cine film sent to you, and also a list of the additional four 35mm black-and-white still films for postcard-size printing for the Kuru files. These are undeveloped films not in cassettes; and thus please warn whoever receives it not to open the film boxes until in a darkroom, for if they do so they will ruin the films. The last cine film has more background material than kuru, but it is important background stuff. With 3400 feet of professional 35mm film being shot on kuru clinically—a good 16mm print of which I shall eventually get off to you once I edit it in Port Moresby—we thought we could use a roll or two on epidemiological and ethnological background material which the 35mm professional film will have none of. Since all the 35mm was shot here at Moke (Okapa) in a few days, using a group of two dozen patients I collected for the purpose, it has no background ethnology, no epidemiologically or ecologically significant shots and—what is most serious—it cannot show the progression of the illness in a single patient. It is by following up patients whom we have previously photographed that we can get this "horizontal" documentation which we are now after; thus I hope more film will be forthcoming. I went and used my own funds to buy a few more reels of cine Kodachrome which I shall shoot on the Kukukuku-Papuan trip and send back to you. It is Australian Kodak with full developing-charges prepaid, which is no longer permitted in the U.S. Thus, it should be developed for you in the U.S. without charging for the developing. Please pass this word on to whomever handles the film, for Kodak anywhere in the world is supposed to recognize this prepayment of developing-charges which they have already collected here in Australia–New Guinea.

In addition I am sending sundry other carbon-copies for the Kuru file. Among them is some identification of spiders eaten mostly by women and children (the usual kuru victims).

95. [instructions regarding handling of enclosed film for shipping to U.S., and a list of supplies to be shipped to Moraei]

96. [The "items": the return of the cine camera and tripod to Okapa; the shipping of exposed film to Smadel at the NIH; the handling of field-notes and case-records that were being sent back with the letter; arrangements to have copies of the field-journal and other notes sent to DCG's mother for safekeeping in Yonkers, N.Y.; and a summary of their plans for the trip south and eventually to the upcoming Kuru Conference in Port Moresby. A list of cine film exposed on the Kukukuku patrol also accompanied this letter.]

97. [a further reminder about shipping off exposed film to the NIH]

98. SIMMONS to DCG, 14 October 1957:

Your letter of 22-9-57. Here are reports on Lots 12–16. I hope your Kukukuku patrol has proved a success and that difficulties were not excessive—I know that the going would be hard.

As the sample cartons arrived, I have handed them to Lois to distribute the portions, as I haven't the time or the staff to handle it. I have suggested to Lois that we would include her in our report as a coauthor, as she has done so much for each of us. Do you agree?

In your letter of 22-9-57 you spoke of 20-odd Yar samples, the result of one week's hard walking. I have not received them in any cartons sent to me.

I do not deserve any repayment for material bought, as I used Wenner-Gren Foundation funds which I have for my research work.

Please confirm that Lot 13 are or are not all kuru. I somehow feel that, with hyphenated tribes etc., the final sorting of my results into tabular form would best be done by you.

99. Enclosed are some additional tags to be used in conjunction with your sending plant material to us, and a copy of the cablegram which I hope you already have received.

100. If, on the other hand, something has gone wrong and my request has gone astray, then select eight of the best pictures—at least five or six of patients and one or two of hospital, background, etc.—and send them instead, since here in Moresby I lack the prints to look through again for choice or the letter-files to find my previous choice. You must have received the original request, however, and I suspect delay is just in photo-processing or in mail. If mail is the matter and there is any chance of loss, can new prints be rushed to Sydney and the editor informed by cable that they are coming and that others are en route (or perhaps lost in transit).

101. Reprints? I did not order from *NEJM* in time. Did your group order enough for Vin and me to get a small stack each? If not, my order sent in today might arrive on time. I hope so. Will NIH order reprints of the Australian and German papers?

102. and have been; and the NFIP grant to me was for only $1000.00 in spite of their misstating that it was for $2000.00 in a recent letter to me re reprints of the articles.

103. I again reiterate that I am in full accord with Visiting Investigator status at NIH if it will mean some salary soon upon my return; a year without salary (I have managed to spend much more than the $1000.00 grant from the NFIP) is about enough, for I cannot swing it much longer. I will be back in the U.S. just after Christmas.

104. Please get a copy of the *NEJM* article off to me by airmail, in order that I may see what appears. Also, Joe, please get the glossy prints off to the Australian journal (*Medical Journal of Australia*), for they *must* publish immediately—being hounded as they are by the lay press.

Enclosed is a carbon-copy of my letter of September 19th, in which I asked for the 12 large glossy prints for the *Medical Journal of Australia*. In case loss or misplacement of the letter has contributed to the delay, this new copy—all I have—may help.

Also enclosed is a carbon-copy of the list of 8 developed-but-unprinted negatives I just mailed off to you. These should be printed, one copy each, in postcard size for the Kuru file in the NIH. They have much pertinent anthropological background material.

105. I shall get the cine camera shipped back to the NIH just prior to my departure from the kuru region. Thank you again for all the support; and thank everyone at the NINDB also, for me.

106. [clarification of specimen sources and numbering]

107. [There follow more than two pages of corrected specimen-lists and questions on labeling of laboratory results.]

108. One bottle of fixed visceral specimens is in trace-metal–free formol-saline and thus may be used as such. Zenker's-fixed specimens are in 70% C_2H_5OH as usual. For Atona, we are sending a good long piece of spinal cord and a good section of long bone also fixed in trace-metal–free formaldehyde. Please cable me whether the first trace-metal–free brain which you have already received was adequate for neurohistology as well. Very restricted quantities of trace-metal–free formaldehyde have made fixing in very small volumes of formaldehyde necessary, and I am worried. The long-bone section may be the only one we shall get. Sorry they did not get it more distally or proximally for the epiphyses.

Cable any further suggestions or results which may give us any ideas, for we are winding up the field-program in the early days of December. Also, list your cable addresses in your next letter or cable. I am wasting a fortune in the long address at international cable rates.

109. I have cabled you requesting selection of another dozen good clinical and epidemiological pictures from the Kuru file for the German article. If large glossy prints of these can be sent off to *Klinische Wochenschrift* and small copies to me so that I may prepare captions, I shall be most grateful. The German paper will be the most definitive and elaborate of the three, and I do not plan to work up any more of our data until I get to see you in Washington.

110. A short cable address: Gajdusek, Okapa, Kainantu, New Guinea. What is a short cable address for you?

111. [requests for more details on specimens that had been sent previously, to facilitate sorting and reporting of results]

112. Can you let us know whether cytochemistry, trace-metal studies, etc. are being arranged for and will be done on the few precious brains we have sent to you, and rush any information—positive or negative—about any unusual element which might be encountered or any pertinent negatives (arsenic, beryllium, mercury, manganese, and copper being already things we must know about)? [CSF and blood specimen-lists follow.]

113. I shall send some more pure hydrochloric acid–preserved 24-hour urine collections for amino acid studies. Can you handle these? BAL (Dimercaprol) has now been used on 10 further cases and to no apparent effect. Let us not dismiss trace elements, but think in terms of their role rather than in terms of toxic amounts thereof.

114. I had not seen it until you sent the pages. The *Medical Journal of Australia* article is out and, except for a few misprints and a rather poor quality of photographic reproduction, is also a very satisfactory article. The journal editor did a mighty job in making all the changes and revisions to bring it up to the 154-case level, and to add all the newer findings which I rushed in the last minute. We had to keep it as up-to-date as *Time*, at least. I am satisfied with what they did.

115. Enclosed find a list of most—I hope—specimens sent to you recently. Some are for our genetic hematological study of the population groups hereabout, others for a comparison of kuru-afflicted with kuru-free populations and of kuru patients with healthy subjects from kuru-affected groups. As I said in my last letter, I intend to send only a few further specimens, those designed to complete adequate sampling of those population groups under investigation. Among the South Fore, I shall send a further set of samples from Agakamatasa—which is one of the most isolated and distant of the South Fore groups and fully kuru-affected, although at the kuru border.

If I can get any further Yar, Kukukuku, and Auiana specimens, I certainly shall do so; and I suspect that I can during the coming week or two. The Kimi collection is about adequate, the Keiagana will be amplified somewhat, and the Fore kuru patients will continue to be sent as specimens are available—for serological survey, chemical studies of serum, hemoglobin and red cell studies.

116. This, then, should make our plans clear. You can still reach me here at Okapa by a prompt reply to this letter. I hope the previous two letters—with explanations of the dreadful mistake made here in sending you the empty containers of glucose-citrate, and with documentation of all the specimens you have been studying—are in your hands.

117. If the article is to be published in German translation and if you have already submitted it to your scientific translator, we believe that he will find very few changes necessary and only small additions here and there. Should this be the case—that is, should it already have been translated—I am enclosing, with the new revised and retyped manuscript, a carbon-copy of the old original manuscript submitted to you on which all changes and revisions have been marked. This could be used by the translator in locating all such changes, additions, and revisions, and thus simplify his task. If no translation has yet been made—or if, of course, the article is to appear in its original English version—then this corrected carbon-copy can be destroyed as can your first original manuscript which is now to be replaced by this corrected, up-to-date, and retyped version.

118. Enclosed also are 12 illustrations, many or all of which we hope you can use. Appropriate captions for these are included on an accompanying sheet. I believe that you have already received a copy of the map locating the kuru region, to be used as the first figure in

the paper. This was to have been mailed from Melbourne to you some time ago. If not, another copy is enclosed herewith. I expect that you will have it redrawn with appropriate German spellings, etc. May I point out that Mt. Michael is spelled incorrectly on the map and should be Mt. Michael, not Mt. Micheal?

Finally, I am enclosing a set of Kodachrome slides; appropriate legends for these are also enclosed on an accompanying sheet. It is our hope that you may consider some worthy of color reproduction in the Journal. If not, perhaps you may find it worthwhile to have black-and-white negatives made of some of them and include them as black-and-white pictures. In any event, please preserve all of these slides for me, and return them to me when you have completed your use of them.

119. To editor: These illustrations are the first and only colored illustrations on the subject of kuru that we have submitted for publication. We realize that we are submitting far more photographic material than your journal can handle; however, we hope that you will select as liberally from what we have submitted as your publishing policy will permit. Kuru *is* a new disease to Western medicine, and it is unique and incredible from many points of view. The two preliminary reports listed in the bibliography to this paper are the only publications on this "new" disease in the world literature thus far. For this reason we feel justified in offering you such an extensive photographic record, in the hope that you will be willing to publish much of it. Please preserve and return the Kodachrome slides to us, both those which you may use and those which you cannot use.

120. Shipping with this letter: cine films, still films (all undeveloped, and thus warn everyone not to "look" at the films until both cine and still are developed); two brains and other tissues from good kuru cases: Waruye, m–6 years, in trace-metal–free 5% formol–saline of limited volume and thus perhaps not adequately fixed (Why not fix it further?); Inare, f–30 years in regular pathological 10% formol–saline in adequate volume and probably well-fixed.

Pituitaries in sellas, peripheral nerves, and cord specimens are in the bottles or packed in the plastic sacks with the brains; do not lose them in unpacking, please.

Enclosed are lists of other specimens off in recent shipments. Yes, I am leaving—in spite of repeated delays. When Professor Robson arrives in a week or so, I am off.

121. [future mailing address c/o Smadel at the NIH]

122. The last of the cine-film rolls, undeveloped (developing prepaid) Kodachrome reels Nos. 29 and 30, are now off en route to you; and these I urge be quickly developed. I shall also enclose with them a few further rolls of developed 35mm still black-and-white film for postcard-size printing for the Kuru Picture file.

123. Further shipments of our kuru-serum file and of our file of normal sera from all 12 surrounding linguistic and cultural groups will be packed off to you from Melbourne in due course. Please find cold-storage space for them and welcome use of aliquots to anyone who has any "bright ideas" which might throw some light onto our problem.

Again, Joe, accept my thanks for all your help and encouragement.

Finally, enclosed are a number of reprint requests to Vin and myself. Since the NFIP sent me a note stating that they have sent you their 500 reprints of this article (since they are forced to photostat it for their collected reprints), will you please ask someone in your office to mail these out for us. Put some reprints aside for me, if you will.

One of the requests is a letter, enclosed herewith, from M. L. Littman at Mount Sinai hospital. Please read it. He offers to do mycological histology on pathological materials. I have not yet answered him. Should there be no one as competent at NIH, it might be an excellent idea to send him material which you have on hand.

124. I shall stop briefly at Kuala Lumpur, and then be off again toward the U.S. If, Joe, you can reach me with copies of the *AMA Archives of Internal Medicine* papers by myself on the AICF test (there should be two, one with Ian MacKay and both in the January issue, I was told), I shall be most grateful.

125. Could you, Joe, ask your secretary to send by airmail some five copies of the *NEJM* reprints to Vin Zigas in Kainantu; another five to Lucy Hamilton Reid (Box 88, Lae, New Guinea—she has just married and is now Mrs. Reid); another five to me at Kuala Lumpur (c/o the Traubs); and finally, five to Dr. Scragg, Director of Public Health, Port Moresby, Territory of Papua and New Guinea. They have ordered their own copies, and some few

hundred will be en route by surface mail to Vin and the P.H.D. in Port Moresby, but thus far no copies are available at all. This will take 20 copies in all. 200 copies of the Australian journal reprints will be sent by Zigas to you shortly. Keep a stack aside for me, if you will.

126. Botanical specimens, some for study at Smithsonian and others for transshipment to appropriately designated taxonomic centers in the U.S., will be arriving sometime in the next month addressed to you. I have left Kevin White and Lucy working at Lae getting them ready to send out, and have organized further local collections.

127. From here I am mailing extensive stacks of maps and cartological data, along with records and books from Dutch New Guinea containing data on New Guinea ecology (which is mighty hard to get). Please file all these in the Kuru file for me.

Correspondents—Writers and Recipients
Titles and Locations in 1957

Dr. Terry K. Abbott. Physician, Territory of Papua and New Guinea Public Health Department, Port Moresby; later Director of Medical Services, Territory of Papua and New Guinea.

Dr. S. Gray Anderson. Virologist, Walter and Eliza Hall Institute of Medical Research, Melbourne.

Dr. Pearce Bailey. Director, National Institute of Neurological Diseases and Blindness, National Institutes of Health, Bethesda, Maryland.

Dr. Carl G. Baker. Assistant to the Associate Director [Smadel], National Institutes of Health, Bethesda, Maryland.

Jack Baker. Patrol Officer, Department of Native Affairs, Territory of Papua and New Guinea; assigned to Okapa Patrol Post, Eastern Highlands, in April 1957. Worked in kuru research continuously from 1957 through 1960.

Dr. Jacques Barrau. Botanist, South Pacific Commission, Noumea, New Caledonia.

John Berkin. European Medical Assistant, Territory of Papua and New Guinea Public Health Department, stationed at Kainantu, Eastern Highlands.

Dr. Ronald M. Berndt. Professor of Anthropology, University of Western Australia, Perth.

Sir Frank Macfarlane Burnet. Director, Walter and Eliza Hall Institute of Medical Research, Melbourne. Nobel laureate in Physiology/Medicine, 1960.

Ian Burnet. Patrol Officer at Lufa, Eastern Highlands, Territory of Papua and New Guinea; son of Sir Frank Macfarlane Burnet.

Brigadier Donald M. Cleland. Administrator of the Territory of Papua and New Guinea, Port Moresby, 1953–1967.

Dr. Cyril Curtain. Research Associate, The Thomas Baker, Alice Baker and Eleanor Shaw Medical Research Institute, Alfred Hospital, Melbourne.

Dr. Perk Lee Davis. Physician, Paoli, Pennsylvania.

Dr. John C. Eccles. Professor of Neurophysiology, John Curtin Graduate School of Medicine, Australian National University, Canberra. Nobel laureate in Physiology/Medicine, 1963.

Dr. Joseph Garland. Editor, *New England Journal of Medicine*, Boston.

Dr. J. Godwin Greenfield. Visiting Scientist, National Institute of Neurological Diseases and Blindness, National Institutes of Health, Bethesda, Maryland.

Dr. John Gunther. Assistant Administrator of the Territory of Papua and New Guinea, Port Moresby. First Director of the Territory of Papua and New Guinea Public Health Department, 1946–1957. First Vice-Chancellor of the University of Papua New Guinea.

Lucy Hamilton. Nutritionist, Territory of Papua and New Guinea Public Health Department. Later Mrs. Jack Reid.

Dr. Henry A. Imus. Assistant to the Director, National Institute of Neurological Diseases and Blindness, National Institutes of Health, Bethesda, Maryland.

Dr. Theodore H. Ingalls. Associate Editor, *New England Journal of Medicine*, Boston.

Dr. Igor Klatzo. Neuropathologist, National Institute of Neurological Diseases and Blindness, National Institutes of Health, Bethesda, Maryland.

Klinische Wochenschrift. Springer Verlag, Heidelberg, Germany.

Dr. Leonard Kurland. Epidemiologist, National Institute of Neurological Diseases and Blindness, National Institutes of Health, Bethesda, Maryland.

Lois Larkin. Microbiological Technician, Walter and Eliza Hall Institute of Medical Research, Melbourne.

Dr. Richard Masland. Assistant Director, National Institute of Neurological Diseases and Blindness, National Institutes of Health, Bethesda, Maryland; later, Director, 1958–1968.

Dr. Marion Morris. Assistant Director of Professional Education, National Foundation for Infantile Paralysis, New York.

Dr. A. V. G. Price. Director, Public Health Laboratory, Territory of Papua and New Guinea Public Health Department, Port Moresby.

Dr. A. L. G. Rees. Assistant Chief, Division of Industrial Chemistry, Commonwealth Scientific and Industrial Research Organization, Melbourne.

Dr. Thomas M. Rivers. Director, National Foundation for Infantile Paralysis, New York.

Prof. Graeme Robertson. Professor of Neurology, School of Medicine, University of Melbourne.

Dr. Roy F. R. Scragg. Director, Territory of Papua and New Guinea Public Health Department, Port Moresby, 1957–1962.

Dr. Roy T. Simmons. Head, Blood Grouping Laboratory, Commonwealth Serum Laboratories, Melbourne.

Dr. Joseph E. Smadel. Associate Director, National Institutes of Health, Bethesda, Maryland.

Dr. P. J. Southern. Soils Chemist, Territory of Papua and New Guinea Department of Agriculture, Stock and Fisheries, Port Moresby.

Dr. Michael M. Wilson. Assistant Director, Public Health Laboratory, Department of Bacteriology, University of Melbourne.

Ronald M. Winton. Editor, *Medical Journal of Australia*, Sydney.

Dr. Ian Wood. Chief, Clinical Research Unit, Walter and Eliza Hall Institute of Medical Research, Melbourne.

Dr. Vincent Zigas. Medical Officer, Territory of Papua and New Guinea Public Health Department, stationed at Kainantu, Eastern Highlands.

Indigenous and Local Terms

Assistant District Officer (A.D.O.)—Territorial Administrative official of the Department of Native Affairs, lower in rank than District Officer and District Commissioner. Often assigned to Sub-district Offices such as Kainantu.

bilum—a net bag made of local materials and used as a carry all.

boi—a local male employee, regardless of age. In this book, a distinction has been made in spelling between the boys (ranging in age from about eight to the mid-teens) who were DCG's constant companions and assistants, and *boi* who were more formalized employees such as *kuk boi* (cooks) and *dokta boi* (Native Medical Assistants); however, anglicized forms such as "cargo-boys" (carriers) are also used for those employees.

bush camp—a temporary campsite in uninhabited forest areas between villages.

census—In 1957 most of the areas into which DCG patrolled were still officially restricted. Part of the process of establishing government control was to take a census of each settlement, determining village composition (from the various hamlets) and recording births, deaths, and migrations. Until the mid-1960s, a "census-book" was kept in each village, under the protection of the local government-appointed official (e.g., *Luluai*, *Tultul*, or Village Constable). Census was the responsibility of patrol officers like Ian Burnet and Jack Baker.

European Medical Assistant (EMA)—a paramedical person, often serving at Administrative outposts instead of a physician, or as an assistant to a physician.

girigiri—small cowrie shells used for personal adornment and as items of exchange in some traditional villages of the time.

haus boi—in this case, a communal dwelling in which initiated but unmarried boys and young men lived, apart from the houses of their mothers and younger siblings and from the separate communal dwelling of the older men.

haus kiap—shelters, called "rest houses," that were built along regular government patrol routes to house Patrol Officers and their entourages on exploratory, census, and medical expeditions.

kaukau—sweet potato, *Ipomea batata*.

kiap—refers to Department of Native Affairs officials from the level of Patrol Officer up to that of the District Commissioner (*kiap namba wan*).

kivung—a meeting of village elders and officials to discuss pressing matters of community concern.

kunai—a tall grass that covers vast tracts of the highlands as an effect of slash-and-burn agriculture.

line (Pidgin: *lain*)—to line up villagers for a census; this standard procedure became so well known in the highlands that in many previously uncensused villages people would line up spontaneously upon the appearance of a European in their midst.

laplap—a waistcloth made of cotton or other trade-fabric, worn by men.

Luluai (L/L)—a village leader, appointed by the Territorial government who was responsible for the safekeeping of the village census-book. This title was chiefly used in the New Guinea side of the Territory, and is no longer in use.

manki—a boy or schoolboy. Sometimes used by DCG to refer to his youthful friends and hangers-on.

mumu—a method of steam-cooking over stones in a covered earth pit; a feast or meal prepared by this method.

Native Medical Assistant (NMA)—a paramedical person, often attached to local "aid posts" where first aid and some emergency medical care could be provided in the absence of a physician.

pitpit—a type of sugar cane cultivated for food and building materials in the Eastern Highlands.

rest house—a campsite with permanent shelters built to accommodate government patrols and other transients; a *haus kiap*.

shutim nil—a form of ritual sorcery and murder practiced by many New Guinea peoples; see *tukabu* below.

singsing—a festival or celebration involving dancing, singing, and often feasting.

skin i ot—fever.

tukabu (tukavu)—Fore name for the form of *shutim nil* as practiced in the kuru region; often conducted in reprisal for kuru sorcery. See Appendix 1.

Tultul—assistant village official, appointed by the Territorial government; subordinate to *Luluai*. The term was used chiefly in the New Guinea side of the Territory and is no longer in use.

Village Constable (V.C.)—a Territorial government-appointed village official; term in use in the Papuan side of the Territory in 1957.

References

Pertaining to Kuru

Alpers, M. P., Gajdusek, D. C., and Ono, S. G. *Bibliography of Kuru*. Third revision. National Institute of Neurological and Communicative Disorders and Stroke, National Institutes of Health, Bethesda, Maryland (March), 220 pp. 1975.

Anonymous. Kuru. Leading article, *Medical Journal of Australia*, 2:21 (November 23), 765–766. 1957.

Bennett, J. H., Rhodes, F. A., and Robson, H. N. Observations on Kuru. I. A possible genetic basis. *Australasian Annals of Medicine*, 7:4 (November), 269–275. 1958.

Berndt, R. M. A "devastating disease syndrome": Kuru sorcery in the Eastern Highlands of New Guinea. *Sociologus*, 8:1 (April), 4–28. 1958.

Fowler, M. and Robertson, E. G. Observations on Kuru. III. Pathological features in five cases. *Australasian Annals of Medicine*, 8:1 (February), 16–26. 1959.

Gajdusek, D. C. Kuru: An appraisal of five years of investigation. With a discussion of the still undiscardable possibility of infectious etiology. Presented at the Tenth Pacific Science Congress, Honolulu, Hawaii, August 21–September 2, 1961. Mimeographed, National Institutes of Health, Bethesda, Maryland, 14 pp. Later published in *Eugenics Quarterly*, 9:1 (March), 69–74. 1962.

Gajdusek, D. C. *Kuru Epidemiological Patrols from the New Guinea Highlands to Papua*, August 21 to November 10, 1957. Monograph, limited edition, with index. National Institute of Neurological Diseases and Blindness, National Institutes of Health, Bethesda, Maryland (August), 223 pp. 1963.

Gajdusek, D. C. Journals 1957–1978. 28 volumes published in limited editions. National Institute of Neurological and Communicative Disorders and Stroke, National Institutes of Health, Bethesda, Maryland. 1963 to 1979.

Gajdusek, D. C. (editor). *Correspondence on the Discovery and Original Investigations on Kuru. Smadel–Gajdusek Correspondence 1955–1958*. Monograph, limited edition. National Institute of Neurological and Communicative Disorders and Stroke, National Institutes of Health, Bethesda, Maryland. Second printing (April), 413 pp + index. 1975.

Gajdusek, D. C., Fetchko, P., Van Wyk, N. J., and Ono, S. G. *Annotated Anga (Kukukuku) Bibliography*. National Institute of Neurological Diseases and Stroke, National Institutes of Health, Bethesda, Maryland (December), 85 pp. 1972.

Gajdusek, D. C. and Gibbs, C. J., Jr. Attempts to demonstrate a transmissible agent in kuru, amyotrophic lateral sclerosis and other subacute and chronic nervous system degenerations of man. *Nature*, 204:4955 (October 17), 257–259. 1964.

Gajdusek, D. C., Gibbs, C. J., Jr., and Alpers, M. (editors). *Slow, Latent and Temperate Virus Infections*. National Institute of Neurological Diseases and Blindness monograph no. 2. Proceedings of the Workshop and Symposium on Slow, Latent, and Temperate Virus Infections, National Institutes of Health, Bethesda, Maryland, December 7–9, 1964. U.S. Government Printing Office, Washington, D.C., 489 pp. 1965.

Gajdusek, D. C., Sorenson, E. R., and Meyer [Farquhar], J. A comprehensive cinema record of disappearing kuru. *Brain*, 93:1 (March), 65–76. 1970.

Gajdusek, D. C. and Zigas, V. Degenerative disease of the central nervous system in New Guinea. The endemic occurrence of "kuru" in the native population. *New England Journal of Medicine*, 257:30 (November 14), 974–978. 1957.

Gajdusek, D. C. and Zigas, V. Untersuchungen über die Pathogenese von Kuru: eine klinische, pathologische, und epidemiologische Untersuchung einer chronischen, progressiven, degenerativen und unter ein Eingeborenen der Eastern Highlands von Neu Guinea epidemische Ausmasse erreichenden Erkrankung des Zentralnervensystems. *Klinische Wochenschrift*, 36:10 (May 15), 445–459. 1958.

Gajdusek, D. C. and Zigas, V. Kuru. Clinical, pathological and epidemiological study of an acute progressive degenerative disease of the central nervous system among natives of the

Eastern Highlands of New Guinea. *American Journal of Medicine*, 26:3 (March), 442–469. 1959.

Glasse, R. M. South Fore society: A preliminary report. Mimeographed. Public Health Department, Territory of Papua and New Guinea, (June) 14 pp. 1962. Reissued, National Institutes of Health, Bethesda, Maryland, 18 pp.

Glasse, R. M. Cannibalism in the kuru region. Mimeographed. Public Health Department, Territory of Papua and New Guinea, 14 pp. 1963. Reissued, National Institutes of Health, Bethesda, Maryland, 14 pp.

Glasse [Lindenbaum], S. The social effects of kuru. Mimeographed (June). Public Health Department, Territory of Papua and New Guinea, 17 pp. Reissued, National Institutes of Health, Bethesda, Maryland, 22 pp. 1962.

Gunther, J. Australia, Kuru and a Nobel Prize. Unpublished manuscript. 1979.

Hadlow, W. J. Scrapie and kuru. *Lancet*, 2:7097 (September 5), 289–290. 1959.

Klatzo, I., Gajdusek, D. C., and Zigas, V. Pathology of kuru. *Laboratory Investigation*, 8:4 (July–August), 799–847. 1959.

Simpson, D. A., Lander, H., and Robson, H. N. Observations on kuru. II. Clinical Features. *Australasian Annals of Medicine*, 8:1 (February), 8–15. 1959.

Zigas, V. and Gajdusek, D. C. Kuru: Clinical study of a new syndrome resembling paralysis agitans in natives of the Eastern Highlands of Australian New Guinea. *Medical Journal of Australia*, 2:21 (November 23), 745–754. 1957.

Other Studies Referred To

Baltazard, M., Bahmanyar, M., Ghodssi, M., Sabeti, A., Gajdusek, C., and Rouzebehi, E. Essai pratique de sérum antirabique chez les mordus par loups enragés. *Bulletin of the World Health Organization*, 13:5, 747–772. 1955.

Gajdusek, D. C. Acute infectious hemorrhagic fevers and mycotoxicoses in the Union of Soviet Socialist Republics. Medical Science Publication No. 2, Army Medical Service Graduate School, Walter Reed Army Medical Center, Washington, D.C. (May), 140 pp. 1953.

Gajdusek, D. C. Hemorrhagic fevers in Asia: A problem in medical ecology. *Geographical Review*, 46:1 (January), 20–42. 1955.

Gajdusek, D. C. Das epidemische hämorrhagische Fieber. *Klinische Wochenschrift*, 34:29–30 (August), 769–777. 1956.

Gajdusek, D. C. An "auto-immune" reaction against human tissue antigens in certain chronic diseases. *Nature*, 179:4561 (March 30), 666–668. 1957.

Gajdusek, D. C. Pneumocystis carinii—etiologic agent of interstitial plasma cell pneumonia of premature and young infants. *Pediatrics*, 19:4 Part I (April), 543–565. 1957.

Gajdusek, D. C. An "auto-immune" reaction against human tissue antigens in certain acute and chronic diseases. I. Serological investigations. *AMA Archives of Internal Medicine*, 101:1 (January), 9–29. 1958.

Gajdusek, D. C. and Bahmanyar, M. Sur la Q fever en Iran. *Bulletin de la Société de Pathologie Exotique*, 48:1 (January–February), 31–32. 1955.

Gajdusek, D. C. and Rogers, N. G. Specific serum antibodies to infectious disease agents in Tarahumara Indian adolescents of northwestern Mexico. *Pediatrics*, 16:6 (December), 819–835. 1955.

Gajdusek, D. C., Rogers, N. G. and Bankhead, A. S. Serological survey of viral and rickettsial diseases among jungle inhabitants of the upper Amazon Basin. Presence during infancy, childhood and adolescence of antibodies to rickettsiae, to viruses of poliomyelitis, mumps, herpes, PLV group, yellow fever and to the group B arthropod-borne viruses. *Pediatrics*, 23:1 Part 1 (January), 121–131. 1959.

Mackay, I. R. and Gajdusek, D. C. An "auto-immune" reaction against human tissue antigens in certain acute and chronic diseases. II. Clinical correlations. *AMA Archives of Internal Medicine*, 101:1 (January), 30–46. 1958.

Schaeffer, M., Gajdusek, D. C., Brown Lema, A., and Eichenwald, H. Epidemic jungle fevers among Okinawan colonists in the Bolivian rain forest. I. Epidemiology. *American Journal of Tropical Medicine and Hygiene*, 8:3 (May), 372–396. 1959.

Schmidt, J. R., Gajdusek, D. C., Schaeffer, M., and Gorrie, R. H. Epidemic jungle fevers among Okinawan colonists in the Bolivain rain forest. II. Isolation and characterization of Uruma virus, a newly recognized human pathogen. *American Journal of Tropical Medicine and Hygiene*, 8:4 (July), 479–487. 1959.

Simmons, R. T., Gajdusek, D. C. and Larkin, L. L. A blood group genetical survey in New Britain. *American Journal of Physical Anthropology*, 18:2 (June), 101–108. 1960.

Simmons, R. T., Graydon, J. J. and Gajdusek, D. C. A blood group genetical survey in Australian aboriginal children of the Cape York Peninsula. *American Journal of Physical Anthropology*, 16:1 (March), 59 78. 1958.

Wilson, K., Zigas, V. and Gajdusek, D. C. New tremor syndromes occurring sporadically in natives of the Wabag-Laiagam-Kundep region of the Western Highlands of Australian New Guinea. *Lancet*, 2:7097 (October 31), 699–702. 1959.

Index

Abbott, Terry K., 74,84,87,321
Abomatasa village, 175;*16*
Abonai village, 127-129,131-132,142,293
Aborigines (Australian). *See* Cape York aborigines
Ada, Gordon L., 46
Addiction Treatment Center, Lexington, Ky., 228
Adelaide University, 253,262,265,268,275,292
Adelaide, Australia, 232,237,267,271-272,276-277,302; *21*
Afakanu. *See* Apekono
AFIP. *See* Armed Forces Institute of Pathology
Aga village, 170;*21*
Agakamatasa village, *xxviii,* 133,176,180,197,201,222, 314,318;*5,22,23,31*
Agamusa. *See* Agamusei village
Agamusei village, 178-181,187;*24,38,47*
Aganaga, 315
Agayagusa village, *xxviii,* 148;*5*
Agurio, *xxviii,* 126,133,314
Aipos (Pe'i), 127,129-130,205,314
Aiwantaia, 284
Aiyura Patrol Post, 218,221
AJPA. See American Journal of Physical Anthropology
Alfred Hospital, Melbourne, 321
Allen, A. D., 313
Alpers, Michael P., *xxvii,* 325
Amakiora, 92,125;*37*
Amboina, Indonesia, 171
Amenetu clan, Mani village, 149
American Journal of Physical Anthropology (AJPA), 59,103,122,257,308,310
American Philosophical Society, *xvii*
Amora village, 282,285-287
Amoraba village, 178
Amusa rest house, 135
Amusa village, 136,138,140,144,282
Amuwaiompa, 105
Anderson, Mrs. Gray, 2,27
Anderson, S. Gray, *xxiii,* 1-3,5,7-8,14-15,17-18,20, 23-24,26-28,31-32,38-40,42-44,46-59,62,69,71-73, 75-77,83-84,91,93-94,97-98,103-104,109-111, 118-119,153,163-164,166,170,229,236,250,252,264, 296,298-301,305,307-308,311,313,316,321
Andi, 176,178,195,206,222;*4*
Aneti, 126,149
Anga. *See* Kukukuku linguistic groups
Anjapte hamlet, 196-197
Anji village, 187,189;*39,53*
Anona, 153
ANU. *See* Australian National University
Anua, *xxviii,* 108,134-140,142,149-150,176,195;*5,10*
Anuma, 126-127,130-131,142
Anumpa village, 282
Aoga, 125
Apa:Nati hamlet, 283

Apekono, 147;*21*
Ara Vainapinto hamlet, 283
Aranaka, 95-97,155
Arawe linguistic group, 10
Archive for the Study of Child Growth and Development and Disease Patterns in Primitive Cultures (NIH), *xix*
Arinoguti hamlet, 283
Armed Forces Institute of Pathology (AFIP), U.S.A., *xxv,* 65
Aroia, *xxviii*
Arora village, *xxviii;5*
Asarumba, 211
Ashton brothers (gold prospectors), 293
Asia, 242
Asisi, 277
Aso, 125
Asoi, *xxviii,* 189,202-203,206,218,222
Asomeia, 33,61,85,97
Asunauveri village, 313-314
Athens, Greece, 142
Atigina region, 125-126,281-286
Atogori, 153
Atona, 100,317
Au'ia, 286-287
Auiana linguistic group, *xxviii,* 43,55,58,137,258,318
Aurieto, *xxviii*
Aurika, *xxviii,* 134
Auroga Kukukuku region, 181-189,192,194,201; *38,43,51,52,56*
Australia (Australian), *xviii,xix,xxii,* 6,18,26-27,30, 37-38,42-43,48,51-54,56,63,71,78,85,91,94,104-105, 133,143,145,158,168,171,187,214,216,231-232, 234-236,238-243,246,251-253,256,265-268,270,272, 275-276,278-279,291,293,298,310,316-317;*23,41*
"Australia, Kuru, and a Nobel Prize" (Gunther), 1
Australian Academy of Science, 217
Australian Commonwealth Film Unit, 239,291;*9*
Australian Medical Journal. See Medical Journal of Australia
Australian Museum, Sydney, 170
Australian National University (ANU), Canberra, 232, 235,292,321
Australian New Guinea Administrative Unit (ANGAU), 293
Australian School of Pacific Administration, Mosman, 4
Aveli linguistic group, 36
Awa linguistic group, 29,35,58,176-178,180,184,188, 295;*17,24,38,47,49*
Awande village, 293
Awande Lutheran Mission Kuru Hospital, 295
Awarosa village, 176,178,180,282,293
Aziana River, 179-181;*38*
AMA Archives of Internal Medicine, 50,88,104,106,121, 166,277,299,301,303,306,309,312,319

329

Bahmanyar, M., 326
Bailey, Pearce, 66,70,109,121,154,232,235,240,289, 306–307,321
Baimuru, 216
Baker, Carl G., 78,98–99,109,157,263,290,310,315,321
Baker Institute. *See* Thomas Baker, Alice Baker and Eleanor Shaw Medical Research Institute
Baker, Jack, *xvi–xxviii,*21–22,34–35,86,125,127–128, 131,133,136,138–139,147–149,167,175–180,183, 187–189,191–192,194,196–206,208,210–211,213–216, 218–220,222,225,236,266,296,321,323;*3,4,6,7,17,18, 40,44,53*
Baker, Lois. *See* Larkin, Lois
Baltazard, M., 326
Bangkok, Thailand, 78,97,124,160,234
Bankhead, A. S., 326
Barata, 286
Baro Ianai. *See* Yani River
Baroi River, 214
Barrau, Jacques, 154,156,321
Barrett, John, 218,221
Barua linguistic group, 192
Baseden, S. C., 55–56,69–70
Beara Government Station, 129–130,149,205–206,210, 214–216,223,313–314
Beethoven, Ludwig von, 197
Belgium, 279
Beltsville, Md., 154
Bena Subdistrict, 89
BenaBena linguistic group, 293–294
Bennett, J. H., 325
Berkin, John, 127,132–133,136–137,139,141–145,148, 150–151,219–223,240–241,321;*6,7,14*
Berlin, Germany, 226
Berndt, Catherine, 89–90,104,294
Berndt, Ronald M., 88,90,104,225–226,236,241,294,321, 325
Bethesda, Md., 99,110,123,156,235,272,278,315,321–322
Beyersdorfer, Mrs. (NINDB staff), 256,266
Biak, Netherlands New Guinea, 237
Blood Grouping Laboratory, CSIRO, 322
Bondar, Father, 145
Bo'oru, 210
Borae. *See* Fore
Bore'e people. *See* Fore
Borneo, 55,280,308
Boston, Mass., 15,41,239,305,321
Brain, xix
Brisbane, Australia, 6,77,94,276
British Journal of Industrial Medicine, 56
British Medical Association, 302
British Medical Journal, 110
Brown Lema, A., 326
Brussels, Belgium, 232
Burnet, Ian, *xxii,*4–5,26,39,41,46–47,63,74,138, 142–144,147–151,164,171,300,321,323
Burnet, Lady, *xxvii*
Burnet, Sir Frank Macfarlane, *xvii,xxii–xxiii,xxvii,*2–5, 8–9,14–15,17–18,23–28,31–33,35–44,46–52,54,55,57, 60–64,70–72,74,79,88,103–105,110,119,121,153–154, 163,165–167,171–172,225–226,230,236,250–253,261, 264,275,277,296–302,305–308,310,321;*1*

Busarasa village. *See* Pusarasa village

Cahill, Murray, 300,302
Cahill, Ross, 297–298,302
Cairns, Australia, 297
Cairns, H. J. F., 28
California Institute of Technology, *xxii*
Calvert, Lyn, 215
Calvert, Peter F., 215
Campbell, Charles, 53
Canberra, Australia, *xix,*28,32,38,217,239,246,276,296, 321
Cantoni, G. L., 263
Cape York Aborigines (Australia), 36,59,77,83,94,103, 133,243,248,306,308–311
Caroline Islands, 118
Casey, Miss (Roy Simmons' assistant), 94
Cat On a Hot Tin Roof (Williams), 302
Central America, 213
Central Highlands District, 60
Central Nakanai linguistic group, 36
Chamorro linguistic group, Guam, *xxvi,*118
Chandler, R. L., *xxvi*
Changai village, 190
Chemogo village, 184
Chimbu linguistic group, 51,60,295
China, 63
Chuave region, Chimbu, 51
Cleland, Brigadier Donald M., 52,54
Cleland, Prof. (Adelaide University), 268,270,272,321
Clinical Investigation, 309
Clinical Research Unit, WEHIMR, 6–7,322
Coleman, John, *xix,*4,24,27,129,178–179,194
Commonwealth Scientific and Industrial Research Organization (CSIRO), 69,72,93–95,102,153,167,225, 233,237,322
Commonwealth Serum Laboratories (CSL), 62,81,92, 97,103,153,257,315,322
Compton, England, *xxv*
Conrad, Joseph, 142
Cooper, G. N., 242
Copper Metabolism: A Symposium on Animal, Plant, and Soil Relationships (McElroy & Glass), 264
Correspondence on the Discovery and Original Investigations on Kuru (DCG), *xvii*
Council, Burt, 214
Cowling, David, 56,58,71
Crabill, R. E., 228
Creutzfeldt, Hans G., 155
Crown Law Office, TPNG, 305
CSIRO. *See* Commonwealth Scientific and Industrial
CSL. *See* Commonwealth Serum Laboratories
Curtain, Cyril, 59,61,77,81,83,88,92–94,100,104–105, 119,122,171–172,230–231,242–245,248,252,257–259, 262,276,298,300–303,311–312
"Cycas Nuts and Amyotrophic Lateral Sclerosis" (Whiting), 313

D Major Concerto (Beethoven), 197

INDEX

Daru Island, 128–129,216,299
Dastur, D. K., 290
Davidson (Waghi valley coffee planter), 299
Davis, Perk Lee, 256, 321
"Degenerative Disease of the Central Nervous System in New Guinea" (DCG & Zigas), 256
Delbrück, Max, *xxii*
Denny-Brown, Derek, 232,239
Department of Agriculture, Ottawa, 218
Department of Agriculture, Stock, and Fisheries, TPNG, 43,55,93,107,307
Department of Agriculture, U.S.A., 154,156,170,228, 279,280
Department of Bacteriology, Univ. of Melbourne, 322
Department of Native Affairs, Canberra, 2
Department of Native Affairs, Port Moresby, 46,321, 323
Department of Territories, Canberra, 39,52,56
De Re Medica (Lilly), 76
Detzner, Hermann, 293
Dineen, John K., 46,297,301
Diseases of the Nervous System in Infancy, Childhood, and Adolescence (Ford), 16,26
Division of Industrial Chemistry, CSIRO, 322
DuShane, Graham, 310
Dutch New Guinea. See Netherlands New Guinea
Dwyer, Mick, 293
Dwyer, L. (TPNG Dept. of Agriculture), 55

Eastern Highlands District, TPNG, *xv,xviii,xxii,xxiv*, 8,24,29,41,46,60,84,88,133,142,145,225–226,236,274, 281,293,305,307,309,321–232;*13*
Eccles, John C., 217,232,235,237–239,271,292,321
Edinburgh, Scotland, *xxv*
Edsall, Geoffrey, 55
Eichenwald, H., 326
Ela Beach, 68
Enders, John, *xxii,*41
Engel, W. King, 306
England, *xxvi,*279
English language, 156,216,309,314,318
Eposi. See Aipos (Pe'i)
Ereio, 298–299
Esita, 134
Etagi, 281
Europe (European), *xvi,xix,xxiii–xxiv,*31,37,41,46–47,63, 67,74,97–98,100,113,157,192,196,204,213–214,230, 239,242,248–249,251,276–278,308,313,323
Evesa, *xxviii,*108

Fairfield Hospital, Melbourne, 26,38,44,71,167
Farquhar, J. See Meyer, J.
Fazekas de St. Groth, Stephen, 28
Fenner, Frank J., 28,63,277,296
Fetchko, P., 325
Finland, 37
Fitzgerald, F. Scott, 142
Flying Doctors Service, Australia, 133
Fly River, 215

Foley, Michael, *xxviii,*127,131,149,222
Fopota, 83
Ford, Edward, 108
Ford, E. M., 104,106
Ford Foundation, 36,42
Fore-Keiagana border area, *15*
Fore linguistic group, *xviii,xxviii,*2,4,8,13–15,28,30,33, 35–36,43,46,50,55,57–58,65,68–70,72–73,78,80–81, 83–84,86,88–90,92,100–103,106–108,113,116,121, 126–131,136–138,140,144,149–150,157,159,161–167, 169–170,176–178,180,184–186,188,195–198,204–205, 207,210–211,218–220,226–229,231,238–240,244–245, 247,253–254,258–260,262–263,265,267,275–276, 278–279,282–285,287–288,291–293,295,313–314,318, 324;*5,15,16,22,23,26,27,31,35,36,44,46*
Forestry Department, Lae, 279
Fort Meade, Md., 256
Fowler, M., 325
France (French), 137,156,197
French, Eric, 20,46,63,298
Frigano region, 144,149

Garland, Joseph, 111,123,256,311,321
Garry, B. (Klatzo's assistant), 155
Geneva, Switzerland, 63–64,77,113,308
German language, 232,266,272,276–277,301,318–319
Germany (German), 44,166,279,302,321
Ghodssi, M., 326
Gibbs, C. J., Jr., *xxiii,xxvii,*325
Gide, André, 133
Gimi. See Kimi linguistic group
Gladwin, Thomas, 290
Glass, Bentley, 264–265
Glasse, Robert M., *xxv,*326
Glasse (Lindenbaum), Shirley, 326
Goilala Patrol Post, 175,293
Goldberger (pellagra researcher), 118
Gono rest house and mission, 140–141,143,150
Gordon Smith, C. E., 104,108
Gordon, W. S., *xxv–xxvi*
Gorogwanga village, 193
Goroka, *xxii,*4–7,11,15,28,46–47,49,51–52,58,89,138, 143–145,148,150,185,197,199,222,261,294,298
Gorrie, R. H., 327
Gottschalk, Alfred, 46,106,166
Grand Canyon, 101
Graydon, J. J., 308,327
Great Britain (British), *xix,xxvi,*37,47,278,301
Greece, 279
Greenfield, J. Godwin, 66,114,229,240,306,321
Greenfield, Mrs. J. G., 230
Growing Up in New Guinea (Mead), 298
Guam, 43,118,313
Guide to the Medicinal and Poisonous Plants of Queensland (Webb), 238–239
Gulf District, TPNG, 149,216,313
Gunther, C. E. M., 24
Gunther, John T., *xvii,xxii,xxvii,*1–5,9,13,15,17,28,32, 38–42,46,48–49,51–52,86–87,110,199,217,250,275,296, 306,321,326

Hadlow, William, *xxiv–xxvi,*326
Haida'abo, 100
Hall Institute. *See* Walter and Eliza Hall Institute of Medical Research
Hamilton, Edith, 142
Hamilton, Lucy, *xix,*53,85,108,136,138–139,147–150, 157,161–162,164,166–167,169–171,216–217,219–223, 226–229,231,233,236,238,239,247,253–254,263,266, 279,291,319–321;*6,14*
Haneo, 144,311
Hari, 313
Harvard School of Public Health, 118
Harvard University, *xxii–xxiii,*36,38
Haymaker, Webb, *xxv,*114
Healy, Clary, 216,313–314
Hegel, Georg, 41
Hegeteru village, 140,149–150
Heidelberg, Germany, 268,321
Henegaru village, *xxviii,*138,149;*14,20*
Henganofi Patrol Post, *xxii,*165
Hepavina village, *xxviii*
Herebebi, 211
Hicks, J. D., 62,73
Highland Road, TPNG, *xxii*
Hilleman, Maurice, 13
Hoi'uru'a'a. *See* Ho'urua village
Hoiuru'a'a. *See* Ho'urua village
Holden, Henry F., 50,56,69,75–77,94,98,171
Holland, George P., 218
Hollandia, Netherlands New Guinea, 67,237,277,279, 306
Holmes, Margaret, 108–109
Homeguei, *xxviii,*129,132,151,220–221
Hong Kong, 63
Honolulu, Hawaii, *xxv*
Hopi Indians, 41
Horning, Evan, 154–155,228–229,256
Ho'urua village, 130–131,313–314
Hughes, Arthur, 308

Iagaria linguistic group, 136,258
Iagusa village, 282
Iam, 287
Iar. *See* Yar Pawaiian
Iaragi, 288
Iasui local grouping, 281,284–285
Iate linguistic group, 88–90,161,258
Ibunarai hamlet, *34*
Iceland, *xxvi*
Iduabada Technical Training School, Port Moresby, 217–218
Ifufurapa region, 282
Igaga, *23*
Iginauri hamlet, 130
Igopiji hamlet, 184;*51*
Ilafo village, 176
Ilagi. *See* Ivuti (Ilagi)
Imus, Henry A., 109–110,117,121,123–124,157,227,290, 321
Inare, 319

India, 153,156,306
Indian Journal of Medical Sciences, 153
Indonesia, 154,156
Ingalls, Theodore H., 111,122,168,316,321
Inome, 155
Institute of Medical Research, Kuala Lumpur, 86,166
Inunkaia, 285
Iokoio, 83
Irakeia village, *49*
Iran, 65
Iresa region, 282
Isoisi, 119
Italy, 113,279
Ivaki village, 125,133,195
Ivuti (Ilagi) village, 82–83,171
Iwane village, 189;*50*

Jackson, Elizabeth, *xxvii,*306
Jakob, Alfons, 155
JAMA. See Journal of the American Medical Association
James, John, 220
James, Mrs. (WEHIMR experimental subject), 296,299
Japanese, 293
Jate. *See* Iate linguistic group
Java, Indonesia, 171
Jimmy River, 299
John Curtin Graduate School of Medicine, ANU, 321
*Journal of the American Medical Association (JAMA),*9, 50,297,299
Journal of Experimental Medicine, 309
Journal of Laboratory and Clinical Medicine, 62
Julius, Charles, 4–5,24,236,288,294

Kadaga, 51
Kafka, Franz, 298
Kagu village, 149
Kaiguanbi hamlet, 190–193
Kainantu, *xxii–xxiii,xxviii,*1–3,5–6,10,13,15–18,25–26,29, 31–33,36,39,43,49,51,55,57,69,90,101,132,136,138,148, 150,176,216,218,220,236–237,241,247–249,255,262, 269,273–274,280,284,288,291,293–294,306–309, 318–319,321–323;*13*
Kainantu Hospital, *xxiii;13*
Kainantu Subdistrict, 41,281,305
Kairimai village, 214
Kako, 82–83,86
Kamano linguistic group, 88–90,127,131,147,177,258, 284,293,295
Kamata village, 282,287;*36*
Kameia, 51
Kami village, 5
Kamira village, 282–283,293
Kamoneku village, 51
Kanite linguistic group, 43,161,165,170,258,294
Kapuna village, 214–215
Karamui Patrol Post, 164
Kasarai village, 126–127,130–132,134,139,210
Kasokana village, 175;*44*
Kasokandi hamlet, 126

INDEX

Kasoru village, 282,288
Kataramapinti hamlet, 195-196,198-199
Kataramunga River, 200
Kavarakea, 287
Kavu, Tommy, 215
Kavuiompa, 153
Kawaina-Nambaira region, Western Highlands, 253
Kaza River, 195
Keakasa. *See* Keiakasa village
Ke'efu rest house, *21*
Ke'efu village, *xxviii,*138,176;*15,21*
Keefu village, 147; *see also* Ke'efu
Keiagana linguistic group, *xxviii,*8,14,28,30,35-36,43, 68,80-81,84,90,127,129,131,136,144,147,150,161,170, 177,185-186,207,258,261,293-294,318;*14,15,20,46*
Keiakasa rest house, *21*
Keiakasa village, 283,286,294
Ke'jagana. *See* Keiagana linguistic group
Keldur, Iceland, *xxv*
Kemiu village, 154
Kemo "native area," 282
Kerema, 128-129,314
Ketabi-Purosa (Ketabi) village, *xxviii,*130,134;*5*
Ketebi "native area," 282
Kety, Seymour, 228
Keugeu, 315
Kiarupamenti hamlet, 283
Kiebu, 284
Kikori Government Station, 128-129,214,216,314
Kima, 287
Kimi linguistic group, *xviii,xix,xxviii,*8,14,28,30,35-36, 43,55,68,78,80-81,84-85,127-132,134-135,137-140, 144,147,149-151,161-164,170,177,185-186,204-205, 207,244,258,281-287,294,314,318;*7,15,25*
Kinao, 88,96,155,245
Klatzo, Igor, 112,114,117,123,154-156,160,170,264,272, 290,321,326
Klinische Wochenschrift, 120,232,266,268,270,274,276, 280,312,318-319,321
Kogu village, 88-90
Kokominjan language, 306,309
Konebada, Port Moresby, 218
Korea, 103
Koriki-Iai linguistic group, 215
Kosinto, *xxviii,*132,141,143-144,147,176,191,195,222; *14,15*
Kuagatnunga village, 202-204,207
Kuala Lumpur, Malaya, 55,78,124,237,274,277,280, 305-306, 319
Kuanimugu hamlet, 184;*51*
Kuata River, 201
Kukukuku linguistic groups (Anga), *xviii,xix,*4,24,29, 35,53,79,81,97,101,119,126,130,132-133,137,143, 147-148,159,161,163,169,175,177,179,181-205, 207-211,213,218-223,231,240-242,245-246,250,258, 266,277-279,293,295,302-303,310,313-318;*5,18,19,22, 38-43,47,48,50-56*
Kume village, 282-283
Kumm, H. W., 99
Kuna, 285
Kundenti hamlet, 283

Kundiawa, 172
Kunimara Patrol Post, 294
Kuraripinti village, 201
Kurland, Leonard T., 117,121,123-124,157-158,227,321
Kuroeva, 176
Kuru (Jack Baker's dog), *17*
Kuru Epidemiological Patrols from the New Guinea Highlands to Papua, 1957 (DCG), *xvii,*190
Kuru Research Hospital. *See* Okapa Kuru Center
Kutena'visa "clan," 283

La Billardiere (18th century Pacific explorer), 171
Lae, 6,15,51-52,67,104,148,216-218,223,236-237,248, 251,254,266,277,279,309-311,319-320
Lamari River, 28-29,34-35,53,58,101-102,126,128-130, 170,176-177,179,195,197,201,205-206,211-213,219, 231,233,241,245,250,314;*22,44,49*
Lamari Valley, 293;*44*
Lancet, 77
Lander, Harry, 268,270-272,276,326;*21*
Larkin, Lois, *v,xxiii,xxvii,*20,46,51,59,61-62,74,77,79, 82-83,88,104,108,122,153,163,242,244-245,257,261, 271,277-278,293,296-297,298-303,306,313,315,317,321, 327;*7,14,15*
Laughlin, Mr. & Mrs. (Gono missionaries), 141,152
Laye (Lalye) village, 201
League of Nations, *xix*
Leahy, Mick J., 293
Lederberg, Joshua, 154,166,299,301-302
Lee, H. T., 280
Lenska, Rula, 27
Life, 231,246,302
Liklik, *xxviii,* 219
Lind, Pat, 46,302
Little Theatre, South Yarra, 302
Littman, M. L., 319
Lois. *See* Larkin, Lois
London, England, 52,137
London Missionary Society Hospital, Kapuna, 214-216
Look Back in Anger (Osborne), 297
Lufa Patrol Post, *xxii,*3-5,11,26-27,39,41,46-49,53,58, 74,132,136-138,140,142,144-145,147-150,161-162, 164-165,175,208,247,255,306,321
Lufa-Goroka Road, 295
Luther, Martin, 302

Macaulay, Thomas B., 142
Mackay, Ian R., 44,46,56,59,106,277,296-297,299-302, 306,319,326
Madame Bovary (Flaubert), 181
Mage village, 294
Magic Flute, The (Mozart), 297
Magorti "native area," 282
Maiara village (Beara Patrol Post), 314
Maiguanga village, 193,195
Malaya, 55,154,306,308
Malekor, *xxviii,*175-176,187,191,201,210;*48*
Mamusi linguistic group, 10,36
Mande, 100,125

Manevu, 281
Mania village, 282
Mani rest house, 137–138,149
Mani village, 132,136,138–140,147,313,315
Manto, 33,86,127,139
Manus Island, 5
Maprik, 172
Mara, 83
Markham Valley, 236,293
Maryland, 321–322
Masasa, *xxviii,*108,176,207,212,222;*5,6,12,19*
Masland, Richard, 109–110,121,256,263–264,270,274, 290,321
May, Dr., 27
McElroy, William D., 264
Mead, Margaret, 300
*Medical Journal of Australia (MJA), xxiii,*24,59,95,104, 106,110,120,159–160,225,232,235,241,243,245,249, 251,256–257,261,264,267,272,275,311,313,315, 317–318,320,322
Medical Service, Public Health Department, TPNG, 321
Melanesia, *xxiii,*100,115,137,156–157,165,248,252
Melbourne, Australia, *xxii,xxvi,*1,4–7,9,11,13,16–18,21, 26,31–32,36–38,41,43–44,46–47,50–52,54–55,60,62, 66–67,69–73,75–76,78–80,82–83,87–88,91,97,100–103, 110,120,132,158,165,168,217,232–233,237,240–241, 244–245,255,259,262–263,266,270–271,276,278,289, 296,301,305,307–308,315,319,321–322;*15*
Melbourne University, 94
Melbourne University Medical School, 299
Mendi Patrol Post, 306
Mengino village, 140–141,150
Menyamya Patrol Post, 277–278
Meyer (Farquhar), J., 325
Miarasa village, *xxviii,*293
Misapi village, 127,129–130,132,134–136,313
MJA. See Medical Journal of Australia
Mobutasa village, 177–181,187,198,219;*49*
Mobutasa Rest House, 178
Moke–Pintogori. *See* Pintogori site
Moke village, *xviii,xxviii,*1,8,10,13,16,21,28–29,34,36, 43–44,49,60,85,88–90,107,116,121,125,127–128, 131–132,135–136,138–139,145,147–150,154,162–163, 165,175–178,193,208,218–219,222–224,226,238–239, 241,250,254,275,291–292,299,307,316;*12,26*
Mononi Kukukuku region, 203–204,209–210
Moraei Kukukuku region, *xxviii,*130,133,143,177,181, 192–193,195–198,200–201,203–204,210,219–223, 314–316;*5*
Moredun Institute, Edinburgh, *xxv*
Morieto, *xxviii; 5*
Morobe District, TPNG, *xviii*
Moroccan, 56,97,101,106
Morris, Marion, 27,42,321
Motu language, 127,205,314
Mozart, Wolfgang Amadeus, 297
Mt. Hagen, 60
Mt. Michael, *xxii,*4,135,138,162,175,195,293,302,319
Mt. Sinai Hospital, 319
Mugaiamuti Seventh Day Adventist Mission, 294

Mugaiamuti village, 204,294;*22*
Mu'i, 314; *see also* Wi'ir village
Mulinapa, 95,155
Muniri Kukukuku region, 192–197,201
Munroe, Eugene, 218
Muriso, *xxviii*
Musgrave, Mr. (Australian Museum entomologist), 171
Myrianthopoulos, Ntinos, 118

Namaina village, 213
Namaorufa, 248
Namugoi'ia, 286–287
Nantati, 285–286
Nasumbi, 282
Nata, 97,100,125;*9*
Natameia, 155
National Foundation for Infantile Paralysis (NFIP), 27,42,60–61,74,103,120,306–308,310,317,319,321–322
National Institute of Mental Health, 8,228
National Institute of Neurological Diseases and Blindness, 77,98–99,102,114,117–118,121,123–124, 157,159,162,227,229,233,235,240,245,256–257, 263–265,272–274,306,313,317,321
National Institutes of Health (NIH), U.S.A., *xvii,xix, xxii,xxv,xxvii,*40,43,61,67,91,97–100,102–103,116,120, 155–156,162,164–166,170,227–229,234,239,245, 253–254,256,265,271–274,276,279,289–290,305, 310–313,315–317,319,321–322
National Library of Medicine, Bethesda, Md., *xix*
*Nature, xxiii,*9,51,59,70,74,77,104,121,297,299,306, 308–309
Negibi New Tribes Mission, 294
Negibi village, 294
NEJM. See New England Journal of Medicine
Neni, 288
Neomelanesian Pidgin, *xxviii,*6,7,9,13,89,127,238,286 294,301,314,323;*15*
Ness, Robert K., 114,290
Netherlands New Guinea, 17,37,55,122,277,279, 305–306,308,320
New Britain, *xxiii,*36,41,59,81,103,122,125,133,137,193, 243–244,248,261,301,303,313
New Caledonia, 171,321
New Crops Research Branch, U.S. Dept. of Agriculture, 156
New England Journal of Medicine (NEJM), 95–97,99, 101,111,120,122,124,154,157,159,168,231,241–242, 245,249,251,256,260,266,272,310–312,315–317,319, 321
New Guinea. *See* Territory of Papua and New Guinea
New South Wales, Australia, 262
New York, 71,137,321–322
New Zealand, 187,215
NFIP. *See* National Foundation for Infantile Paralysis
Niaia, 61–62
Nicholas, Marta, *xix*
Nicholson, Ray, 220
NIH. *See* National Institutes of Health
NINDB. *See* National Institute of Neurological Diseases and Blindness

INDEX

1984 (Orwell), 181,197
Nondugl Subdistrict, 60
North Fore linguistic group, *xix,xxviii,*175,282,293-294; *5,16,26,36,44*
Nossal, Gustav J. V., 302
Nottingham, England, 307
Noumea, New Caledonia, 321
Nuffield Foundation, 172
Numpuru rest house, 144

O'Dea, John, 59,62
Oceania, 60,89-90,226,236
Ogia, 220-221
Oiya, *xxviii*
Okapa Kuru Center (Hospital), *xvi,xviii,xxviii,*29,35,57, 79,85,100,102,107-108,112,120,143,161,164,170,226, 246-247,255,265,294;*8,9,12,15,31*
Okapa Patrol Post, *xvii-xix,xxii,xxvii-xxviii,*1-5,7-8,13, 18,21-22,26-28,39,41,47-49,51-53,59,63,69,83,85,88, 93,95,97,104,107,110,131,133,140,144-145,165-167, 175-177,192,198,202,208,216-220,223,231-234, 236-237,241-242,245,249,268,274,279,283-284, 286-288,291-292,294-296,302,305-306,309,316,318, 321;*3-6,8,9,15,17,31,45,46,48*
Okapa Pine forest, *See* Okasa Pine forest
Okapa Primary School, 105
Okasa (Okapa) Pine Forest, 195,293
Okasa (Okapa) rest house, 175
Okasa village, 282-283,287,312
Oldar village, 52
Oma, 126
Onarunka Lutheran Mission, 293
Oni, 83
Ono, S. G., *xix,*325
Oriei rest house, 125-126,133
Oriei village, 313
Orikaiva linguistic group, 36,300,303
Ormond College, University of Melbourne, 302
Orwell, George, 197
Owen, John, 94,172

P.H.D. *See* Public Health Department, TPNG
Paiti village, 126-130,132-134,208
Pallson, Pall, *xxv*
Pamir Mountains, USSR, 298
Paoli, Pennsylvania, 256,321
Papua, *xix,*18,36,63,79,125-131,149,151,162-164,169, 175,192-193,197,199,204-205,208,210-212,214,223, 231,241,245-246,248,250,267,269,273,293,301,314, 316,324;*18; see also* Territory of Papua New Guinea
Papua New Guinea, *xv*
Paretai, 203
Paris, France, 137
Pattison, I. H., *xxvi*
Pauling, Linus, *xxii*
Pawaiian. *See* Yar Pawaiian linguistic group
Pe'i. *See* Aipos (Pe'i)
Peak Molotov, 298
Peak Pravda, 298

Peak Stalin, 298
Pediatrics, 58,70,299,305-306,308-309
Peking, China, 63
Pennsylvania, 145,256,321
People (Australian tabloid), 218
Perth, Australia, 104,321
Pidgin English. *See* Neomelanesian Pidgin
Pile, 52
Pintogori site, 71,116,133,135,138,177,195-197,241,250, 294;*8,12,15,31; see also* Okapa Kuru Center, Okapa Patrol Post
Pittsburgh, Pennsylvania, 145
Poona, India, 153
Port Moresby, *xix,xxii,*3,5-7,13,15,17-18,20,22-23, 25-26,28,30-33,36-37,39,42-44,46-48,51-52,55-56, 59,64,67,69,71-73,75-76,78-79,81,85-86,93,95,102, 104,106-107,110-111,119-120,128-129,159,161, 163-165,167,172,185,187,198-199,216-218,221-226, 231,233-234,237,239,240-241,245-246,248-253,266, 272,274,291,294,298-299,301,305-309,311,314, 316-317,319-322
Port Romilly, 216
Price, A. V. G., 26,44,56,68,70,76-77,84,104,106,111, 119,167,217,308,311,321
Proceedings, Seventh International Grassland Conference, 280
Public Health Department, Netherlands New Guinea, 306
Public Health Department, TPNG, *xvii,*3,7-8,13,21, 30,43,46,48,52-53,60,104,106-107,111,167,216-217, 230,236,253,275,288,291,298,305-306,309,311, 319-322
Public Health Laboratories, University of Melbourne, 242,302,311,322
Public Health Laboratory, Port Moresby, 321
Public Health Service, U.S.A., 9,36
Purari River, 169,192,212-215,231,241,245,250,293,313
Purosa region, *xxviii,*120,129-131,133-134,176,195,220, 282,294;*5,28*
Purosa rest house, 125;*28*
Purosa Road, 262,294
Purosa World Mission station, 294
Puruya River, *18,50*
Pusarasa village, *xxviii,*89-90

Queensland, Australia, 143

R.M.H. *See* Royal Melbourne Hospital
Rabaul, New Britain, 216-217,244,284,294,307
Radio Indonesia, 187
Raipinka Lutheran Mission, 293
Raro clan, Kimi, 140
Rees, A. L. G., 75,93-95,229,236,264,322
Reid, Jack, 148
Reid, Lucy Hamilton. *See* Hamilton, Lucy
Revesavipi hamlet, 151
Rhodes, F. A., 325
Rivers, Thomas M., 27,40-41,60-61,63,70,78,80,88, 99-100,103,120,307-308,322

Robertson, E. Graeme, 62,75,87,114,164,167,171,225, 229-230,240,245-246,252,266,271-273,322,325
Robson, H. N., 217,232,234,239,268,270,272,274-275, 277,292,319,325-326
Rockefeller Foundation, 36
Rocky Mountain Laboratory, U.S. Public Health Service, 113,123
Rogers, Nancy G., 306,326
Rome, Italy, 137
Rouzebehi, E., 326
Rowlands, Ned, 293
Royal Children's Hospital, Melbourne, 153
Royal Melbourne Hospital, 13-15,23,54,56,75-77,102, 165
Royal Papua and New Guinea Constabulary, *xix,xxviii,* 176,294;*21*
Runman, John Paul, *xix*

Saave, 295
Sabeti, A., 326
Saipan, Caroline Islands, 118
Sama, 221,223-224
Sarason, Dr. (NIH), 290
Savoy Theatre, Melbourne, 297
Schaeffer, M., 326-327
Schmidt, J. R., 327
School of Medicine, University of Melbourne, 322
Schubert, Bernice, 154
Science, 73,76,95,101,310
Scotland, *xxvi*
Scragg, R. F. R., *xxii-xxiii,*3-6,10,13,17,21,23-28, 31-33,35,38-41,44,46,48,51-52,55,59,63,87-88,94-95, 97,104,110-111,114,164,166-167,172-173,216-217, 222-223,225-226,230,240,250,275,291,306,319,322
Senau, 313
Seventh Day Adventists, 294
Shanghai, China, 63
Shaw (Baker Institute), 298
Sherrington, Sir Charles, 235
Shy, Milton, 109,114,123,160,290
Sigurdsson, Bjorn, *xxvi*
Simbari Kukukuku region, *xxviii,*177,189-195,199,201; *5,18,50*
Simmons, Roy T., 56,59,61,73,77,80,82-84,92-94,97, 103-105,118-119,122,153,162,240-245,249,257-259, 261-262,267,298,300-303,308,310,312-313,315,317, 322,327
Simpson, Colin, 300
Simpson, Donald A., 268-272,326;*21*
Sinclair, Alex, 102
Sinoko, *xxviii,*141,145,150-151,315
Sir Mac. *See* Burnet, Sir Frank Macfarlane
Skinner (1947 Eastern Highlands Patrol Officer), 293
Slovak, 145
Smadel, Joseph E., *xvii,xxii,xxviii,*8-10,27,29,31,40-43, 54,60-61,64-66,68,70-71,74,77-80,87,90,95-104,109, 111-116,120-124,154-157,160,162-163,166,168-169, 171-172,220-224,226-229,231,234-236,239-240,246, 253,256-257,259,261-262,264,266,268-270,273-277, 279-280,289,306,308,310-312,315-316,319,322;*2*

Smadel, Elizabeth, 160,306
Smithsonian Institution, 162,170,228,254,320
Smythe, William E., *xxvii,*53,65,89-90,157,161,167, 216-217,220,231,244,307,312
Sociologus, 226
Soiyanadudau ground, Yar Pawaiian, 314
Sojotu village. *See* So'o
Solomon Islands, 172
Somai village, 130,133
So'o village, 127,129-132,149,205-206,210-215,221, 313-314
Sorobi, 126-129
South America, 92,213,228
South Fore linguistic group, *xix,xxviii,*28-29,52, 119-120,133,180,281-282,293-295,318;*5,22,23, 27,31,34*
South Pacific Commission, Noumea, New Caledonia, 154,156,321
Southeast Asia, 46
Southern, P. J., 69-70,93-94,229,322
Spector, Wally G., 46
Springer Verlag, Heidelberg, Germany, 321
St. John, Prof. (botanist, University of Hawaii), 279
Stamp, J. T., *xxv*
Stockholm, Sweden, 297
Subu River, 205,209,211-213,245,250,313; *see also* Lamari River
Suguia, 282
Summer Institute of Linguistics, 294
Sunday News, 125
Sunday Times, Perth, 225
Sunderland, Sydney, 217,232-233,239,270,292
Sweden, 37
Sydney, Australia, 197,217,232,251,276,302,305,317,322
Sydney Blood Bank, 172
Sydney Herald, 217,246,252
Sydney University, 276
Symes, William, 5,11
Szent-Ivany, J. J. H., 217

Ta'ati'a, 313
Taft, L. I., 50
Tainoraba village, 178
Taka, *xxviii,*108,176,207,212,222;*5,12,19,21*
Takai village, 129-131,293
Taketo, 286
Tamogavisa village, 315
Tarabo village, 139,147,294
Tarabo Lutheran Mission, 293
Tarangau, *xxviii,*176,188,195,222;*16*
Tarato, 312
Tari Patrol Post, 47
Tasiko, 316
Tasmania, 133
Tau language, 314
Tawasa, 125
Taylor, James L., 293
Tchaiorogoro camp, *42*
Tchaiorogoro hamlet, 185-187;*39,41,50,54,55*
Tchetchai village, 190-191

Te'hei language, 131,313–314; *see also* Yar Pawaiian
Telefomin, 286
Territory of New Guinea, *xix*
Territory of Papua, *xix,*314
Territory of Papua and New Guinea, *xv–xix,xxii–xxiii,* 2,4,7–9,14,25,27,32,36–42,46–47,52–53,58–60,62–63, 71,73–75,77–78,80–81,84,88–89,101,103,108–109,118, 121,123,125,128–129,133,135,137,143,145,154,156, 160,171–172,187,193,197,208,213–216,218,225,228, 230,235,236,239–240,244–248,256–257,261,263, 274–277,279,281,284,286,289,291,295–301,303, 305–307,309–311,315,318–324
Thailand, 228
Thomas Baker, Alice Baker and Eleanor Shaw Medical Research Institute, 59,100,257,321
Tiarana village, 315
Time, 142,218,231,239,246,252,257,266,302,318
Tiu, *xxviii,*108,176,180,195–196,201,207,212,222;*5,6,16, 18,19,21*
Tobosa-Lamari junction, 211–212
Tobosa River, 211–212
Tolai linguistic group, 284
Tomasetti, Bill, 177;*7*
Toneaso, 92
Tori, 179
Tosetnam, *xxviii;10*
Tove, 312
Tovepi, 153
Tracer Elements, Human and Animal Nutrition (Underwood), 97
Traub, Robert, 86,166,218,274,277,319
Tu River, 202
Tübingen, Germany, 106
Tu'efa, 90
Tufanel, Father, 293
Tunuku village, 176

Uarevana village, 313
Ubank, Ted, 293
Ulysses (Joyce), 298
Umi River, 105
Underwood, E. J., 97
Ungar, Sue, 77
Union of Soviet Socialist Republics (Russia), *xxvi,*137, 166,187,297,300–302
Union Theatre, Melbourne, 297,302
United Nations, 291
United States of America (American), 9,17,24,27, 30–31,36–38,41–42,46–48,60–61,64–65,71,74–75, 77–78,81,88,97,100,103–104,118,134,141–142, 144–145,152,160,162,164–167,171,187,197,217–218, 221,223,225,231–234,236–237,239,244–246,250–251, 256–257,261,268,270,272,275–276,279,291,299–302, 305,308–310,316–317,319–320
University of Hawaii, 279
University of Melbourne, 108,114,167,172,242,292,322
University of Papua New Guinea, 321
University of Sydney, 279
University of Western Australia, Perth, 90,321
Urahau, 130,205,208,210,213,313–314

Urai, 140,144
Urai River, 134
Urai village, 126;*34*
Ureba village, 183–184,186;*38,43,52*
Uri village, 213
U.S. Academy of Pediatrics, 38
Usurufa linguistic group, 72,80,88–90,161,170,258, 293–294
Uvae, 82–83
Uvai rest house, 136,315
Uvai village, 132,135–136,149,295,315
Uwami village, 293

Vailala River, 195
Vallee, Bert, 263
Van Wyk, N. J., 325
Vienna, Austria, 19
Voice of America, 187
Voors, Antonie W., 280

Wa'abowa village, 130–131,214,314
Waghi Valley, 299–300
Wagiri hamlet, 283
Wai'oti, 283
Waiajeke (Haus Kapa), *xxviii,*133,176,181–182,186,189, 191,193,195–196,222,314;*5,18,21*
Waibibi. *See* So'o village
"Walk Into Paradise" (film), 125
Walsh, Robert J., 28,172,244
Walter and Eliza Hall Institute of Medical Research, *xxii–xxiii,xxvii,*1–2,8–9,15,21,25,30,37–40,42–43, 46–48,51,53,63,69,73,81–82,93,108,110,119,149,154, 158,164–168,171,229–230,236,244–245,257,264,275, 277,296–297,299,308,315,321–322;*1,15*
Walter Reed Army Institute of Research, 274,306
Walter Reed Army Medical Center, *xxii,*305
Wame River, 214
Wane, 285
Wanevi, *xxviii,*108,134,176–177,180,189,192,195,202, 206,212,214,218,222;*6,11,19*
Wanikanto village, 282,286
Wanitabi village. *See* Wanitavi village
Wanitavi village, 281–282,285,287–288;*22*
Wanovi (Wanovibi) hamlet, 180
Wantekia Kukukuku region, 187–189,192,194;*39,41,42, 50,53–55*
Wanto, 134
Wanto'abarasa region, 281–283,285
Ward, Hugh, 28
Waruata, 285
Waruye, 319
Wasanamuti hamlet, Moke village, *12*
Washburn, S. L., 77,94,310
Washington, D.C., 40–41,60,63,71,137,156,254,267,271, 306,318
Watcheramapinti hamlet, 196–200,204
Watkins, W. W., 305
Wau, 298
Wa'u'wa, 313

Wedding Preparations (Kafka), 298
WEHIMR. *See* Walter and Eliza Hall Institute of Medical Research
Weiden, Sara, 50,244
Wellington, New Zealand, 280
Weme village, 202-204,206-213,216,221;*57*
Wenabi village, 186-187
Wenner-Gren Foundation, 94,317
West Australian, Perth, 225
West Nakanai linguistic group, 36
West, Harry, 179,194
Western District, TPNG, 215
Western Highlands District, TPNG, 172,244
Western Reserve University, Cleveland, 36
White, F. O., *xxvii*
White, Kevin, 217,254,279,320
Whiting, Marjorie Grant, 118,124,313
Wi'ir village, 125,127-128,205,208,210,221,314
Williams, Tennessee, 302
Wilson, K., 327
Wilson, Michael M., 108,311,322
Winton, Ronald M., 160,232,236,249-250,322
Wipf, Eckert, 77
Women's Hospital, Melbourne, 76,97,252
Womersley, John S., 53
Wood, Ian, 5-6,8,14-15,17-18,44,46,49,56,62,104,302, 306,308,322
Wood, Mrs. Ian, 51
Woodward, Ted, 2
Works Department, TPNG, 34
Work, Telford, 153,165
Wugamuwa River, *39,40*

Yabaiotu, 88,112,155
Yagusa Highland Christian Mission, 295
Yagusa village, 295
Yakeia village, 176-179
Yakurimba, *31-33*
Yani, 97,100;*36*
Yani River, 85,126-131,133,136,201,205,233,314
Yar. *See* Yar Pawaiian linguistic group
Yarevana, 151,313,314; *see also* Yar Pawaiian
Yar Pawaiian linguistic group, 126-131,136,142,145, 149,162-163,169,192,194,200-212,215,231,241,294, 302,313-315,317-318;*37*
Yate linguistic group, 294
Yearbook of Neurology, 26
Yir-Yoront language, 306,309
Yoeia, 155
Yonkers, New York, 316
Yosetme, 176

Zaritsky, Raul A., *xix*
Zigas, Gloria, *xxvii,*51,216-217,239,301
Zigas, Vincent, *xvi-xvii,xxii-xxiii,xxvii-xxviii,*1,3-4,6-9, 11,13-18,21-26,28-31,33-37,39-40,42-44,46-48,51, 53-57,63-64,67-68,70,72-73,75,79,81-82,84,86-87,91, 97,103,106-107,110,116,122,125,127,131-132,136, 138-139,145,147-150,161,164,167-169,172,175-176, 198-199,216-219,221-223,225,231,234-237,239,241, 246,249-250,252,254,259,266,269-270,273,275-276, 278-280,291,300-301,307-308,315,317,319-320,322, 325-327;*3,8,13*